Handbook of Circuit Mathematics for Technical Engineers

Handbook of Circuit Mathematics for Technical Engineers

Robert L. Libbey

CRC Press
Boca Raton Ann Arbor Boston

Library of Congress Catalog Card Number 90-86414

This book represents information obtained from authentic and highly regarded sources. Reprinted material is quoted with permission, and sources are indicated. A wide variety of references are listed. Every reasonable effort has been made to give reliable data and information, but the author and the publisher cannot assume responsibility for the validity of all materials or for the consequences of their use.

Direct all inquiries to CRC Press, Inc., 2000 Corporate Blvd. N. W., Boca Raton, FL 33431.

International Standard Book Number 0-8493-7404-9
Printed in the United States of America

To Rob

1959-1987

Contents

Preface

This book is about the use of the concepts and the calculations of mathematics to aid the student and the practicing engineer in understanding and performing circuit analysis. From its beginning, the technology of electronics has been a happy and exciting mixture of both the theoretical and the applied. Lee De Forest's invention of the electron tube (circa 1900) started the branch of science that has expanded, at a feverish rate, to produce products that provide pleasure, information, control and security. As with most areas of science, mathematics provides the rigor, the structure, and the insights to bridge the separation between the theoretical and the practical. The brilliant, creative experimental work of many pioneers has been confirmed and classified using the techniques of mathematics. Mathematics augments both general circuit theory and practical experience. A knowledge of how to use fundamental mathematics provides both the new student and the experienced practitioner with a key that allows them to open the doors of past, present and future technology.

This book is designed to aid both the first-time student and the working professional in learning and reviewing the principles, rules and laws, and techniques that are the underlying collection of information termed circuit analysis. It is also designed to help the reader establish more efficient learning techniques and habits.

In learning, updating, or reviewing most any subject, one of the major avenues to more efficient learning is some insight into the learning process itself. Without delving into the details of the theories of learning, it is quite well established that some people learn best from the printed word, some from symbols such as those used in mathematics, and some from pictures, diagrams, or charts—that is, graphics. This book is structured to aid and reinforce the knowledge acquisition of different types of learners. It presents each idea or technique in three different ways. These learning aids or learning models include a Text Model (TM), a Mathematical Model (MM), and a Graphic Model (GM). Each new topic is introduced using English text. Next, the pertinent equations or mathematical principles for the new topic are introduced with an MM. The general outline of this MM parallels the progression of the ideas that were just presented in the TM. Finally, the topic is illustrated with a GM, using a diagram, a plot, or a chart, to present the topic from a third perspective and to reinforce the relationships between the written word and the appropriate equations.

The use of modern computer circuit analysis tools is another major feature of the book. Typically, a topic is presented using classical, non-automated techniques, and this is followed by examples of how the analysis can be done with up-to-date personal computer or work station programs such as PSpice© or MathCAD©, Mathematica©, and Theorist©. Not only do contemporary computer programs give the user the ability to do complex analog and digital circuit analysis, they provide the ability to solve both simple and difficult circuit-related equations and to plot the results so the investigator can see the results in graphic form. The ability to graph the results of a circuit analysis or a mathematical solution is one of the most important tools for learning circuit analysis. The numerous uses of these analysis, computing, and graphing tools establishes a learning pattern that will aid the reader for years to come.

Robert L. Libbey
Cherry Hill, NJ

Acknowledgments

The experience of producing this book has, once again, brought to mind the wonderful competence and friendship that colleagues have shared with me throughout the years.

To Murlin Corrington, who gave me a glimpse of what world-class mathematics is about, and to John Reisner, who gave me a glimpse of what world-class science is about, a most warm thank you. Steve Barboun, Lue Bazin, Mike Carroll, John Sank and Dan White were extremely helpful in my research as was Natalie Mamchur of the Moorestown General Electric library. Jon Dattoro of ENSONIQ Corporation was most helpful with references on digital audio. I am especially grateful to Bob Hurst and Tom Murakami for allowing me to adapt some of their work. Dean John TenBrook of Camden County College was most supportive.

Alan Rose of Intertext Publications and Joel Claypool of CRC Publications always were there to give the guidance and continuing encouragement that was needed throughout the writing process.

Arch Luther contributed his proven talents as an engineer, scientist, and author to provide almost limitless insights and support; this book would not exist without him.

And to my dear wife Helen—whose courage, patience, leadership, and above all, her sense of humor, made it possible for me to reach this goal—all my love.

1

Fundamental Circuit Analysis Models: Resistor Networks

1.1 AN INTRODUCTION

An electric circuit can first be considered as an input generator, a transmission or signal modification medium, and an output receiver or load. These circuit units can be very simple such as a battery, conductors and a switch, and a light bulb. Likewise, the concept of a circuit can be expanded into a network where there may be several elements that can act as input generators, several paths that can transmit and/or modify the input signals and several other elements that can receive the output signal. There are two major approaches to circuit analysis—the white box and the black box. The clear or white box lets one see all of the circuit, including the generator, the mid-section or transmission medium, and the output. The black box approach includes the input generator, the output load and a black box that represents all the components of the transmission medium.

The first major requisite of circuit analysis is to find the magnitude of the currents and voltages. Next, it is usually necessary to find the time (phase and/or delay) and frequency relationships between these currents and voltages. Circuit analysis is based upon physical relationships (laws) that often originated with careful experimentation and were confirmed by the techniques and rigors of mathematics. All fundamental circuit analysis, be it manual or by computer, is based upon *Ohm's law, Thevenin's theorem* and *Kirchhoff's current and voltage laws*. These fundamental principles have been expanded or modified to include such rules (laws) and techniques as Norton's theorem, the superposition theorem, and the convention of network geometry, including *branches, nodes, loops* (defined later) and their resulting circuit equations. All these rules and laws are defined in this chapter.

The formal process of circuit analysis is based upon some combination of these three steps:

1. Create current paths in the given circuit using the above laws.
2. Write a series of equations for each proposed circuit current using the values of the components. Each component will have one or more currents flowing through it. At this point in the analysis process, the designated currents may only be mathematical quantities and not necessarily the actual, physical currents.
3. Solve these equations to produce the actual, physical currents using some of several techniques to be described. Once the true currents are obtained, the component voltage drops can be calculated. It is also possible to obtain voltage drops first and then calculate component currents. Above all, it should be remembered that it is extremely important, in the final analysis, to relate actual, physical operation to any mathematical calculations.

GM A FUNDAMENTAL ELECTRIC CIRCUIT MODEL

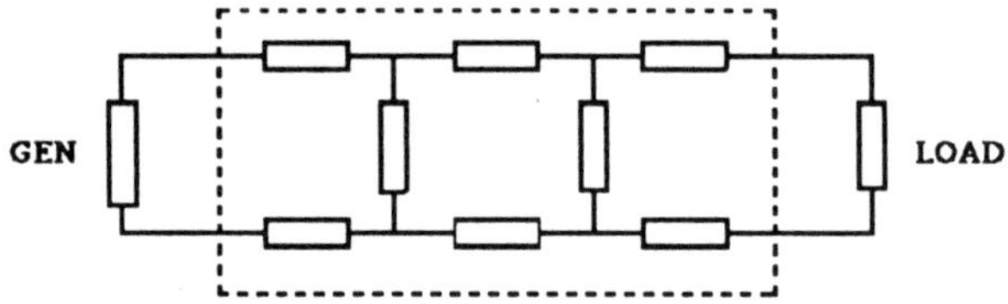

FIGURE 1.1 A GENERIC NETWORK The circuit elements inside the dashed line are the transmission medium and would be opaque using "Black Box" circuit analysis.

1.2 FUNDAMENTAL NETWORKS AND NETWORK LAWS

In circuit analysis, perhaps the most important problem is to find the current, the voltage, and the relationship between these two for any system, sub-system, or component. The following laws (rules) and theorems state the general relationships for dc and resistive circuits. They will be expanded in later chapters for ac and transient conditions using reactive components (capacitors and inductors) as well as resistors.

TM OHM'S LAW

In any resistor (an energy-dissipating circuit element), the current in the resistor is proportional to the voltage applied across it and inversely proportional to its magnitude resistance.

MM OHM'S LAW

$$I = \frac{E}{R} \text{ and, by algebra, } E = I\,R \qquad \text{(E 1.1)}$$

KIRCHHOFF'S CURRENT LAW

The point where two or more components are connected together in an electric circuit is termed a *node*. In a simple series circuit, the same current flows in all components. As a CONVENIENT CONVENTION, the current is assumed to leave the first component and enter the next one at the connecting node. If there were three components connected at a node, current might leave the first component and (dividing) enter the other two.

TM KIRCHHOFF'S CURRENT LAW

Kirchhoff's current law states that the ALGEBRAIC SUM of the currents entering and leaving a node is zero. Using different words, the sum of the currents entering a node must equal the sum of the currents leaving a node.

MM KIRCHHOFF'S CURRENT LAW

$$\sum_{k=1}^{n} i_k = 0 \qquad \text{(E 1.2)}$$

GM KIRCHHOFF'S CURRENT LAW

Branches, loops, and nodes are defined as follows using Figure 1.2:

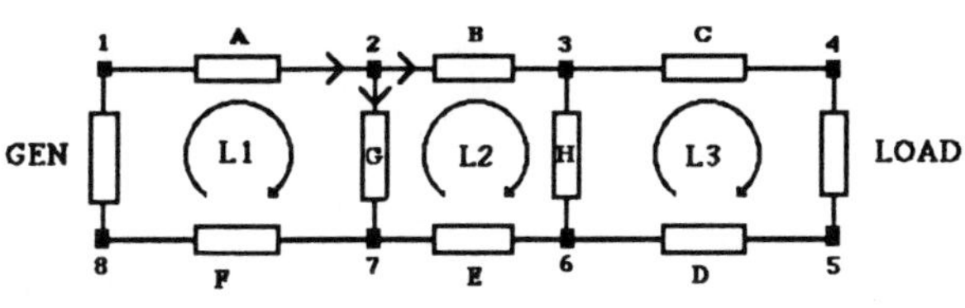

FIGURE 1.2 Branches, loops, and nodes.

An *element* is one single part of a circuit. An element is usually (but not necessarily) one component such as a resistor.

A *branch* is any portion of the circuit that has one or more circuit elements in series. A branch current is unique to the branch elements.

A *node* is the junction or connection of two or more elements.

A *loop* (sometimes called a *mesh*) is any portion of a circuit that closes back on itself. Loop currents are unique to that particular loop.

Figure 1.2 shows three loop currents. The circuit has eight nodes and three loops. Figure 1.2 is also used to illustrate how branch currents (at node 2) would be assigned.

KIRCHHOFF'S VOLTAGE LAW

TM KIRCHHOFF'S VOLTAGE LAW

The *algebraic sum* of the voltage rises and/or drops around a closed loop is zero.

MM KIRCHHOFF'S VOLTAGE LAW

$$\sum_{k=1}^{n} v_k = 0 \qquad \text{(E 1.3)}$$

As a convention, the voltage (arrow) rises in a power source – to + and drops in a resistor + to –. If the voltage drop across a resistor is entered as positive for a loop current or branch path, then voltage source (– to +) will be considered negative. Likewise, if the direction of the loop I_1 is considered positive in a *down* direction, and if a current I_2 is going *up*, then I_2 should be considered as going in a negative direction.

GM KIRCHHOFF'S VOLTAGE LAW

Branch voltage rises and drops:

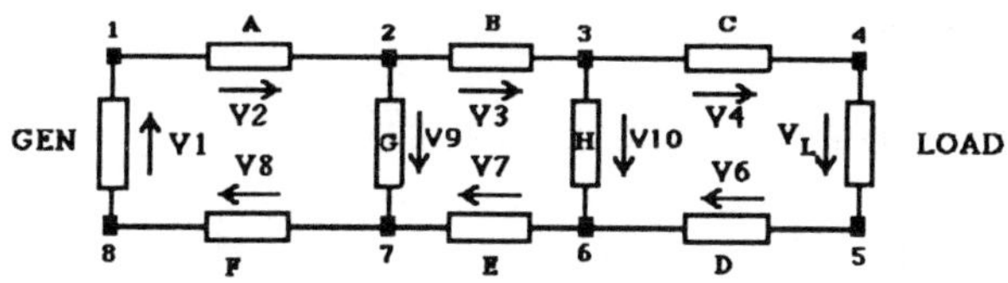

FIGURE 1.3 Element voltages.

The direction of the arrows is arbitrary but must be consistent with the above convention. The load voltage was called V_L for clarity. Most computer programs would call it V_5.

1.3 CIRCUIT EXAMPLES SHOWING LOOP AND BRANCH CURRENTS

The following examples show ONE WAY to configure loop and branch currents and to solve the equations that result from these configurations. It is sometimes difficult to know what and when to simplify. Note that (1) the current directions are arbitrary, but must ALWAYS follow some convention (such as given above); and (2) the resulting loop or branch equations can be solved using a number of mathematical techniques (algorithms). In fact, most of the chapters that follow will present different mathematical or computer techniques for solving network equations.

USING LOOP CURRENTS

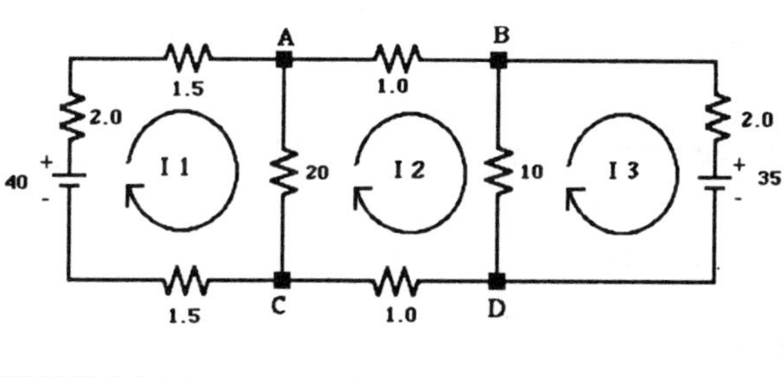

FIGURE 1.4 Loop current calculations.

For Loop 1: $(2 + 1.5 + 20 + 1.5)\ I_1 - 20\ I_2 - 40 = 0$
or $25\ I_1 - 20\ I_2 + 0 = 40$ (E 1.4)

For Loop 2: $-\ 20\ I_1 + (1 + 10 + 1 + 20)\ I_2 - 10\ I_3 = 0$
or $-20\ I_1 + 32\ I_2 - 10\ I_3 = 0$ (E 1.5)

For Loop 3: $0 - 10\ I_2 + 12\ I_3 + 35 = 0$ or $0 - 10\ I_2 + 12\ I_3 = -35$ (E 1.6)

For Loop 1: $I_1 = \dfrac{40 + 20I_2}{25}$ (E 1.7)

For Loop 3: $I_2 = \dfrac{10I_2 - 35}{12}$ and substituting these values
for I_1 and I_3, (E 1.8)

For Loop 2: $20\left(\dfrac{40 + 20I_2}{25}\right) + 32I_2 - 10\left(\dfrac{10I_2 - 35}{12}\right) = 0$ or
$7.67I_2 = 2.8\quad I_2 = \dfrac{2.8}{7.67} = 0.37$ amps. (E 1.9)

For Loop 1 $I_1 = \dfrac{40 + 20 * 0.365}{25} = 1.9$ amps. (E 1.10)

For Loop 3 $I_3 = \dfrac{(10 * 0.365) - 35}{12} = -2.6$ amps (E 1.11)

USING BRANCH CURRENTS (BRANCH PATHS)

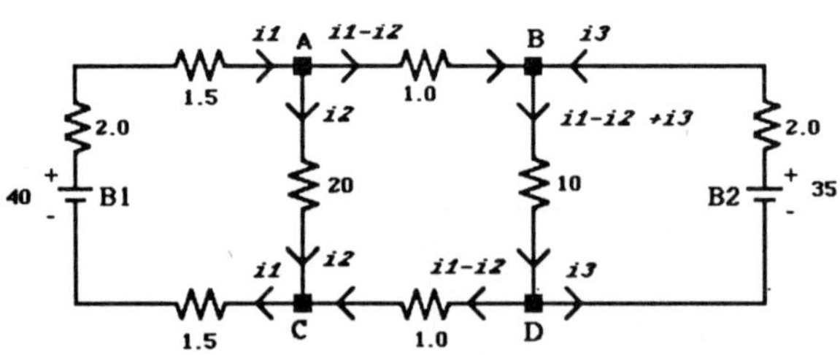

FIGURE 1.5 Branch current calculations.

For the B1AC Path: $2i1 + 1.5i1 + 20i2 + 1.5i1 - 40 = 0$
$5i1 + 20i2 = 40$ (E 1.12)

For the ABDC Path: $1(i1 - i2) + 10(i1 - i2 + i3) + 1(i1 - i2) - 20i2 = 0$
$12i1 - 32i2 + 10i3 = 0$ (E 1.13)

For the BB2D Path: $0 + 10(i1 - i2 + i3) + 2i3 - 35 = 0$
$10i1 - 10i2 + 12i3 = 35$ (E 1.14)

There are now three equations and three unknowns—*i1*, *i2*, and *i3*.

For the ABDC Path: $i3 = -1.2i1 + 3.2i2$ (E 1.15)

Substitute in E 1.13: $10i1 - 10i2 - 14.4i1 + 38.4i2 = 35$
$-4.4i1 + 28.4i2 = 35$ (E 1.16)

Solving E 1.11 and E 1.15: $i1 = 8 - 4i2$ (E 1.17)

From E 1.15: $35.2 - 17.6i2 + 28.4i2 = 35$

$$i2 = \frac{35 + 35.2}{17.6 + 28.4} = 1.5 \text{ amps.} \qquad \text{(E 1.18)}$$

From E 1.16 & 1.14: $i1 = 8 - (4 * 1.53) \approx 1.9$ amps. (E 1.19)

From E 1.14: $i3 = 2.6$ amps. (E 1.20)

1.4 SOLVING NETWORK EQUATIONS USING DETERMINANTS

In the preceding two examples, the simultaneous equations were solved by a process of substitution and addition or subtraction. As has been stated, there are many ways to solve simultaneous equations. For simple equations, an informal process of inspection and elimination is most obvious. However, for more than two equations, especially if the terms are complex (see Chapter 3), a more orderly process or algorithm is required. The method of determinants, which is presented in college algebra textbooks, is often used for both manual and computer analysis.

TM SOLVING NETWORK EQUATIONS USING DETERMINANTS

The use of determinants requires that the equations be presented in a special array or matrix in terms of the unknown, such as current, voltage, or impedance, and the coefficients for these unknowns. The determinant will contain both a numerator and a denominator. The denominator, or Δ, is simply an ordered array—a matrix—of the coefficients of the equations. The numerator does not contain the coefficients of the unknown in question, but does include the constant terms. Thus, if there are three equations with three unknowns, using determinants requires the composition and solution of four arrays—one for the common denominator and one for each unknown. Once the arrays are formed, the unknowns, such as currents, are found using a set of rules.

MM SOLVING NETWORK EQUATIONS USING DETERMINANTS

The example loop equations, E 1.4, 1.5, and 1.6, will be used to illustrate the determinant model.

The denominator matrix, composed ONLY of the coefficients, is common to all the three arrays and is constructed from the equations. Note that the constants are included so that the numerator arrays can be constructed. These constraints are NOT included in the denominator Δ.

Δ ARRAY

For	I_1	I_2	I_3	Constants
Loop 1:	25	–20	0	40
Loop 2:	–20	32	–10	0
Loop 3:	0	–10	12	–35

Each numerator matrix starts with the three constant terms on its left side and, in turn, adds the “other two” rows of coefficients for the unknowns as:

I_1 ARRAY

For I_1:	40	–20	0
	0	32	–10
	–35	–10	12

I_2 ARRAY

For I_2:	40	25	0
	0	–20	–10
	–35	0	12

I_3 ARRAY

For I_3:	40	25	–20
	0	–20	32
	–35	0	–10

Once these arrays are composed, they can be used to evaluate the three unknown currents. There are two sets of rules for using the matrix arrays—one set involves the common denominator determinant, sometimes called the delta or Δ, and one set for each numerator.

To construct the common denominator or Δ matrix:

1. Write out the coefficient terms in columns as above—do NOT use the constant terms.
2. Add n–1 more columns by repeating the first n–1 columns where n is the number of original columns (three in this example).
3. Starting with the top left coefficient, draw n (in this example, three) downward diagonal lines. The first diagonal will include the first coefficient in the first column, the second coefficient in the second column, etc. The second diagonal will include the first term in the

second column, the second term in the third column, etc. The third downward diagonal will start with the first term in the third column and go down in the same manner as the others. These diagonals are used to indicate the coefficients and their order in the process that will be used to evaluate the unknown currents.

4. In a similar manner, draw three upward diagonals starting with the bottom coefficient in the first column.
5. The value of any matrix (determinant) is the algebraic sum of the products of the coefficients on the downward diagonals less the algebraic sum of the coefficients of the upward diagonals.

To construct any numerator matrix:

1. For the first column, use the constant terms from the current equations—see Array I. Then add the coefficients for the currents NOT being evaluated. Thus, do NOT use the I_1 column in the I_1 matrix.
2. Add n–1 more columns by repeating the first n–1 columns where n is the number of original columns.
3. Starting with the top left coefficient, draw n (three) downward diagonal lines. The first diagonal will include the first coefficient in the first column, the second coefficient in the second column, etc. The second diagonal will include the first term in the second column, the second term in the third column, etc. The third downward diagonal will start with the first term in the third column and go down in the same manner as the others. These diagonals are used to indicate the coefficients, and their order in the process will be used to evaluate the unknown currents.
4. In a similar manner, draw three upward diagonals starting with the bottom coefficient in the first column.
5. The value of any matrix (determinant) is the algebraic sum of the products of the coefficients on the downward diagonals less the algebraic sum of the coefficients of the upward diagonals.

GM THE DENOMINATOR MATRIX MODEL

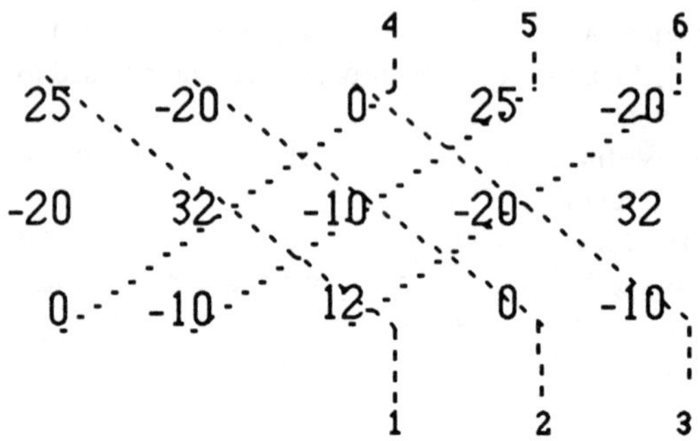

FIGURE 1.6 Using diagonals to evaluate the denominator in the example problem (see Array I).

MM TO EVALUATE THE DENOMINATOR MATRIX EQUATION

(25 ∗ 32 ∗ 12) + (–20 ∗ –10 ∗ 0) + (0 ∗ –20 ∗ –10) – (0 ∗ 32 ∗ 0) –
(–10 ∗ –10 ∗ 25) – (12 ∗ –20 ∗ –20)

9600 + 0 + 0 – 0 – 2500 – 4800

$\Delta = 2300$ the value for the denominator (E 1.21)

GM THE I_1 NUMERATOR MATRIX MODEL

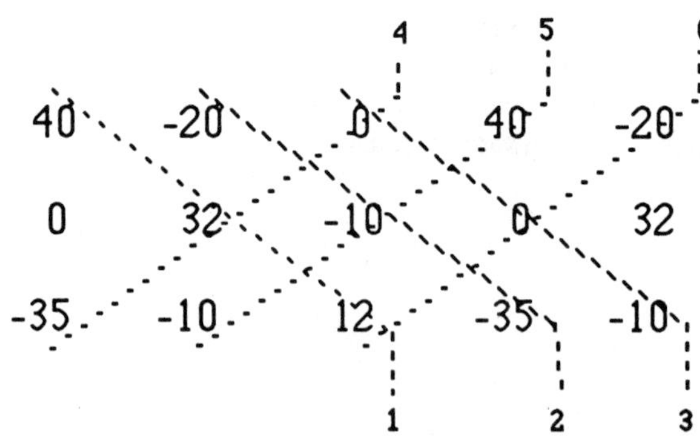

FIGURE 1.7 Using diagonals to evaluate the numerator for I_1 in the example problem (see Array II).

MM **TO EVALUATE THE I_1 MATRIX EQUATION**

$(40 * 32 * 12) + (-20 * -10 * -35) + (0 * 0 * -10) - (-35 * 32 * 0) -$
$(-10 * -10 * 40) - (12 * 0 * -20)$

$15360 + (-7000) + 0 - 0 - 4000 - 0$

$= 4360$ for the I_1 numerator (E 1.22)

MM **TO EVALUATE THE I_2 MATRIX EQUATION (SEE ARRAY III)**

$(40 * -20 * 12) + (25 * -10 * -35) + (0) - (-35 * -20 * 0) -$
$(0 * -10 * 40) - (12 * 0 * 25)$

$-9600 + 8750 + 0 - 0 - 0 - 0$

$= -850$ for the I_2 numerator (E 1.23)

MM **TO EVALUATE THE I_3 MATRIX EQUATION (SEE ARRAY IV)**

$(40 * -20 * -10) + (25 * 32 * -35) + (0) - (-35 * -20 * -20) - (0) - (0)$

$8000 + (-28000) + 0 - 14000 - 0 - 0$

$= -6000$ for the I_3 numerator (E 1.24)

MM **THE EVALUATION OF THE FINAL LOOP CURRENTS USING DETERMINANTS**

$$I_1 = \frac{\text{numerator}}{\text{common denominator}} = \frac{4360}{2300} = 1.9 \text{ amps.} \quad \text{(E 1.25)}$$

$$I_2 = \frac{\text{numerator}}{\text{common denominator}} = \frac{-850}{2300} = -0.37 \text{ amps.} \quad \text{(E 1.26)}$$

$$I_3 = \frac{\text{numerator}}{\text{common denominator}} = \frac{-6000}{2300} = -2.6 \text{ amps.} \quad \text{(E 1.27)}$$

1.5 NETWORK THEOREMS

TM THEVENIN'S THEOREM

Any active linear dc network with two output terminals can be replaced by an equivalent circuit consisting of a voltage source and a series resistor. Thevenin's theorem is usually said to obtain the "looking back" equivalent. The equivalent Thevenin voltage is the sum of all voltage sources, and the Thevenin resistance is the sum of all the series resistors. The symbol Σ is used to indicate "the sum of."

MM

$$E_{Thev.} = \Sigma\,(E_1 + E_2 + \cdots) \text{ and } R_{Thev.} = \Sigma\,(R_1 + R_2 + \cdots) \qquad \text{(E 1.28)}$$

GM THEVENIN'S THEOREM

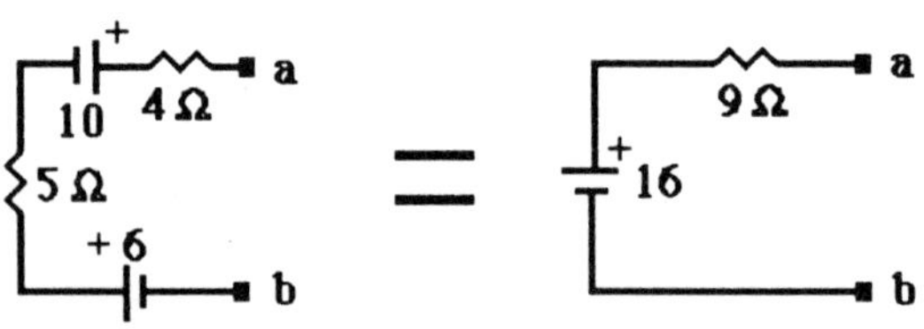

FIGURE 1.8 The Thevenin equivalent.

TM NORTON'S THEOREM

Any two-terminal linear dc network can be replaced by an equivalent circuit consisting of a current source and a parallel resistor. The Norton equivalent is not as obvious as the Thevenin equivalent because even simple circuits usually require several steps to convert to the Norton equivalent—see the next section on parallel resistors.

A circuit example of Norton's theorem showing the conversion steps required:

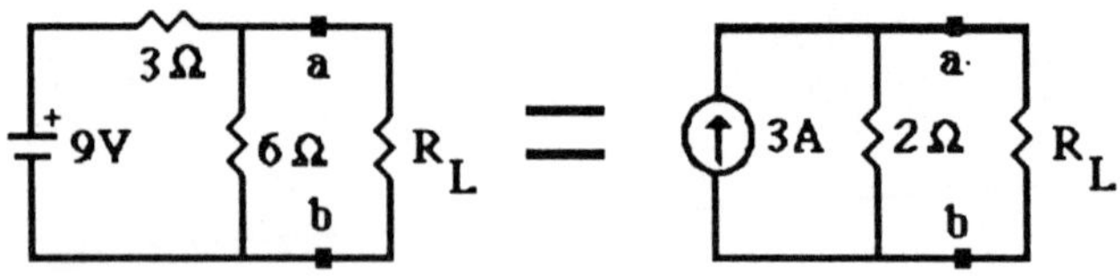

FIGURE 1.9 Norton's theorem.

MM NORTON'S THEOREM—A NORTON'S THEOREM EXAMPLE—SEE FIGURE 1.9

Step 1: Temporarily remove the load resistor R_L at points a and b in the left-hand portion of Figure 1.9 leaving an open circuit. Likewise, assume the 9V battery is a short circuit.

Step 2: R_N (R_{Norton}) will be the Thevenin equivalent of R_1 and R2

$$R_N = R_1 \parallel R2 = \frac{(3)(6)}{3+6} = \frac{18}{9} = 2\,\Omega \qquad \text{(E 1.29)}$$

Step 3: Add a temporary short circuit from a to b. The 9V battery voltage (without the short circuit) is now only across the 3 Ω resistor and its current $I_N = \frac{V}{R} = \frac{9}{3} = 3$ A.

The final Norton equivalent circuit is shown in the right-hand portion of Figure 1.9 with a current generator = 3 A and its shunt Norton equivalent resistance = 2 Ω.

RESISTORS IN SERIES

When two or more resistors are connected end to end, the total resistance is the sum of the individual resistances.

$$R_{Total} = R_1 + R_2 + R_3 + \ldots \qquad \text{(E 1.30)}$$

RESISTORS IN PARALLEL

When two or more resistors are connected in parallel, the total resistance is the reciprocal of the sum of the RECIPROCALS of all the resistances.

$$\frac{1}{R_{Total}} = \frac{1}{R_1} + \frac{1}{R_2} + \frac{1}{R_3} + \ldots \qquad \text{(E 1.31)}$$

Also, using algebra, the resistance of TWO resistors in parallel can be simplified to:

$$R_{Total} = \frac{R_1 R_2}{R_1 + R_2} \qquad \text{(E 1.32)}$$

CONDUCTORS (CONDUCTANCE) IN SERIES

In some circuit analysis techniques, it is convenient to sum the *conductance* G (the ability to conduct) of resistors rather than the resistance. Conductance is defined as the reciprocal of resistance and has the units of mho or siemen.

For conductors in series, $\frac{1}{G_{Total}} = \frac{1}{G_1} + \frac{1}{G_2} + \frac{1}{G_3} + \ldots$ (E 1.33)

CONDUCTORS IN PARALLEL

Conductors in parallel act as resistors in series, so:

$$G_1 = G_1 + G_2 + G_3 + \ldots \quad \text{(E 1.34)}$$

SUPERPOSITION THEOREM

When there are two or more voltage sources (batteries, power supplies, or ac generators) in a network, the voltage or current of any other one component is the result of energy coming from each source. The frequency of the source generator is not a limiting factor. For instance, an amplifier will have a dc power supply and probably an ac input. The principle of *superposition* will allow analysis by shorting the ac generators and finding the dc current and voltage and then shorting the dc generators and finding the ac current and voltage. It is assumed that the voltage sources represent their Thevenin equivalents.

TM

When a network has two or more voltage sources, the current and/or voltage of any network component is the algebraic sum of the current or voltage caused by each source acting independently.

GM

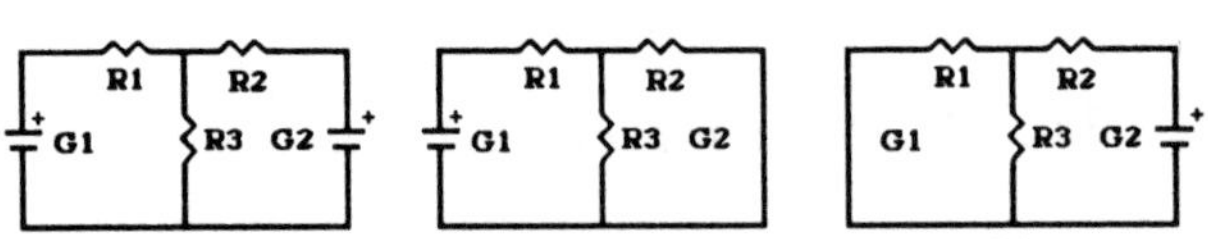

(a) (b) (c)

FIGURE 1.10 A SUPERPOSITION EXAMPLE (a) A network with two voltage sources (these sources could be dc or ac generators); (b) The network with G2 shorted for calculation; (c) The network with G1 shorted for calculation.

1.6 THE ANALYSIS OF TWO-PORT, FOUR-TERMINAL BLACK BOX NETWORKS

TM GENERAL TWO-PORT NETWORKS

The Fundamental Electric Circuit Model, shown in Section 1.1, can be presented as a black box with only input and output terminals—an input and output port. This conceptual model, called a two-port network, is used quite often in practical circuit analysis. Often transmission lines, wave guides, and antennas are analyzed using this model. Likewise, high-frequency components and amplifiers adapt well to this type of analysis as do most transistors. Modern test instruments such as network analyzers, many of which are computer controlled with data recording and readout, make this approach to analysis, test, and maintenance very attractive.

There are several ways to analyze two-port networks. The first three to be described, *impedance*, *admittance*, and *hybrid*, are fundamental and can be applied to many types of networks. Two other specialized analysis parameters will also be described. In the following formulas and definitions, the short-circuited parameters are assumed to be *ac* short circuits only.

GM GENERAL TWO-PORT NETWORKS

FIGURE 1.11 The general two-port black box network.

TM and MM IMPEDANCE (Z) PARAMETERS

This method of network analysis and these parameters determine the impedance (resistance and reactance), looking forward from the source into the network and looking back from the load.

$$E_1 = z_{11} I_1 - z_{12} I_2 \qquad \text{(E 1.35)}$$

$$E_2 = z_{12} I_1 - z_{12} I_2 \qquad \text{(E 1.36)}$$

Where:

z_{11} = the input impedance between terminals 1–1' with terminals 2–2' open circuited = E_1 / I_1 when $I_2 = 0$ (E 1.37)

z_{22} = the output (looking back) impedance between terminals 2–2' with 1–1' open circuited = $-E_2 / I_2$ when $I_1 = 0$ (E 1.38)

z_{12} = the open-circuit impedance with terminals 1–1' open circuited is called the forward-transfer impedance = $-E_1 / I_2$ when $I_1 = 0$ (E 1.39)

z_{21} = the open-circuit impedance with terminals 2–2' open circuited is called the reverse-transfer impedance = E_2 / I_1 when $I_2 = 0$ (E 1.40)

TM ADMITTANCE (Y) PARAMETERS

$$I_1 = y_{11} E_1 - y_{12} E_2 \qquad \text{(E 1.41)}$$

$$I_2 = y_{21} E_1 - y_{22} E_2 \qquad \text{(E 1.42)}$$

Where:

y_{11} = the input admittance between terminals 1–1' with terminals 2–2' short circuited = I_1 / E_1 when $E_2 = 0$ (E 1.43)

y_{22} = the output (looking back) admittance between terminals 2–2' with 1–1' short circuited = $-I_2 / E_2$ when $E_1 = 0$ (E 1.44)

y_{12}= the short-circuit, reverse-transfer admittance with terminals 1–1' short circuited = $-I_1 / E_2$ where $E_1 = 0$ (E 1.45)

y_{21} = the short-circuit, forward-transfer admittance with terminals 2–2' short circuited = I_2 / E_1 where $E_2 = 0$ (E 1.46)

TM HYBRID (H) PARAMETERS

$$E_1 = h_{11} I_1 + h_{12} E_2 \qquad \text{(E 1.47)}$$

$I_2 = h_{21} I_1 + h_{22} E_2$ (E 1.48)

Where:

h_{11} = the input impedance between terminals 1–1' with terminals 2–2' short circuited = E_1 / I_1 where $E_2 = 0$ (E 1.49)

h_{22} = the output (looking back) admittance between terminals 2–2' with 1–1' open circuited = I_1 / E_2 where $I_1 = 0$ (E 1.50)

h_{12} = the open-circuit, reverse-transfer voltage transfer with terminals 1–1' open = E_1 / E_2 where $I_1 = 0$ (E 1.51)

h_{21} = the open-circuit, forward-transfer current ratio with terminals 2–2' short circuited = I_2 / I_1 where $E_2 = 0$ (E 1.52)

TM IMAGE IMPEDANCE PARAMETERS

The concept of image impedance is a specialized circuit technology used for transmission lines and filters. It is included in this section for background and completeness—see Chapter 5. A two-port network, such as a transmission line, is operating with image impedance if the generator impedance equals the input impedance of the network and the load impedance equals the output impedance of the network. While it is true that the input and output impedances of most transmission-line type networks are usually alike, this is not necessary to fulfill the image-impedance criterion. The relationship between the image coefficients and other network coefficients is as follows:

$Z_{I1} = \sqrt{Z_{oc1} Z_{sc1}}$ (E 1.53)

$Z_{I2} = \sqrt{Z_{oc2} Z_{sc2}}$ (E 1.54)

Where:

Z_{I1} = the image impedance looking into the input terminals and is equal to the generator impedance

Z_{I2} = the image impedance looking into the output terminals and is equal to the load impedance

Z_{oc1} = the input impedance with the output terminals open circuited

Z_{sc1} = the input impedance with the output terminals short circuited

Z_{oc2} = the output impedance with the input terminals open circuited

Z_{sc2} = the output impedance with the input terminals short circuited

TM S OR SCATTERING PARAMETERS

S parameters represent specialized measurements of incident and reflected waves in transmission lines, special filters and high-frequency transistors. The measurement of these parameters is similar but not the same as the measurements described above. They are included for comparison and completeness.

$$a_1 = \frac{1}{2}\left(\frac{V_i}{\sqrt{Z_0}} + \sqrt{Z_0}\, I_i\right) \text{ a is the incident (input)wave} \quad \text{(E 1.55)}$$

$$b_1 = \frac{1}{2}\left(\frac{V_i}{\sqrt{Z_0}} - \sqrt{Z_0}\, I_i\right) \text{ b is the reflected wave} \quad \text{(E 1.56)}$$

$$S_{11} = \frac{b_1}{a_1} = \text{the input reflection coefficient when } Z_L = Z_0$$
and thus $a_2 = 0$ (E 1.57)

$$S_{22} = \frac{b_2}{a_2} = \text{the output reflection coefficient when } Z_L = Z_0$$
and thus $a_1 = 0$ (E 1.58)

$$S_{12} = \frac{b_2}{a_1} = \text{the reverse transmission gain when } Z_L = Z_0$$
and thus $a_1 = 0$ (E 1.59)

$$S_{21} = \frac{b_2}{a_1} = \text{the forward transmission gain } Z_L = Z_0$$
and thus $a_2 = 0$ (E 1.60)

GM

FIGURE 1.12 The S-parameter two-port configuration.

1.7 THE ROLE PLAYED BY COMPUTER CIRCUIT ANALYSIS

This chapter has presented the fundamental techniques and mathematics for analyzing component and network electric circuits. While the chapter concentrated on simple resistive models, the last section did, for completeness, extend the techniques to include resistance and reactance (complex) circuits and concluded with parameters used in quite sophisticated high-frequency transmission lines and transistors. Computers, in one form or another, can both detect and calculate all the dc, ac and high-frequency parameters. Computer analysis can be divided into three areas—test and maintenance, circuit design analysis, and mathematical analysis. For test and maintenance, many contemporary test instruments not only determine the circuit parameters but show them in a convenient display, record them for future use and do (programmed) processing based upon the user's needs. For design analysis, there are programs to calculate dc, ac, network, filter, high-frequency, transient, statistics, and worst-case parameters. Some of these programs are simple, but can still give the student a depth of understanding that was difficult to obtain in years past, while others have the complexity necessary to analyze elaborate systems. One of the most exciting aspects of modern circuit analysis is the development of very powerful, but usable, mathematical analysis programs. Even with a personal computer, such programs as *Mathematica* and *MathCAD* give the student, the circuit and system designer, and the test and maintenance technician the power to understand and do a better and more complete job than in the past.

2

Fundamental Models: Batteries, Sine Waves, and Other Nice Inputs

2.1 ABOUT THIS CHAPTER

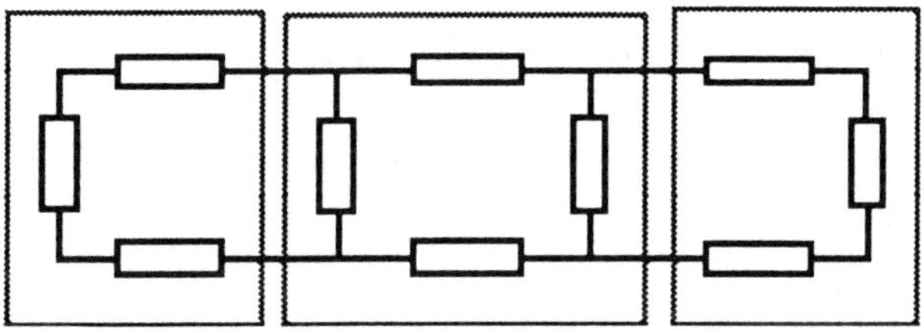

FIGURE 2.1 A generic transmission system.

The graphic in Figure 2.1 shows a fundamental model for a complete transmission system, which is an expansion of the generic network model shown in Figure 1.1. Note that both the generator and the load have been represented as multi-element circuits. Not only does this indicate that, in the general case, both the generator or the load may contain more than one component, it should also introduce the possibility of changing the definition of a generator (section) or a load (section) as the problem dictates.

Electronic generators, in general, produce two types of outputs. First, and most important, the (input) generator is the source information for the system. Second, special generators produce input test signals, which are used to simulate the original source information. These test signal generators are used to design and maintain electronic systems. Different systems require different test signals for analysis and maintenance. The following are "standard wave shapes" that can be combined to build almost any signal one would desire for test purposes.

2.2 CONSTANT OUTPUT GENERATORS

It is usually convenient (and often necessary) to separate an output generator into two parts. These two parts should completely define the generator and its output for both theoretically perfect conditions and for practical or "real-world" conditions.

TM THE CONSTANT VOLTAGE GENERATOR

An ideal *constant voltage generator* will produce a stated output voltage (or wave shape) regardless of the value of the load placed across the generator. Thus, whether in the no-load case (an open circuit), in the short-circuit case, or any intermediate load case, the theoretically ideal constant-voltage generator will exactly output its stated voltage.

TM THE THEVENIN EQUIVALENT VOLTAGE GENERATOR

For real-life, physical systems, the actual voltage generator can contain (1) a truly constant voltage generator and (2) a series resistor. This series resistor is either the internal resistor of the generator or a Thevenin equivalent of the combination of the internal resistance and one or more external resistances that are considered part of the voltage generator system (see Figure 1.8). An automobile battery is an example of a near-perfect dc voltage generator.

TM THE CONSTANT CURRENT GENERATOR

An ideal *constant current generator* will produce a stated output current regardless of the load. The current will be the stated value (including the wave shape) for loads of infinity (an open circuit), for loads of zero (short circuit), or for any intermediate load value.

TM THE NORTON EQUIVALENT GENERATOR

For real-life, physical systems, the actual current generator can contain (1) a constant current generator and (2) a shunt resistor. The value of the shunt resistor will be the Norton equivalent of the generator internal resistance and any other external resistors that are considered part of the current generating system (see Figure 1.9). The collector output of a transistor, especially an FET, is a near-perfect current generator.

TM THE SINE OR COSINE WAVE GENERATOR

A generator that produces an output vs. time, that can be described in the mathematical terms of a sine (angle) function or a rotating vector function, is termed a sine (or cosine) wave generator. BY DEFINITION, the output of a sine-wave generator is a (mathematically) *complex function.* It inherently should contain both amplitude and phase information.

MM THE SINE OR COSINE WAVE GENERATOR

$$v(t) = V_{max} \sin(\omega t + \phi) \text{ or } V_{max} \cos(\omega t + \phi) \qquad \text{(E 2.1)}$$

Where $\omega = 2\pi f = 2\pi/T$ (radians/second) and ϕ is the phase shift ± from zero° or zero radians.

GM THE SINE OR COSINE WAVE GENERATOR

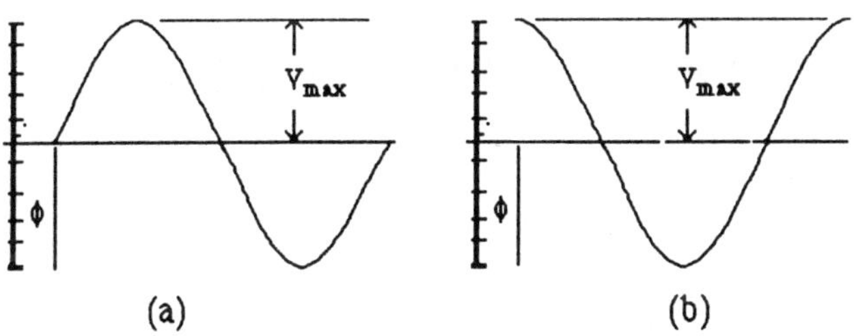

FIGURE 2.2 Sine (a) and cosine (b) wave shape models. The ϕ is used to indicate any phase shift from the vertical reference axis.

The sine/cosine-wave generator is, by far, the most common wave-form generator, partially because its wave shape is very easy to simulate and analyze and also because the sine wave is the building block for many motions found in nature. Unfortunately, especially in the past, many engineers have been woefully remiss by specifying networks with just their simple sine-wave response and have not given proper emphasis to their transient (and phase) response.

TM THE SQUARE-WAVE GENERATOR

A square-wave generator, again usually, but not necessarily, a Thevenin Equivalent voltage generator, produces a pulse output with a fifty percent duty cycle (half on and half off). A square wave can be described as in Figure 2.3 or it can be described mathematically by using a combination (a series) of sine and cosine waveforms. Such a combination of sine and cosine components is termed a *Fourier Series.*

GM THE SQUARE-WAVE GENERATOR

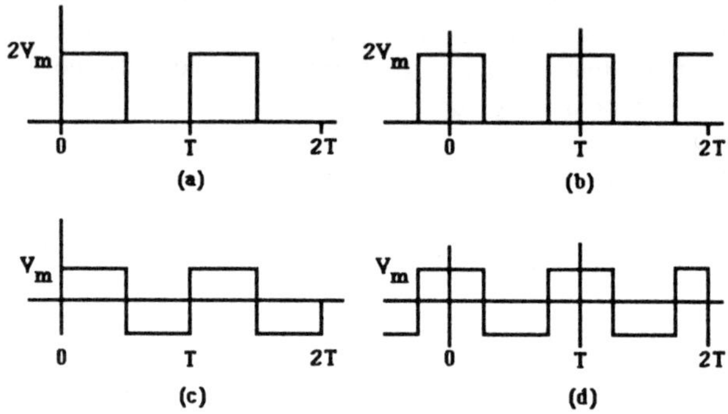

FIGURE 2.3 Model square waves with different axis shifts (see Chapter 10).

MM THE FOURIER SERIES EQUATION FOR EACH SQUARE WAVE IN FIGURE 2.3

$$\text{(a) } V_m + \frac{4}{\pi} V_m \left(\sin \omega t + \frac{1}{3} \sin 3\omega t + \frac{1}{5} \sin 5\omega t + \cdots\cdots\right) \quad \text{(E 2.1)}$$

$$\text{(b) } V_m + \frac{4}{\pi} V_m \left(\cos \omega t + \frac{1}{3} \cos 3\omega t + \frac{1}{5} \cos 5\omega t + \cdots\cdots\right) \quad \text{(E 2.2)}$$

$$\text{(c) } \frac{4}{\pi} V_m + \left(\sin \omega t + \frac{1}{3} \sin 3\omega t + \frac{1}{5} \text{sine } 5\omega t + \cdots\cdots\right) \quad \text{(E 2.3)}$$

$$\text{(d) } \frac{4}{\pi} V_m + \left(\cos \omega t + \frac{1}{3} \cos 3\omega t + \frac{1}{5} \cos 5\omega t + \cdots\cdots\right) \quad \text{(E 2.4)}$$

Historically, a square wave has several advantages. It is symmetrical and easy to synchronize to ocilloscopes. Likewise, it gives SOME insight into the characteristics of a network by indicating both its transient and frequency response. The vertical output's rise and fall times roughly indicate the circuit's transient characteristics and the horizontal rise or fall (droop) roughly indicates its frequency response. It should again be emphasized that the square wave is far from the ideal test signal. The prudent investigator only uses square waves as qualitative indicators and relies on more sophisticated test signals for quantitative information.

TYPICAL NETWORK OUTPUTS USING A SQUARE-WAVE INPUT TEST SIGNAL

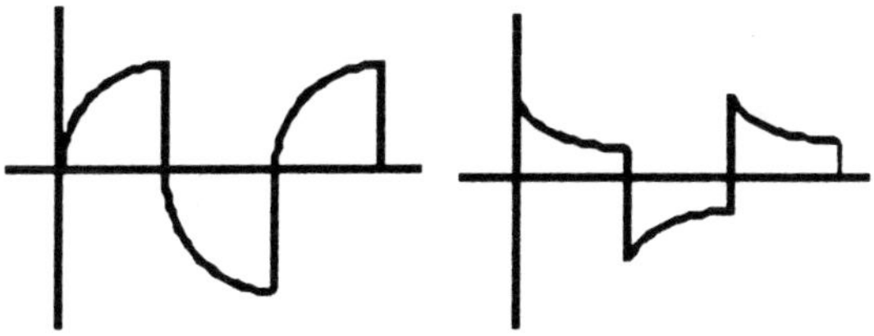

FIGURES 2.4–2.5 The outputs of networks with (left) slow rise time (poor high - frequency response) and (right) fast rise time (poor low-frequency response).

TM THE SAWTOOTH GENERATOR

A voltage (or current) that starts at (usually) zero and rises linearly with time to a preset value and then returns to the zero with an infinite fall time and then repeats itself is termed a sawtooth. The sawtooth wave form has both theoretical—see Chapter 10—and practical use. It is especially useful in checking the linearity of analog-to-digital converters.

MM THE SAWTOOTH GENERATOR

For a sawtooth starting at zero voltage (or current), its Fourier Series can be written:

$$v(t) = \frac{A}{\pi}\left(\frac{\pi}{2} - \sin\theta - \frac{1}{2}\sin 2\theta - \frac{1}{3}\sin 3\theta - \ldots\right) \qquad \text{(E 2.5)}$$

GM THE SAWTOOTH GENERATOR

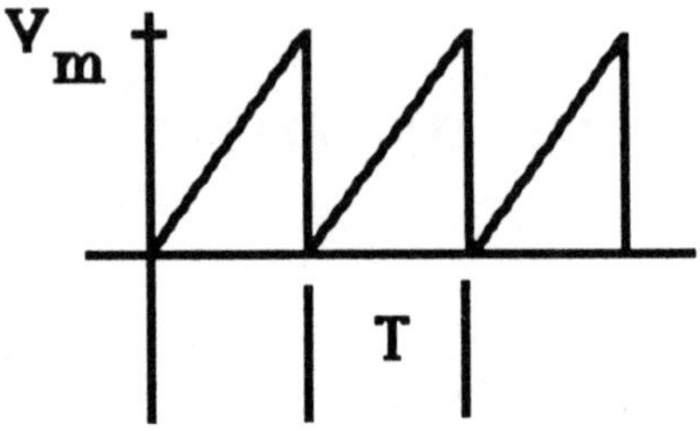

FIGURE 2.6 The model sawtooth wave.

TM THE UNIT STEP GENERATOR

A voltage (or current) that is "turned on" at a designated time, has an infinite (or very fast, compared to any response of the network under test) rise time, and is left on "forever" is termed a unit step. While there are some limitations to the generation of this wave form in practical circuits, it can be simulated reasonably well and has great appeal in theoretical design and analysis. For test purposes, the unit step can be simulated by a very, very long period pulse or square wave (see Chapter 10).

MM THE UNIT STEP GENERATOR

$V_m = 0, t < 0$

$V_m = 1, t > 0$

GM THE UNIT STEP GENERATOR

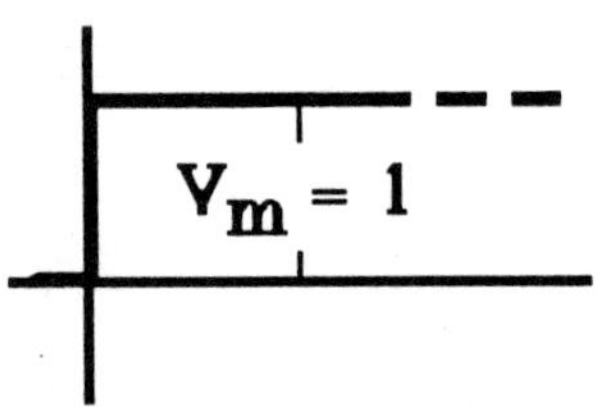

FIGURE 2.7 The model step function.

TM THE UNIT PULSE GENERATOR

A pulse of area K that has zero width but is infinitely high is termed an impulse. If the height times the width (the area) is DEFINED as one or unity, it is termed a *unit impulse*. This unit impulse or unit pulse has great theoretical significance in advanced circuit analysis—see Chapter 11 on the Laplace transform. The concepts of the unit pulse can be used in the laboratory if a pulse is used that is very, very narrow compared to the response in analog circuits or the sampling period in digital circuits. The unit impulse function is also known as the Dirac delta (δ) function.

MM THE UNIT PULSE GENERATOR

$$K = \int_{-\infty}^{\infty} K\,\delta(t)\,dt \quad K\delta(t) = K * 1 = K \text{ when } t = 0 \text{ and } = 0 \text{ when } t \neq 0 \qquad \text{(E 2.6)}$$

GM THE UNIT PULSE GENERATOR

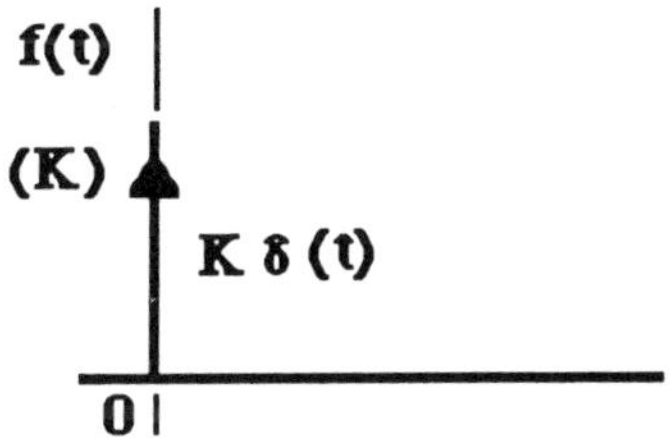

FIGURE 2.8 The graphic model for the unit impulse is the arrow.

3

RC and RL Circuits: An Introduction to AC and Transient Circuit Functions

3.1 ABOUT THIS CHAPTER

The combination of RL and RC elements form the essential building blocks for coupling and filter circuits. The study of L's and C's involves a new dimension in circuit use and analysis—the concept of energy storage. If a circuit contains only resistors, there is no storage of input energy (voltage or current). However, if circuits contain L's and/or C's, the input energy will be temporarily stored in these L and C elements. This ability of L's and C's to store energy, even temporarily, makes these circuits act "funny" when compared to simple resistive circuits. RL and RC combinations produce rapid, unexpected signal changes (transients) that require very special circuit analysis techniques.

3.2 INDUCTORS IN SERIES AND PARALLEL

In addition to resistors, inductors and capacitors are used in most circuits and networks. (Unwanted inductors and capacitors also may exist as stray or unwanted elements around circuit leads, etc.) The simple Thevenin equivalent of these combinations can be visualized in terms of the way it reacts to the input (signal) energy.

TM VOLTAGE, CURRENT, AND ONE INDUCTOR

When a current passes through an inductor (in very simple terms, a coil of wire), it produces a magnetic field or magnetic lines of force. Theoretically, once a current has established a magnetic field, that field will continue unless it is disturbed by more changes in the current. Likewise, this field tends to resist more changes in the current. The field (lines of force) is proportional to the current multiplied by a value called inductance, or L. If the current through the inductance is changing, it produces a voltage across the inductor proportional to the rate-of-change of the current.

MM VOLTAGE, CURRENT, AND ONE INDUCTOR

$$\text{L (in henrys)} = \frac{\text{flux lines}}{\text{current producing flux lines}}\,10^{-8} \qquad \text{(E 3.1)}$$

$$\text{v (across the inductance)} = L\frac{di}{dt}$$

$$\left(\frac{di}{dt} \text{ is the rate-of-change of the current—see Chap. 12.}\right) \qquad \text{(E 3.2)}$$

Note that $v = L\frac{di}{dt}$ (or $v = -L\frac{di}{dt}$) is similar to the Ohm's law model v = RI.

The minus sign indicates that the polarity of the current is opposite from the polarity of the voltage it produces. An inferred plus sign, or a minus sign, will indicate a particular direction of the current flow—see Figure 3.1.

GM VOLTAGE, CURRENT, AND ONE INDUCTOR

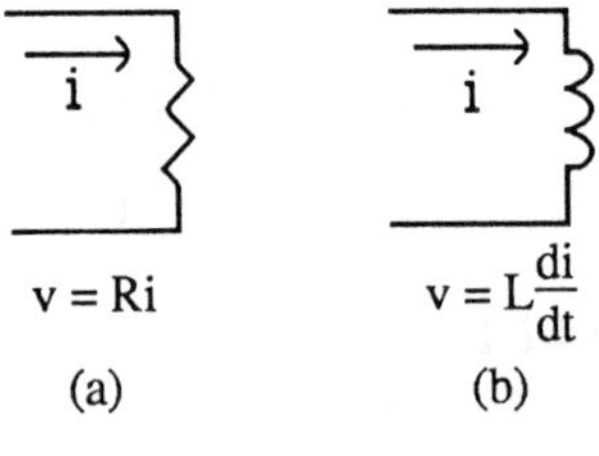

FIGURE 3.1 Voltage and current relationships for a resistor and a single inductor.

TM VOLTAGE, CURRENT, AND INDUCTORS IN SERIES

When two or more inductors are connected in series, the total inductance is equal to the sum of the inductors. Since the current is the same in all components of a series circuit, the voltage across these components must be the sum of the individual voltages.

MM VOLTAGE, CURRENT, AND INDUCTORS IN SERIES

$$v_{total} = v_{L1} + v_{L2} + v_{L3} + \cdots$$

$$= L_1 \frac{di}{dt} + L_2 \frac{di}{dt} + L_3 \frac{di}{dt} + \cdots = (L_1 + L_2 + L_3 \cdots) \frac{di}{dt}$$

$$= L_{series} \frac{di}{dt} \text{ so,}$$

$L_{series} = L_1 + L_2 + L_3 \cdots$ Thus, inductances in series add. (E 3.3)

GM VOLTAGE, CURRENT, AND INDUCTORS IN SERIES

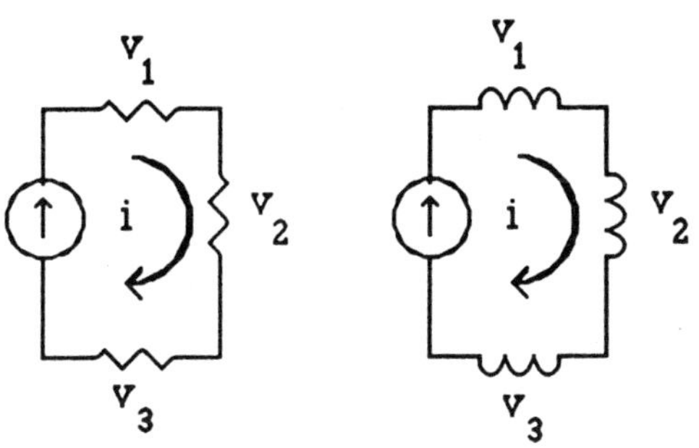

FIGURE 3.2 Resistors and inductors in series.

TM VOLTAGE, CURRENT, AND INDUCTORS IN PARALLEL

When inductors are connected in parallel, the voltage across them is the same and the current in and out of the nodes that connect them is divided among them. This current division is somewhat analogous to the current division in parallel resistors. When inductors are connected in parallel, the reciprocals of their inductances add.

MM VOLTAGE, CURRENT, AND INDUCTORS IN PARALLEL

In section 3.2, it was stated that the voltage across an inductor is $v = L\frac{di}{dt}$.

This equation can be used to find the current in each inductor. Since,

$v_{total} = v_{L1} = v_{L2} = v_{L3} = \cdots$ for one inductor $i = \frac{1}{L}\int v\,dt$.

($\int v\,dt$ is the integral of the changing voltage—see Chapter 14.)

$i_{total} = \frac{1}{L_1}\int v\,dt + \frac{1}{L_2}\int v\,dt + \frac{1}{L_3}\int v\,dt$ for three inductors in parallel and

since $i = i_1 + i_2 + i_3$

$$i_{total} = \left(\frac{1}{L_1} + \frac{1}{L_2} + \frac{1}{L_3}\right)\int v\,dt$$

$$\frac{1}{L_{total}} = \frac{1}{L_1} + \frac{1}{L_2} + \frac{1}{L_3} \cdots \qquad \text{(E 3.4)}$$

GM VOLTAGE, CURRENT, AND INDUCTORS IN PARALLEL

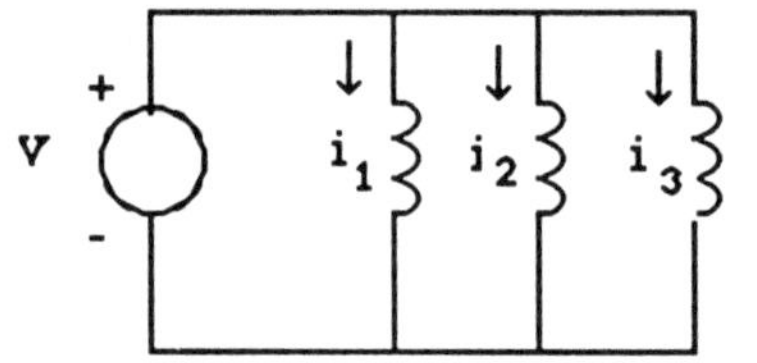

FIGURE 3.3 Inductors in parallel.

3.3 CAPACITORS IN SERIES AND PARALLEL

TM VOLTAGE, CURRENT, AND ONE CAPACITOR

A capacitor is made up of two conducting plates separated by an insulator (air, mica, paper, etc.). When a voltage is applied to the terminals of a capacitor, there is a charge (a collection of electrons) stored on the conductors. If the voltage is changing, the charge will change. However, if the input voltage is suddenly removed, the charge will remain, if it is not disturbed, as stored energy. The ratio of the stored charge (q) to the input voltage (v) is defined as the capacitance (C) of the capacitor.

The unit of charge (q) is the coulomb. A coulomb is approximately 1.6×10^{19} electrons (a large handfull of electrons). Since electric current is defined as "the flow of electrons," or the flow of charge, a current of one ampere is defined as a flow of one coulomb per second.

MM VOLTAGE, CURRENT, AND ONE CAPACITOR

$$C = \frac{q}{v}$$

$$v = \frac{1}{C} q$$

$$\frac{dv}{dt} = \frac{1}{C}\frac{dq}{dt} \quad \text{and} \quad i(t) = \frac{dq(t)}{dt} \quad \text{so,}$$

$$i = C\frac{dv}{dt} \quad \text{and} \qquad \text{(E 3.5)}$$

$$v = \frac{1}{C}\int i\,dt + k \quad \text{(k is a } \textit{constant of integration}\text{)} \qquad \text{(E 3.6)}$$

TM VOLTAGE, CURRENT, AND CAPACITORS IN SERIES

When two or more capacitors are connected in series, the reciprocals of their capacitances add. This is similar to resistors in parallel where the voltages across each element add to equal the input voltage.

MM VOLTAGE, CURRENT, AND CAPACITORS IN SERIES

$$v_{total} = v_{C1} + v_{C2} + v_{C3} + \cdots$$

$$v_{total} = \frac{1}{C_1}\int i\,dt + \frac{1}{C_2}\int i\,dt + \frac{1}{C_3}\int i\,dt + \cdots$$

$$= \left(\frac{1}{C_1} + \frac{1}{C_2} + \frac{1}{C_3} + \cdots\right)\int i\,dt = \frac{1}{C_{ser}}\int i\,dt$$

$$\frac{1}{C_{total}} = \frac{1}{C_1} + \frac{1}{C_2} + \frac{1}{C_3} + \cdots \qquad \text{(E 3.7)}$$

GM VOLTAGE, CURRENT, AND CAPACITORS IN SERIES

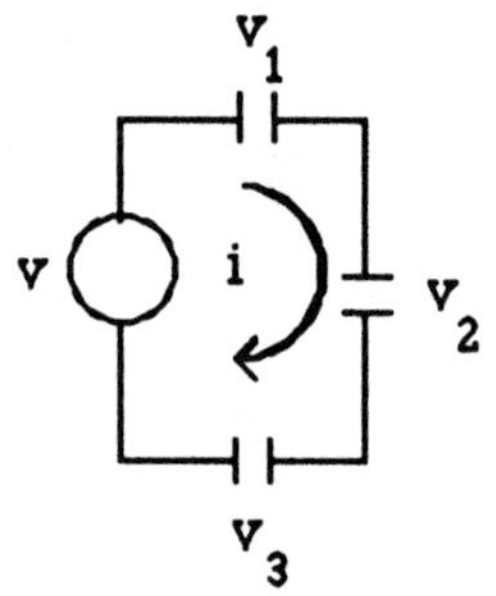

FIGURE 3.4 Capacitors in series.

TM VOLTAGE, CURRENT, AND CAPACITORS IN PARALLEL

When two or more capacitors are connected in parallel, their total capacitance is the sum of the individual units.

MM VOLTAGE, CURRENT, AND CAPACITORS IN PARALLEL

$$v_{total} = v_{C1} = i_{C2} = i_{C3} = \cdots$$

$$i_{total} = i_{C1} + i_{C2} + i_{C3} + \cdots$$

$$i_{total} = C_1 \frac{dv}{dt} + C_2 \frac{dv}{dt} + C_3 \frac{dv}{dt} + \cdots = (C_1 + C_2 + C_3)\frac{dv}{dt}$$

$$C_{total} = C_1 + C_2 + C_3 \qquad \text{(E 3.8)}$$

GM VOLTAGE, CURRENT, AND CAPACITORS IN PARALLEL

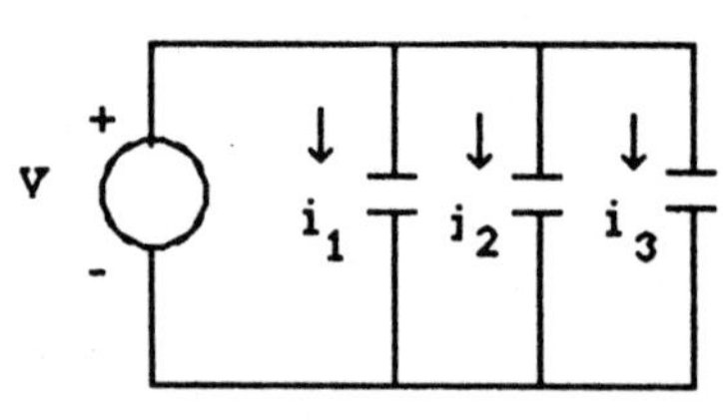

FIGURE 3.5 Capacitors in parallel.

3.4 THE REACTANCE OF INDUCTORS

TM PERFECT INDUCTORS AND CAPACITORS

An ideal or "perfect" inductor, one with no resistance, would be a complete short circuit to direct current. Likewise, a "perfect" capacitor would be a complete open circuit to a direct current. In the practical, real world, inductors usually can be considered to have a very low resistance and capacitors, while not perfect, usually possess a very high shunt or leakage resistance. A capacitor usually approaches the ideal much more closely then an inductor.

TM THE AC CHARACTERISTICS OF A SINGLE INDUCTOR

Inductors are usually modeled using an ideal or "perfect" inductor (L) and a series resistor (R). The L is the measured (or calculated) value of the inductance and the R the measured (or calculated) resistance value of the wire in the inductor coil.

Since the inductor will store energy in a magnetic field and the rate of this storage will depend upon the rate of change of the current in the inductor, the ac characteristics of the inductor will be a function of the rate of change of the ac series current. The *reactance* (X_L) of an inductor is thus given in terms of this changing current (frequency).

MM THE AC CHARACTERISTICS OF A SINGLE INDUCTOR

For a sine wave of frequency f, the reactance of an inductor is:

$$X_L = \omega L = 2\pi f L \qquad \text{(E 3.9)}$$

GM THE AC CHARACTERISTICS OF A SINGLE INDUCTOR

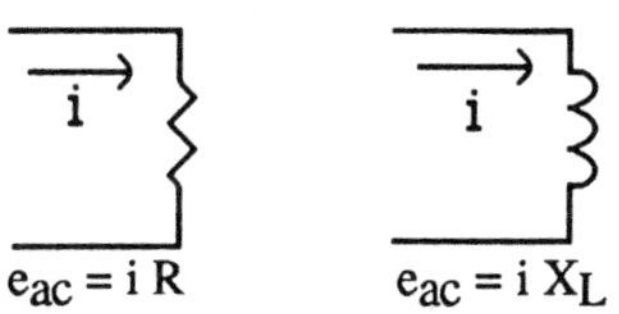

FIGURE 3.6 The ac voltage across a resistor and an inductor.

3.5 THE AC CHARACTERISTICS OF AN INDUCTOR AND A RESISTOR IN SERIES

TM THE AC CHARACTERISTICS OF AN INDUCTOR AND A RESISTOR IN SERIES

Ohm's law for a resistor and an inductor in series states that the combined voltage across the RL combination is equal to the series current times the sum of the resistance and the reactance. However, because of the storage characteristics of the inductor, the reactance of the inductor does not add directly to the resistance but requires a "square root of squares function." In addition, when the reactive component (X_L) is in series with a resistor, there is a phase angle (ϕ) between the voltage across the resistor and the inductor.

When a single resistor and a single inductor are connected in series across an ac voltage generator, the current follows Ohm's law for ac circuits. As with two resistors in series across a voltage, there is a single loop current (i). It is assumed that the representation of the inductor (X_L) is ideal and that the real-world resistance of the inductor is combined with the separate, series resistor.

MM THE AC CHARACTERISTICS OF AN INDUCTOR AND A RESISTOR IN SERIES

$$v_{ac} = \frac{I_{ac}}{Z_{ac}} = \frac{I_{ac}}{\sqrt{R^2 + X_L^2}}$$

$\theta = \tan^{-1}\frac{X_L}{R}$ = the phase angle between the voltage across the resistor and the voltage across the inductor and resistor in series. (E 3.10)

$Z = R + jX_L$ j is a mathematical operator that places the X_L 90° from R—see Chapter 4. (E 3.11)

GM **THE AC CHARACTERISTICS OF AN INDUCTOR AND A RESISTOR IN SERIES**

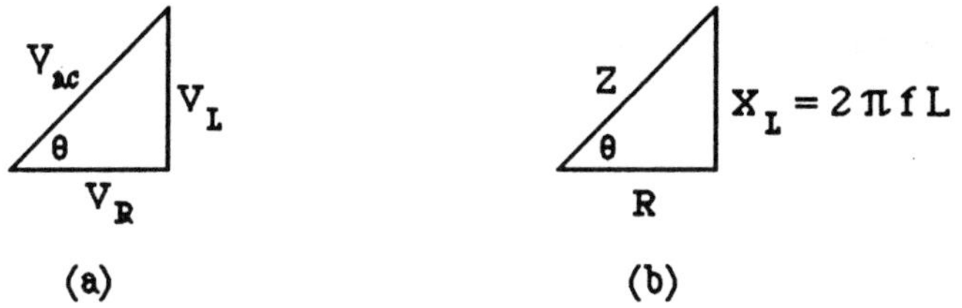

FIGURE 3.7 The phase relations of voltage, resistance, reactance, and impedance in a series RL circuit.

3.6 THE TRANSIENT CHARACTERISTICS OF AN INDUCTOR AND RESISTOR IN SERIES

TM **THE TRANSIENT CHARACTERISTICS OF AN INDUCTOR AND RESISTOR IN SERIES**

One simple approach (and there are many) to the transient response of a circuit is to state its response to a step or (long) pulse function. The output is usually stated as a function of time. A given circuit will have a particular time response, which is often given in units of the circuit time constant. For a series LR circuit, one time constant is equal to the value of the inductor, in henrys, divided by the value of the resistance, in ohms.

Special Note: Forcing a current through an inductor is directly analogous to pushing a very heavy car. It is very difficult to start the car moving but, as it starts to roll, pushing it faster becomes easier. Likewise, once it is moving, it is very difficult to stop it.

MM **THE TRANSIENT CHARACTERISTICS OF AN INDUCTOR AND RESISTOR IN SERIES**

In a series RL circuit with a generator producing a step or wide voltage pulse of E:

$$i_L = I_m(1 - e^{-t/\tau}) = \frac{E}{R}(1 - e^{-t/(L/R)}) = \frac{E}{R}(1 - \frac{1}{e^{t/(L/R)}}) \qquad \text{(E 3.12)}$$

Where t = the time, τ = *time constant* (L/R) and e is the base of natural logarithms = 2.7183 · · ·

GM THE TRANSIENT CHARACTERISTICS OF AN INDUCTOR AND RESISTOR IN SERIES

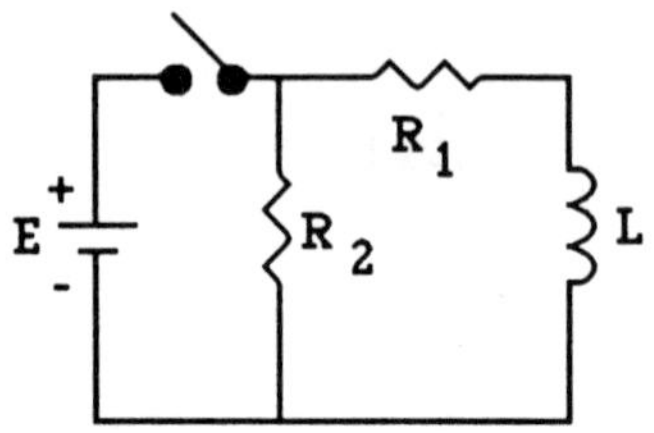

FIGURE 3.8 A circuit model showing the transient current rise in the inductor (when the switch is closed) and the transient flux decay (when the switch is opened).

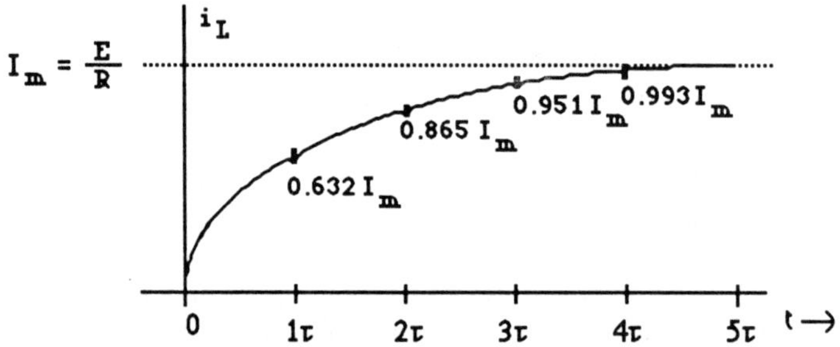

FIGURE 3.9 The transient current rise, with the switch closed, in Figure 3.8 ($\tau_1 = L/R_1$).

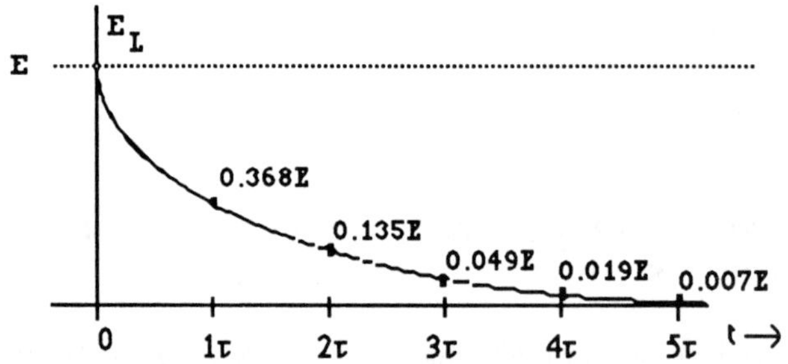

FIGURE 3.10 The current decay, and thus the inductor voltage decay, with the switch opened, in Figure 3.8 ($\tau = \tau_2 = L/(R_1+R_2)$). (Note that the rise or fall is completed in about $5\tau_2$.)

3.7 THE AC CHARACTERISTICS OF AN INDUCTOR AND A RESISTOR IN PARALLEL

TM THE AC CHARACTERISTICS OF AN INDUCTOR AND A RESISTOR IN PARALLEL

When a resistor and an ideal inductor are connected in parallel, the current divides at the node(s) combining them. Since, as with model parallel resistors, the same voltage appears across both the R and L, the current in the resistor will be in phase with this voltage and the current in the inductor will lag the voltage. The derivation of the impedance equation is quite straightforward if the relationship between these parallel currents and the terminal voltage is used. If one starts from the model impedance formula $Z_{tot} = \frac{Z_R Z_L}{Z_R + Z_L}$, the algebra becomes a bit messy (see Chapter 4).

Special Note: In an RL circuit, the voltage (e) will lead the current (i). In an RC circuit, the current will lead the voltage. As a memory aid, remember "eLi the iCe man."

MM THE AC CHARACTERISTICS OF AN INDUCTOR AND A RESISTOR IN PARALLEL

e = the voltage across R and L

$i_R = \frac{e}{R}$ the current in the resistor

$i_L = \frac{e}{X_L}$ the current in the inductor

$i_{tot} = i_R + i_L$

$Z = \frac{e}{i_{tot}}$ or, using the impedance derivation,

$$|Z| = \frac{RX_L}{\sqrt{R^2 + X_L^2}}$$ for a resistor and an inductor in parallel.

(|Z| indicates "absolute" value.) (E 3.13)

$$\theta = \tan^{-1} \frac{X_L}{R} \qquad \text{(E 3.14)}$$

GM THE AC CHARACTERISTICS OF AN INDUCTOR AND A RESISTOR IN PARALLEL

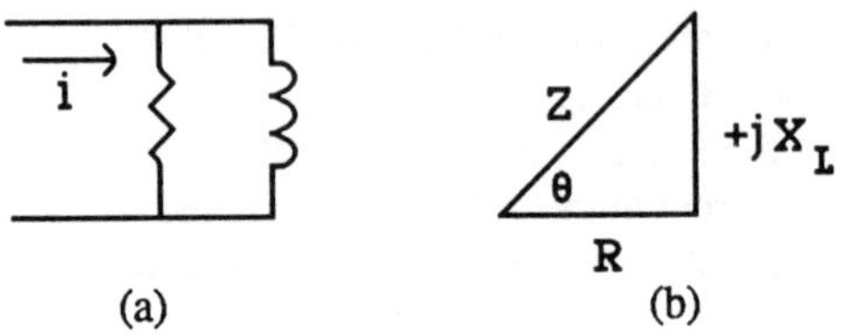

FIGURE 3.11 The phase relations of resistance and an inductance in a parallel RL circuit.

3.8 THE TRANSIENT CHARACTERISTICS OF AN INDUCTOR AND RESISTOR IN PARALLEL

TM THE TRANSIENT CHARACTERISTICS OF AN INDUCTOR AND RESISTOR IN PARALLEL

The transient current and voltage relationships in a parallel LR circuit are similar to those of a series LR circuit, but the derivation of these relationships is more difficult. The series circuit has only one mesh (loop) whereas the parallel circuit has two. These more complex transient circuit relationships and their derivations will be covered in later chapters on transient response.

MM THE TRANSIENT CHARACTERISTICS OF AN INDUCTOR AND RESISTOR IN PARALLEL

For a step function (or the first part of a very long square wave):

$$i_L = I_m = \frac{E}{R}\left(1 - e^{-R/L}\right) \qquad \text{(E 3.15)}$$

$$v_L = L\frac{di}{dt} = E\, e^{-R/L} \qquad \text{(E 3.16)}$$

GM THE TRANSIENT CHARACTERISTICS OF AN INDUCTOR AND RESISTOR IN PARALLEL

The graphs of the current in the inductance are similar to those in Figure 3.7 except the fall time constant for the parallel case is just R/L.

3.9 THE AC CHARACTERISTICS OF A SINGLE CAPACITOR

TM THE AC CHARACTERISTICS OF A SINGLE CAPACITOR

A capacitor C can usually be modeled as if it were an ideal element. The very low series (lead) resistance and the very high (leakage) shunt resistance are only significant at very high frequencies.

A capacitor will store energy in an electrostatic field. This storage will make the current and voltage effects different than those in a resistor. The capacitive reactance (X_C) is thus given in terms of this changing current (frequency).

MM THE AC CHARACTERISTICS OF A SINGLE CAPACITOR

For a sine wave, the reactance of a capacitor at a frequency f is:

$$X_C = \frac{1}{\omega C} = \frac{1}{2\pi f C} \qquad \text{(E 3.17)}$$

GM THE AC CHARACTERISTICS OF A SINGLE CAPACITOR

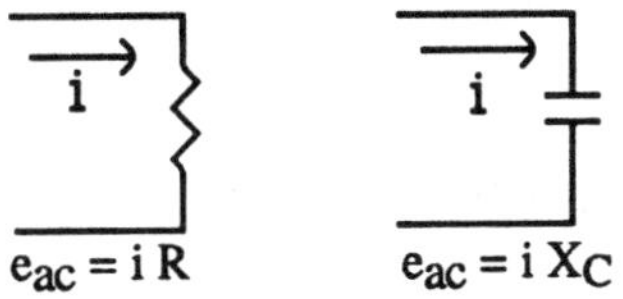

FIGURE 3.12 The ac current in a resistor and a capacitor.

3.10 THE AC CHARACTERISTICS OF A CAPACITOR AND A RESISTOR IN SERIES

TM THE AC CHARACTERISTICS OF A CAPACITOR AND A RESISTOR IN SERIES

Ohm's law for a resistor and a capacitor in series states that the combined voltage across the RC combination is equal to the series current times the sum of the resistance and the reactance. However, because of the storage characteristics of the capacitor, the reactance does NOT add directly to the resistance but requires a "square root of squares function." In addition, when the reactive component (X_C) is in series with a resistor, there is a phase angle (ϕ) between the voltage across the resistor and the voltage across the inductor.

When a single resistor and a single capacitor are connected in series across an ac voltage generator, the current follows Ohm's law for ac circuits. As with two resistors in series across a voltage, there is a single loop current (i). It is assumed that the representation of the capacitor (C) is ideal and that any real-world lead resistance is combined with the separate, series resistor.

MM THE AC CHARACTERISTICS OF A CAPACITOR AND A RESISTOR IN SERIES

$$V_{ac} = \frac{I_{ac}}{Z_{ac}} = \frac{I_{ac}}{\sqrt{R^2 + X_C^2}} \qquad \text{(E 3.18)}$$

$\theta = -\tan^{-1}\frac{X_C}{R}$ = the phase angle between the voltage across the resistor and the voltage across the capacitor. (E 3.19)

$Z = R - jX_C$ $-j$ is a mathematical operator that places the X_C -90° from R—see Chapter 4. (E 3.20)

GM **THE AC CHARACTERISTICS OF A CAPACITOR AND A RESISTOR IN SERIES**

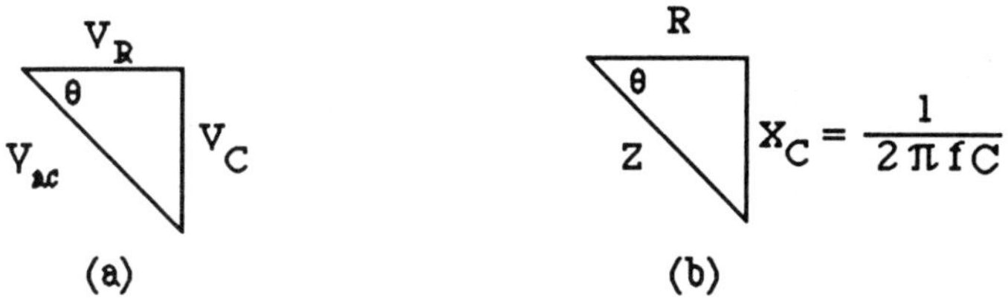

FIGURE 3.13 The phase relations of voltage, resistance, reactance, and impedance in a series RC circuit.

3.11 THE TRANSIENT CHARACTERISTICS OF A CAPACITOR AND RESISTOR IN SERIES

TM **THE TRANSIENT CHARACTERISTICS OF A CAPACITOR AND RESISTOR IN SERIES**

One simple approach (and there are many) to the transient response of a circuit is to state its response to a step or (long) pulse function. The output is usually stated as a function of time.

Often the time is given in units of the circuit time constant. For a series RC circuit, the time constant is equal to the value of the capacitor, in farads, multiplied by the value of the resistance, in ohms.

MM **THE TRANSIENT CHARACTERISTICS OF A CAPACITOR AND RESISTOR IN SERIES**

In a series RC circuit with a generator producing a step or wide voltage pulse of amplitude E:

$$i_C = \frac{E}{R} e^{-t/(RC)} = \frac{E}{R}\frac{1}{e^{t/(RC)}} = \frac{E}{R}\frac{1}{e^{t/\tau}} \qquad \text{(E 3.21)}$$

Where t = the time and τ = time constant (RC).

GM THE TRANSIENT CHARACTERISTICS OF A CAPACITOR AND RESISTOR IN SERIES

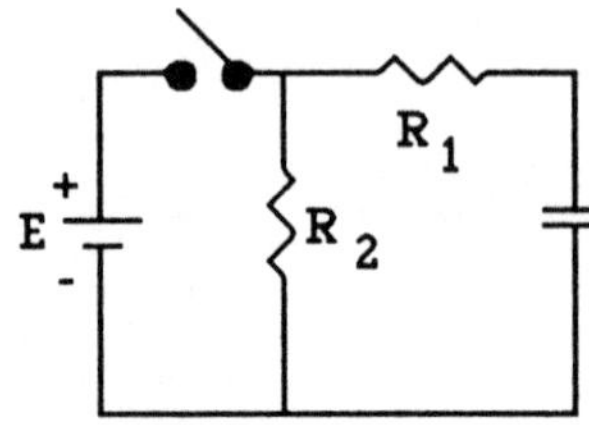

FIGURE 3.14 A circuit model showing the transient charging current in the capacitor when the switch is closed.

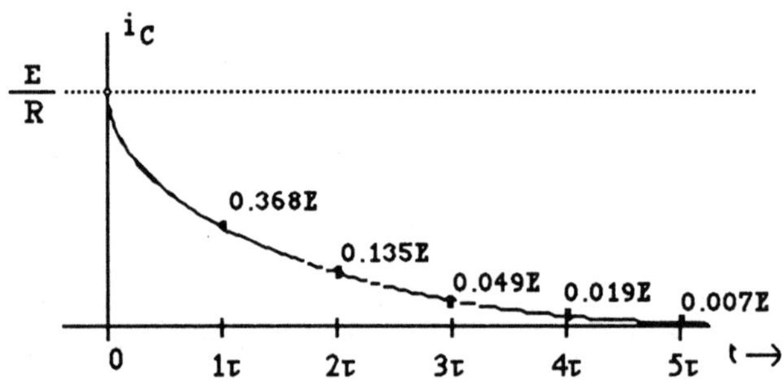

FIGURE 3.15 The charging current in a series RC circuit as a function of time (constants) ($\tau = R_1C$).

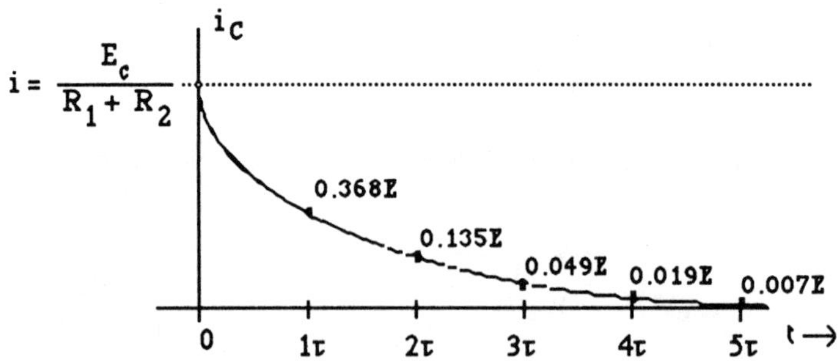

FIGURE 3.16 The decay current when the switch in Figure 3.14 is opened (τ = ((R_1 + R_2)C).

3.12 THE AC CHARACTERISTICS OF A CAPACITOR AND A RESISTOR IN PARALLEL

TM THE AC CHARACTERISTICS OF A CAPACITOR AND A RESISTOR IN PARALLEL

When a resistor and an ideal capacitor are connected in parallel, the current divides at the node(s) combining them. Since, as with model parallel resistors, the same voltage appears across both the R and C, the current in the resistor will be in phase with this voltage and the current in the capacitor will lead the voltage. The derivation of the impedance equation is quite straightforward if the relationship between these parallel currents and the terminal voltage is used. If one starts from the model impedance formula $Z_{tot} = \frac{Z_R |Z_C|}{Z_R + |Z_C|}$, the algebra becomes a bit messy.

Special Note: Using the vertical lines in the term $|Z_C|$ means it is the "absolute value" and disregards the –j in the capacitive reactance (see Chapter 4).

MM THE AC CHARACTERISTICS OF A CAPACITOR AND A RESISTOR IN PARALLEL

e the voltage across R and L is assumed.

$i_R = \frac{e}{R}$ the current in the resistor

$i_L = \frac{e}{X_C}$ the current in the inductor

$i_{tot} = i_R + i_L$

$Z = \frac{e}{i_{tot}}$ or, using the impedance derivation,

$$|Z| = \frac{RX_C}{\sqrt{R^2 + X_C^2}} \quad \text{for a resistor and an inductor in parallel.} \qquad \text{(E 3.22)}$$

$$\theta = -\tan^{-1} \frac{X_C}{R} \qquad \text{(E 3.23)}$$

GM THE AC CHARACTERISTICS OF A CAPACITOR AND A RESISTOR IN PARALLEL

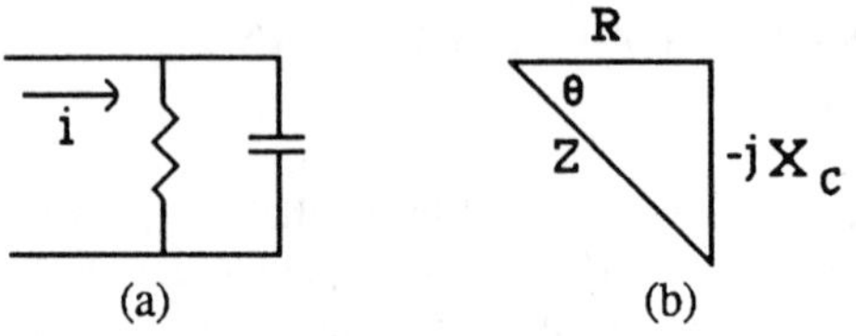

FIGURE 3.17 The phase relations of resistance and a capacitor in a parallel RC circuit.

3.13 THE TRANSIENT CHARACTERISTICS OF A CAPACITOR AND RESISTOR IN PARALLEL

TM THE TRANSIENT CHARACTERISTICS OF A CAPACITOR AND RESISTOR IN PARALLEL

The transient current and voltage relationships in a parallel RC circuit are similar to those of a series RC circuit, but the derivation of these relationships is more difficult. The series circuit has only one mesh (loop) whereas the parallel circuit has two. These more complete transient circuit relationships and their derivations will be covered in later chapters on transient response.

MM THE TRANSIENT CHARACTERISTICS OF A CAPACITOR AND RESISTOR IN PARALLEL

For a step function (or the first part of a very long square wave):

$$v_C = E\left(1 - e^{-RC}\right) \qquad \text{(E 3.24)}$$

$$i_C = \frac{E}{R} e^{-RC} \qquad \text{(E 3.25)}$$

GM THE TRANSIENT CHARACTERISTICS OF A CAPACITOR AND RESISTOR IN PARALLEL

The shape of the graphs is similar to those in Figures 3.15 and 3.16.

3.14 THE VOLTAGE DIVIDER RULE FOR SERIES CIRCUITS

TM THE VOLTAGE DIVIDER RULE

In a series circuit, the voltage across any one circuit element can be obtained by using a simple ratio. The voltage across a single resistor or reactor is the value of the input voltage times the ratio of the value of the given resistance or reactance divided by the total series circuit impedance. If only dc voltages and resistors are involved, all of the numbers are real. If the generator is ac and the circuit elements include reactances, the computations must take phase angles into account.

MM THE VOLTAGE DIVIDER RULE

$$V_x = V_{in} \frac{R_x}{R_1 + R_2 + \cdots R_n} = V_{in} \frac{R_x}{R_{tot}}$$

For series resistors and a dc input (E 3.26)

$$v_x = v_{in} \frac{X_x}{X_1 + X_2 + \cdots X_n} = v_{in} \frac{X_x}{Z_{tot}}$$

For series reactances and an ac input (E 3.27)

GM THE VOLTAGE DIVIDER RULE

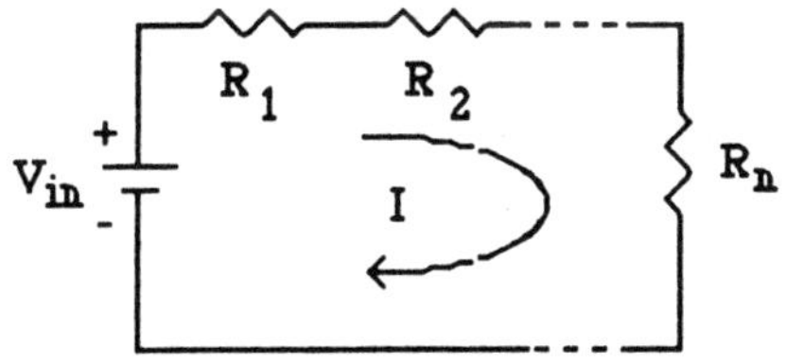

FIGURE 3.18 A series resistive circuit to illustrate the voltage divider rule.

3.15 THE CURRENT DIVIDER RULE FOR PARALLEL CIRCUITS

TM THE CURRENT DIVIDER RULE FOR TWO PARALLEL ELEMENTS

The use of a simple current divider rule is restricted to circuits with only two parallel paths. Although there is a "universal" current divider rule for n parallel branch elements, the calculations required are just about as complicated as those to determine the loop or mesh currents.

In a simple parallel circuit with only two branches or two elements, the current in any one circuit branch can be obtained by using a simple ratio. The current through one resistor or reactor (branch) is the value of the total input current times the ratio of the "other" resistor or reactor branch divided by the total of the two impedances. In simple two-element parallel circuits, each branch is an individual circuit element. However, this current divider rule also applies if each branch has more than one circuit element.

MM THE CURRENT DIVIDER RULE FOR TWO PARALLEL ELEMENTS

$$I_1 = I_{in}\frac{R_2}{R_1 + R_2} \text{ or } I_2 = I_{in}\frac{R_1}{R_1 + R_2} \text{ for resistive circuits} \quad \text{(E 3.28)}$$

Note the use of the numerator R_2 to find I_1 or R_1 to find I_2.

$$i_1 = i_{in}\frac{Z_2}{Z_1 + Z_2} \text{ or } i_2 = i_{in}\frac{Z_1}{Z_1 + Z_2} \text{ for reactive circuits} \quad \text{(E 3.29)}$$

GM THE CURRENT DIVIDER RULE FOR TWO PARALLEL ELEMENTS

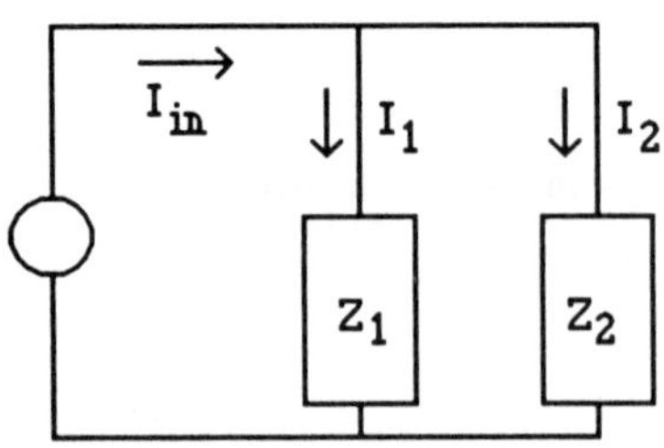

FIGURE 3.19 A two-branch parallel circuit illustrating the current divider rule.

4

Vectors, Phasors, and Complex Numbers: An Introduction to Poles and Zeros

4.1 ABOUT THIS CHAPTER

There are many ways to present or describe a principle or an idea, including words and pictures. The more complex the concept, the more it is necessary to be creative in attempting to successfully transfer the concept from one person or group to another. The graphic capability of contemporary computers has changed the way most printed information is presented. The newspaper *USA Today* revolutionized the print media with its use of color graphics. It is interesting that the use of graphics to help describe principles in electronics has been used for many, many years.

4.2 PHASOR NOTATION WITH SINUSOIDAL AND COSINUSOIDAL WAVE FORMS

TM VECTORS AND PHASORS

"Arrowed" lines, called *vectors* or *phasors*, are one of the fundamental uses of graphics in electronics. The length of the line can be used to indicate, among many things, the magnitude of a quantity such as voltage or current. Likewise, the position of the arrow can indicate a second quantity (dimension) such as relative motion (frequency) or relative position (phase). However, to understand the use of these vectors requires an understanding of their derivation. The following four fundamental principles are always assumed with vectors or phasors:

Vector lines usually indicate TWO dimensions. In electronics, vectors often represent values such as amplitude and frequency, amplitude and phase, or resistance and reactance. When these vectors are used to indicate the phase difference between the (alternating) sine-wave current and voltage or between two circuit voltages, the vectors are termed phasors.

The vector quantity, representing TWO dimensions, is derived from the geometric form of the right triangle using the Pythagorean theorem. (In a right triangle, the square of the length of the vector—the hypotenuse—is equal to the sum of the squares of the other two sides.) The general equation for an abc right triangle is usually written $c^2 = a^2 + b^2$ or $c = \sqrt{a^2 + b^2}$.

When a vector is decomposed into its component parts, i.e., the other two sides of a right triangle, one side—called the opposite side— is usually represented in its "imaginary" form as jb. Although the operator j (i is used by mathematicians) is termed imaginary, in NO WAY does this infer that the j or the quantity jb is fanciful! Here, imaginary is used to denote visualization in more than one dimension. j is defined as $\sqrt{-1}$ and $j^2 = 1$. An expression such as a + jb is termed a complex number.

As a further aid to understanding the physical operation or the mathematics of a circuit, any complex vector form, such as a + jb (this is also called the rectangular form), is really derived from the fundamental trigonometric form $c(\cos\theta + j\sin\theta)$. The vector form can also be converted to the polar form $c\angle\theta$ or to the exponential form $c\,e^{j\theta}$.

MM VECTORS AND PHASORS

A graph of a sine wave is a plot of the ratio of two of the three sides of a right triangle. This ratio is called sin x (or sin θ). For an abc right triangle:

$$\sin x \text{ (or } \sin\theta) = \frac{\text{opposite}}{\text{hypotenuse}} = \frac{b}{c} \text{ and } \cos\theta = \frac{\text{adjacent}}{\text{hypotenuse}} = \frac{a}{c} \quad \text{(E 4.1)}$$

A plot of the sin of a series of angles, such a 0°, 15°, 30°, ... vs. that angle itself (usually in radians), produces the familiar sine wave graph. Likewise, when using a circle with its 360° or 2π radians and a rotating radius vector, similar information about amplitude and angle or frequency can be displayed.

For a right triangle:

$$(\text{hypotenuse})^2 = (\text{adjacent})^2 + (\text{opposite})^2 \text{ or } c^2 = a^2 + b^2 \quad \text{(E 4.2)}$$

GM VECTORS AND PHASORS

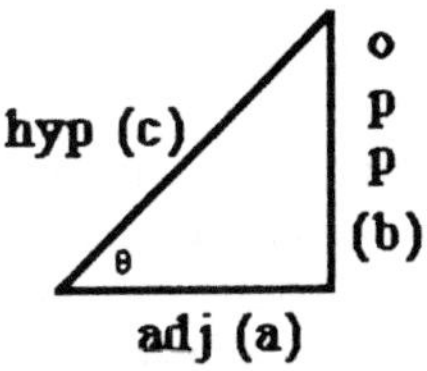

FIGURE 4.1 A prototype abc right triangle.

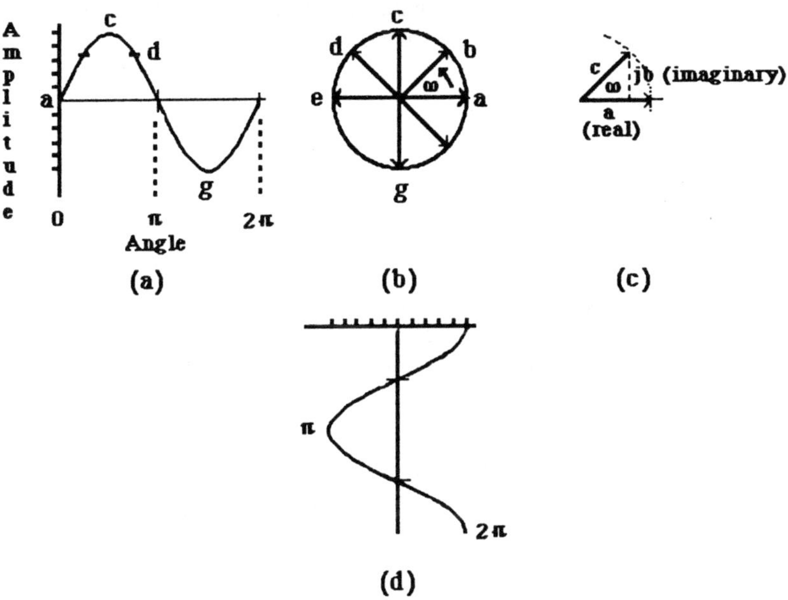

FIGURE 4.2 Vector representation of a sine wave form: (a) A sine wave with amplitude on the vertical axis and angle (or time) on the horizontal axes; (b) A circle diagram showing a radius vector rotating counterclockwise at a rate of ω radians per second. The vectors at positions a, b, c, d, e, f, and g are shown at about 0.0π (the reference position), 0.25π, 0.5π, 0.75π, 1.0π, 1.5π, and 1.75π radians; (c) A segment, removed from the (b) circle, that shows a fundamental abc triangle inferred by the vector at position b. Note the addition of the dashed side jb which could, of course, be constructed at any vector position in the circle (b) (see Section 4.2 above); (d) A cosine wave with amplitude on the horizontal (x) axis and angle on the vertical (y) axis.

4.3 COMPLEX NUMBERS

TM THE VECTOR CIRCLE

For convenience and clarity, the circle in Figure 4.2 (b) can be divided into four separate zones, or quadrants. This is something like splitting a clock face in half both horizontally and vertically. Going counterclockwise, the two top quadrants are defined as one and two, and the lower quadrants are three and four. A prototype abc right triangle can be constructed in any of these four quadrants. When abc triangles are constructed using j, the sign of j is + in the two upper quadrants and – in the lower two quadrants. Likewise, the sign of a is + in the first and fourth quadrants and – in the second and third.

MM THE VECTOR CIRCLE

The general equations will be

$$c^2 = a^2 \pm b^2 \text{ or } c = \sqrt{a^2 \pm b^2} \text{ (the Pythagorean theorem)} \qquad \text{(E 4.3)}$$

where the choice of + or – will depend upon the quadrant of the triangle. Using the operator j, the abc triangle will be ±a ± jb where the sign of a or b will depend upon the quadrant. BY DEFINITION, the radius vector c (the hypotenuse) is always positive. Also, some books use the arrow above or $\overrightarrow{c} = \sqrt{a^2 \pm b^2}$ to indicate a vector quantity.

GM THE VECTOR CIRCLE

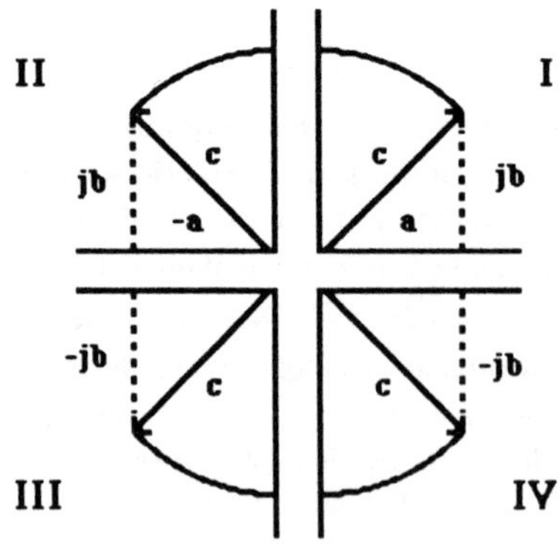

FIGURE 4.3 The prototype a + jb triangle shown in four quadrants.

4.4 THE J OPERATOR

TM THE J OPERATOR AND VECTOR ADDITION

A vector represents a (complex) number. This number is the result of two other numbers. Likewise, two vectors can add to give another third resultant vector which, of course, is really the final result of the four numbers that were combined to form the original two vectors. Vectors can be added mathematically or graphically. In each case, special rules for the vector addition must be applied. Specific rules apply to the operator j, including:

(a) By definition, j is equal to the square root of minus one and, therefore, j squared will be equal to minus one. j cubed equals j squared (-1) times the square root of minus one, which equals minus the square root of minus one (or –j).
(b) j quantities add and subtract arithmetically.
(c) j quantities multiply and divide arithmetically.

MM THE J OPERATOR AND VECTOR ADDITION

(a) $j = \sqrt{-1}$, $j^2 = -1$, $j^3 = j^2 \cdot j = -1 \cdot \sqrt{-1} = -\sqrt{-1} = -j$,
$j^4 = j^2 \cdot j^2 = -1 \cdot -1 = 1$

(b) $j6 + j4 = j10$ and $j10 - j6 = j4$

(c) $j2 \cdot j6 = j^2 \cdot 12 = -12$ but $j12 \div j4 = 3$ since $j \div j = 1$ (E 4.4)

GM THE J OPERATOR AND VECTOR ADDITION

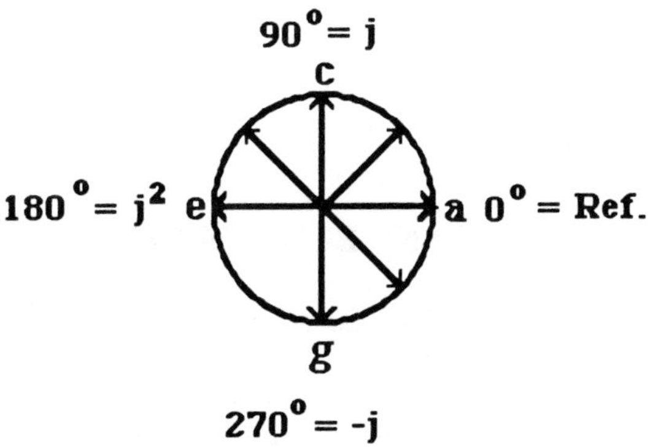

FIGURE 4.4 A vector circle showing the use of the operator j (see Figure 4.2).

4.5 THE POLAR FORM

The fundamental abc triangle and its trigonometric form c ($\cos\theta + j\sin\theta$) can be modified into a more convenient form for electronic circuits using the operator j. This j or rectangular form can again be modified into a simpler polar form.

TM THE POLAR FORM

The polar form of the right triangle uses the vector c (the hypotenuse) and the base angle θ. With these two numbers and fundamental trigonometry, the other two sides and the other angle can be derived.

MM THE POLAR FORM

$c\angle\theta$ is derived from $a + jb$ and $5\angle 53.1$ is derived from $3 + j4$ (E 4.5)

GM THE POLAR FORM

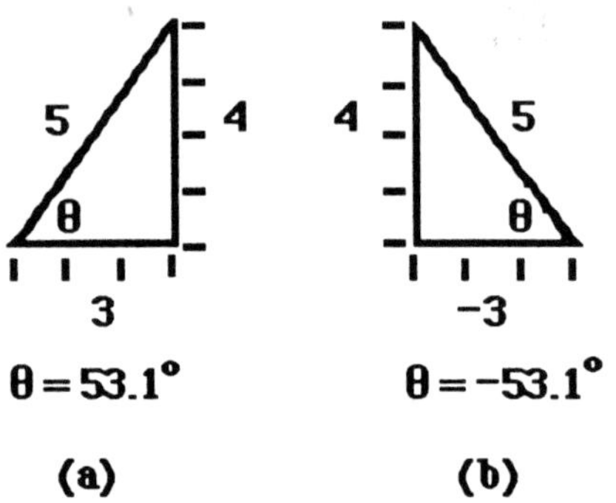

FIGURE 4.5 The abc triangles derived from polar equations: (a) The polar equation $5\angle 53.1$; (b) The polar equation $5\angle{-53.1}$.

4.6 THE EXPONENTIAL FORM AND EULER'S IDENTITY

Complex vector numbers can also be put in a form based upon sines and cosines and their derivation from number series.

Through the centuries, mathematicians have found it fascinating and very intellectually rewarding to reduce numbers to a related *series* of lesser numbers. Many numbers or constants used in circuit analysis can be broken into a series of numbers, including Pi, e, and the trigonometric functions such as sine and cosine.

TM EULER'S IDENTITY

Starting with the trigonometric form c (cos θ + j sin θ), Euler's identity theorem states that, using the appropriate number series, the trigonometric form is equal to a simpler exponential form $e^{j\theta}$.

MM EULER'S IDENTITY

$$\cos\theta = 1 - \frac{\theta^2}{2!} + \frac{\theta^4}{4!} - \frac{\theta^6}{6!} \cdots \text{ the cosine series expansion} \quad \text{(E 4.6)}$$

$$\sin\theta = \theta - \frac{\theta^3}{3!} + \frac{\theta^5}{5!} - \frac{\theta^7}{7!} + \cdots \text{ the sine series expansion} \quad \text{(E 4.7)}$$

$$j\sin\theta = j\theta - j\frac{\theta^3}{3!} + j\frac{\theta^5}{5!} - j\frac{\theta^7}{7!} + j\cdots \text{ and, using another series} \quad \text{(E 4.8)}$$

$$e^x = e^\theta = 1 + \theta + \frac{\theta^2}{2!} + \frac{\theta^3}{3!} + \frac{\theta^4}{4!} + \frac{\theta^5}{5!} + \cdots$$

(factorial 2 = 2! = 1 * 2, 3! = 1 * 2 * 3 etc.) (E 4.9)

$$e^{j\theta} = 1 + j\theta + j^2\frac{\theta^2}{2!} + j^3\frac{\theta^3}{3!} + j^4\frac{\theta^4}{4!} + j^5\frac{\theta^5}{5!} + \cdots \text{ where } j^2 = -1,$$

$$j^3 = -j, \quad j^4 = +1, \quad j^5 = -j \cdots \quad \text{(E 4.10)}$$

$$e^x = e^{j\theta} = 1 + j\theta - \frac{\theta^2}{2!} - j\frac{\theta^3}{3!} + \frac{\theta^4}{4!} + j\frac{\theta^5}{5!} + \cdots \text{ then} \quad \text{(E 4.11)}$$

$$\cos\theta + j\sin\theta = 1 + j\theta - \frac{\theta^2}{2!} - j\frac{\theta^3}{3!} + \frac{\theta^4}{4!} + j\frac{\theta^5}{5!} - \cdots = e^{j\theta} \quad \text{(E 4.12)}$$

4.7 COMPLEX NUMBER MATHEMATICAL OPERATIONS

USEFUL IDENTITIES

(a) $j = \sqrt{-1}$, $j^2 = -1$ where j is defined as an arithmetic operator or 90° CCW rotator

(b) $-j = \frac{1}{j}$ where $\frac{1}{j} = (1)\left(\frac{1}{j}\right) = \left(\frac{j}{j}\right)\left(\frac{1}{j}\right)$
$= \left(\frac{j}{j^2}\right) = \left(\frac{j}{-1}\right) = \left(\frac{-1}{-1}\right)\left(\frac{j}{-1}\right) = \left(\frac{-j}{1}\right) = -j$

(c) $n = a + jb$ where $a + jb$ is the rectangular or Cartesian form and n is any complex number

(d) $n = c\angle\theta$ where $c\angle\theta$ is the polar form, c is magnitude and θ is the angle

(e) $n = ce^{j\theta}$ where $ce^{j\theta}$ is the exponential form, c is the magnitude, θ is the angle and e is the base of the natural logarithm (2.7182 ···)

(f) $\sqrt{-a} = \sqrt{-1 \cdot a} = \sqrt{-1} \cdot \sqrt{a} = j\sqrt{a}$

(g) $e^{j\theta} = \cos\theta + j\sin\theta$ where $\cos\theta$ = the real part of $e^{j\theta} = \text{Re}\{e^{j\theta}\}$ and $\sin\theta$ = the imaginary part $= \text{Im}\{e^{j\theta}\}$

(h) $e^{-j\theta} = \cos\theta - j\sin\theta$

(i) $e^{\pm j\pi} = -1$

(j) $e^{\pm j\pi/2} = \pm j$

(k) $c^* = 2 - j3$ is the complex conjugate of the original $2 + j3$. The asterisk "*" is the sign of a complex conjugate.

(l) $\text{Log}_b N = n$ and $b^n = N$ such as $\text{Log}_{10} 100 = 2$ or $10^2 = 100$ or $\text{Log}_e 10 = 2.30$ and $e^{2.30} = 10$

(m) $\text{Log}_b MN = \text{Log}_b M + \text{Log}_b N$

RECTANGULAR-POLAR CONVERSIONS

When $c = a + jb$ in the rectangular form:

Then $C = \sqrt{a^2 \pm b^2}$ in the polar form

And $\theta = \tan^{-1} \frac{b}{a}$ in the polar form

When $C = c \angle\theta$ in the polar form:

Then $a = C \cos\theta$ in the rectangular form

And $b = C \sin\theta$ in the rectangular form

THE LAWS OF EXPONENTS

(n) $a^n = a \cdot a \cdot a \cdot a \cdots$ for n factors and $a^0 = 1$

(o) $a^{-n} = \frac{1}{a^n}$

(p) $a^n + b^n = (a \cdot a \cdot a \cdots) + (b \cdot b \cdot b \cdots)$

(q) $a^n - b^n = (a \cdot a \cdot a \cdots) - (b \cdot b \cdot b \cdots)$

(r) $a^m \cdot a^n = a^{(m+n)}$

(s) $(a^m)^n = a^{mn}$

(t) $\frac{a^m}{a^n} = a^{m-n}$

(u) $a^{m/n} = \sqrt[n]{a^m}$

TM ADDITION OF COMPLEX NUMBERS

To add two or more complex numbers, add the real and imaginary parts separately as $c_{new} = \pm a_1 + \pm a_2 + \pm a_3 + \cdots + \pm jb_1 + \pm jb_2 + \pm jb_3 + \cdots$

MM ADDITION OF COMPLEX NUMBERS

$$\begin{array}{r} 2+j3 \\ \underline{3+j1} \\ 5+j4 \end{array} \qquad\qquad \begin{array}{r} 3+j6 \\ \underline{-6+j3} \\ -3+j9 \end{array} \qquad\qquad \text{(E 4.13)}$$

GM ADDITION OF COMPLEX NUMBERS

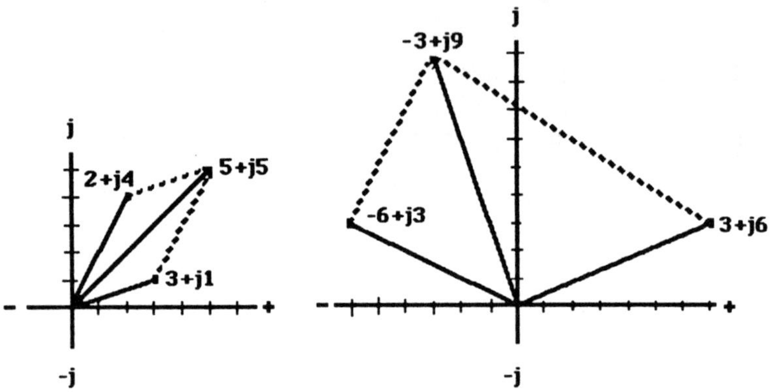

FIGURE 4.6 Addition of complex numbers.

TM SUBTRACTION OF COMPLEX NUMBERS

To subtract two or more complex numbers, subtract the real and imaginary parts separately as $c_{new} = \pm a_1 - \pm a_2 - \pm a_3 - \cdots - \pm jb_1 - \pm jb_2 - \pm jb_3 - \cdots$

MM SUBTRACTION OF COMPLEX NUMBERS

$$\begin{array}{rl} c_1 = & 2+j3 \\ c_2 = & -\underline{(3+j1)} \\ c_3 = & -1+j2 \end{array} \qquad\qquad \begin{array}{rl} c_1 = & 3+j6 \\ c_2 = & -\underline{(6+j3)} \\ c_3 = & -3+j3 \end{array} \qquad\qquad \text{(E 4.14)}$$

GM SUBTRACTION OF COMPLEX NUMBERS

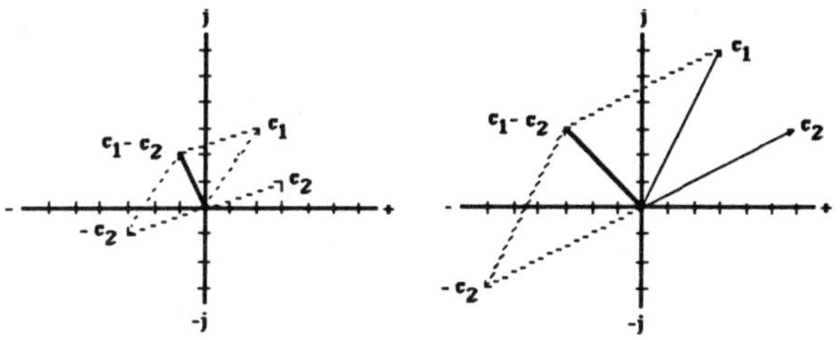

FIGURE 4.7 Subtraction of complex numbers.

TM MULTIPLICATION OF COMPLEX NUMBERS

Multiplication of these numbers can be accomplished in rectangular or polar form. Multiplication in the polar form uses the laws of exponents.

MM MULTIPLICATION OF COMPLEX NUMBERS

Rectangular:

	2 + j3		3 + j6
(x)	3 + j1	(x)	6 + j3
	j2 $\quad j^2 3$		j9 $\quad j^2 18$
	6 $\quad$ j9		18 $\quad$ j36
	6 + j11 − 3 = 3 + j11		18 + j45 − 18 = j45 (E 4.15)

Polar:

$$c_1 \angle\theta_1 * c_2 \angle\theta_2 = c_1 * c_2 \angle\theta_1 \pm \theta_2 \text{ or}$$

$$= c_3 \angle\theta_1 \pm \theta_2 = c_3 \angle\theta_3 \quad \text{(E 4.16)}$$

$$3.6 \angle 56.30 * 3.16 \angle 18.43 = 11.38 \angle 74.73 \; 45 \angle 90 \quad \text{(E 4.17)}$$

TM DIVISION OF COMPLEX NUMBERS

Division of two complex numbers in rectangular form is performed by multiplying the numerator and denominator by the conjugate of the denominator. Division in the polar form uses the laws of exponents.

MM DIVISION OF COMPLEX NUMBERS

Rectangular:

$$\frac{c_1}{c_2} = \frac{a_1 + jb_1}{a_2 + jb_2} \cdot \frac{(a_2 - b_2)}{(a_2 - b_2)} = \frac{2 + j3}{3 + j1} \cdot \frac{3 - j1}{3 - j1} = \frac{6 - j2 + j9 - j^2 3}{9 - j3 + j3 - j^2} = \frac{9 + j7}{10}$$

$$= 0.9 + j0.7 \qquad \text{(E 4.18)}$$

Polar:

$$\frac{c_1}{c_2} = \frac{c_1 \angle \theta_1}{c_2 \angle \theta_2} = \frac{3.6 \angle 56.30}{3.16 \angle 18.43} = \frac{3.6}{3.16} \angle 56.30 - 18.43 = 1.14 \angle 37.87 \qquad \text{(E 4.19)}$$

TM LOGARITHMS OF COMPLEX NUMBERS

The natural logarithm of a complex number can be found most easily by converting the number to its exponential form and using the laws of exponents and logarithms.

MM LOGARITHMS OF COMPLEX NUMBERS

$$\text{Log}_\varepsilon\, x = \text{Log}_\varepsilon\, Ae^{j\theta} = \text{Log}_\varepsilon\, A + \text{Log}_\varepsilon\, e^{j\theta} \text{ where } \varepsilon \equiv e \qquad \text{(E 4.20)}$$

$$\text{Log}_\varepsilon\, 4e^{j1.5\pi} = \text{Log}_\varepsilon\, 4 + \text{Log}_\varepsilon\, e^{j1.5\pi} = 1.386 + j1.5\pi = 1.36 + j4.71 \qquad \text{(E 4.21)}$$

TM INTEGER POWERS OF COMPLEX NUMBERS

A complex number can be raised to an integer power k using its polar or rectangular form and the laws of exponents.

MM INTEGER POWERS OF COMPLEX NUMBERS

$$(ce^{j\theta})^k = c^k e^{jk\theta} \text{ or } c^k (\cos k\theta + j \sin k\theta) \qquad \text{(E 4.22)}$$

$$(2e^{j1/5\pi})^5 = 2^5 e^{j5/5\pi} = 32\, e^{j\pi} = 32 - 1 = 31 \qquad \text{(E 4.23)}$$

$$(3 + j4)^4 = (5\, e^{j53.13^\circ})^4 = 625\, e^{j212.52^\circ} \text{ or, using } e^{j\theta} = \cos\theta + j \sin\theta$$

$= 625\ (\cos 212.52° + j \sin 212.52°) = -527 - j336$ (E 4.24)

TM ROOTS OF COMPLEX NUMBERS

A complex number $ce^{j\theta}$ may be written $ce^{j\theta + 2\pi n}$ where $n = 0, \pm 1, \pm 2, \pm 3, \cdots$ or, similarly, $c \angle\theta$ may be written $c \angle\theta + n360°$. The kth root (such as the square root, the cube root, etc.) of these numbers may be extracted in the exponential or polar form using the laws of exponents.

MM ROOTS OF COMPLEX NUMBERS

For the exponential form, the kth root will be:

$\sqrt[k]{c}\, e^{j(\theta + 2\pi n)/k}$ where $n = 0, 1, 2, 3, \cdots$

To find the five fifth roots of $1 = 1e^{j2\pi n}$, the roots $= \sqrt[5]{1e^{j2\pi n/5}}$

For $n = 0$, the root is $1 \angle 0°$ because $e^{j0} = 0°$ and any root of 1 is 1

For $n = 1$, the root is $1e^{j2\pi 1/5} = 1 \angle 72°$

For $n = 2$, the root is $1e^{j2\pi 2/5} = 1 \angle 144°$

For $n = 3$, the root is $1e^{j2\pi 3/5} = 1 \angle 216°$

For $n = 4$, the root is $1e^{j2\pi 4/5} = 1 \angle 288°$ (E 4.25)

For the polar form, the kth root will be:

$\sqrt[k]{c} \angle(\theta + n360°)/k.$ (E 4.26)

To find the three cube roots of $8 \angle 60° = \sqrt[3]{8} \angle(60° + n360°)/3$

For $n = 0$, the root is $2 \angle(60° + 0°)/3 = 2 \angle 20°$

For $n = 1$, the foot is $2 \angle(60° + 360°)/3 = 2 \angle 140°$

For $n = 2$, the foot is $2 \angle(60° + 720°)/3 = 2 \angle 260°$ (E 4.27)

4.8 THE COMPLEX VARIABLE OR S PLANE—POLES AND ZEROS

TM GEOMETRIC PLANES USED IN CIRCUIT ANALYSIS

Chapter 1 introduced a graphical technique for aligning equation coefficients to facilitate the solution of those equations by determinants. Likewise, Chapter 3 detailed the use of the impedance triangle z-plane graph to display and analyze complex voltage, current and resistance-reactance relationships. Also, this chapter introduced the concept of complex numbers and introduced another graphical display technique using a two-dimensional plane. The complex variable plane is still another graphical technique that is often used in circuit analysis. This complex variable plane presents an overview or "big picture" of the roots of equations such as those of a transfer function (see Section 5.2). By plotting a special property of the *roots* of these equations, called poles and zeros, it is possible immediately to see the general CHARACTER of the function. Many times, a more complete solution is unnecessary. Both the complex number (z) plane and the complex variable (s) plane are logical extensions of elementary plane geometry and its common x-y plane.

4.9 TM GEOMETRIC PLANES AND THE COMPLEX VARIABLE—A BRIEF INTRODUCTION TO THE COMPLEX PLANE

A ZERO is the factored root of the numerator—zeros can be simple (real) or complex numbers.

A POLE is the factored root of the denominator—poles can be real or complex numbers.

The $j\omega$ axis is a *frequency axis* since $\omega = 2\pi f$. $s = j\omega$ for, AND ONLY FOR, steady-state sine waves.

The α axis is the *damping axis*. The damping controls the time-response of the circuit.

If the poles of a transfer function lie anywhere to the right of the $j\omega$ axis, the circuit, the amplifier, network, etc., is very, very unstable—its operation cannot be defined. This area indicates a theoretically ever-increasing oscillation.

If the poles of a transfer function lie directly on the $j\omega$ axis, the circuit, the network, etc., will oscillate. This is excellent for oscillators, but very bad for amplifiers.

If the poles of a transfer function lie to the left of the $j\omega$ axis, the system is stable—it will not oscillate.

The distance to the left of the $j\omega$ axis indicates the amount of damping in the system. Thus, this distance gives an indication of the amount of transient overshoot.

A pole that lies on the $-\alpha$ axis is critically damped. (See an explanation of the significance and use of poles and zeros in Chapters 6, 9, and 11.)

GM GEOMETRIC PLANES USED IN CIRCUIT ANALYSIS

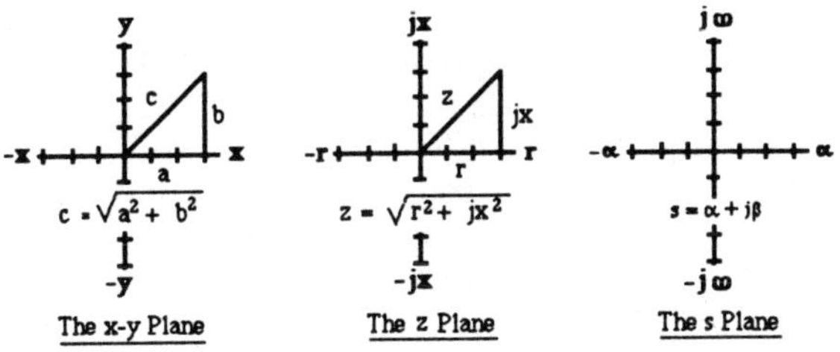

FIGURE 4.8 A comparison of three planes used in circuit analysis.

MM DEFINITIONS USED WITH THE COMPLEX-VARIABLE PLANE

The General s-Form of a *Transfer Function*:

$$F(s) = \frac{A(s)}{B(s)} = \frac{a_m s^m + a_{m-1} s^{m-1} + \cdots\cdots + a_1 s + a_0}{s^n + b_{n-1} s^{n-1} + \cdots\cdots b_1 + b_0} \qquad \text{(E 4.28)}$$

The Factored or Root Form of a Transfer Function:

$$F(s) = \frac{A(s)}{B(s)} = \frac{A\,(s-z_1)\,(s-z_2)\cdots\cdots(s-z_n)}{(s-p_1)\,(s-p_2)\cdots\cdots(s-p_n)} = A\,\frac{\text{zeros}}{\text{poles}} \qquad \text{(E 4.29)}$$

5

Filter Models

5.1 ABOUT THIS CHAPTER

A filter is a device that controls the amount of material or information that will pass through. A paper coffee filter will block the coffee grounds, but will allow the water to pass. The size of the openings in the filter has been predetermined to stop the grounds. Likewise, the tone controls in a stereo amplifier are used to reject some frequency components and pass or enhance others. Electric circuit (analog) filters can be created from combinations of capacitors, inductors, and resistors to form *passive or lumped constant* filters. Likewise, these circuit elements can be combined with high-gain amplifiers to create *active filters*. The preliminary design of a filter is dependent upon the frequencies that it is required to pass and stop and the phase relationship of these frequencies in both the pass and stop bands. Later design parameters usually include the (group) time delay and the transient response required as well as the power handling capacity, the operating temperature, size limitations, etc.

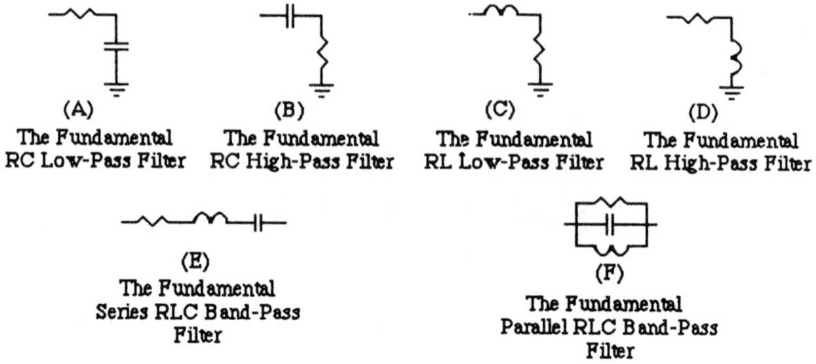

FIGURE 5.1 The simplest analog filter configurations.

Figure 5.1 shows the basic building blocks for passive analog filter design. Some variations of these elementary filter configurations are also called equalizers. An equalizer is a very specific filter configuration that is used to correct for or insert a special frequency and/or phase characteristic in a transmission system. Examples of these specialized filters include line equalizers used by telephone companies and pre-emphasis and de-emphasis equalizers used by the broadcasting and recording industries.

5.2 THE TRANSFER FUNCTION

The transfer function is one of several ways to describe the relationship between the output and input of circuits such as amplifiers or filters. In its simplest form, the transfer function is used to show how the input parameter is modified to form the output. This output-from-input change function can be put in the form of ratio numbers. This modifying (transfer) function can be (complex) gain, voltage, current, impedance, integration, etc.

TM THE TRANSFER FUNCTION

The complex transfer function is composed of two parts, the modulus or amplitude $|H|$ and the phase shift $|\theta|$. The two parts are usually stated as (functions of) being in some complex plane such as $|H(j\omega)|$ and $|\theta(j\omega)|$ or $|H(s)|$ and $|\theta(s)|$. Symbols such as $|G(j\omega)|$ or $|A(j\omega)|$ are also used to indicate the modulus or real part of the transfer function. *Special Note*: Some, but not all, books use the vertical lines to indicate absolute values (assume a plus value) for the modulus and phase shift.

MM THE TRANSFER FUNCTION

$H_{oi}\ (j\omega) = \frac{V_o}{V_i}$ the voltage transfer function where the subscripts o and i indicate output and input.

This is the most common form of the transfer function, but others include the following:

$K_{oi}\ (j\omega) = \frac{I_o}{I_i}$ the current transfer function

$Z_{oi}\ (j\omega) = \frac{Z_o}{Z_i}$ the impedance transfer function

$Y_{oi}\ (j\omega) = \frac{I_o}{V_i}$ the transfer admittance transfer function (E 5.1)

GM THE TRANSFER FUNCTION

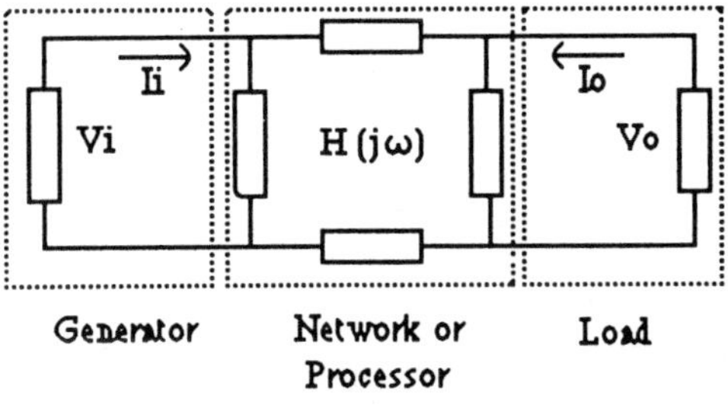

FIGURE 5.2 A generalized transmission system showing a transfer function in the center block.

5.3 TM THE BODE PLOT

H. W. Bode (circa 1945) of the Bell Telephone Laboratories created a graphical technique to display both the output-to-input (transfer) sinusoidal frequency response and the output-to-input phase response. The voltage ratio is presented in decibels, the phase shift in degrees, and the frequency is on a semi-logarithmic scale. The cutoff (sometimes called break point or corner) frequency represents transitions from the pass bands to the attenuation or stop bands. These cutoff points are due to time constants (RC, L/R, etc.) which also can be represented as poles or zeros in the complex frequency or s plane. The plots of the corner frequencies are the rounded transitions coming and going from straight-line asymptotes. The asymptotic rise or fall in the frequency response is usually at a rate of 6 dB per octave (20 dB per decade), 12 dB per octave (40 dB per decade), etc. Six dB represents a voltage output change of about two to one and 12 dB is an output change of about four to one. An octave represents a two-to-one change in frequency and a decade represents a ten-to-one change in frequency.

MM THE BODE PLOT

$H(j\omega) = \dfrac{V_o\angle out}{V_i\angle in}$ is one form of a network or filter transfer function

where $V_o\angle out$ and $V_i\angle in$ are considered complex numbers (E 5.2)

$\text{Gain} = 20 \log \dfrac{V_o}{V_i}$ is the Bode magnitude (amplitude) plot (E 5.3)

$\text{Phase} = \angle out - \angle in$ is the Bode phase angle plot (E 5.4)

GM THE BODE PLOT

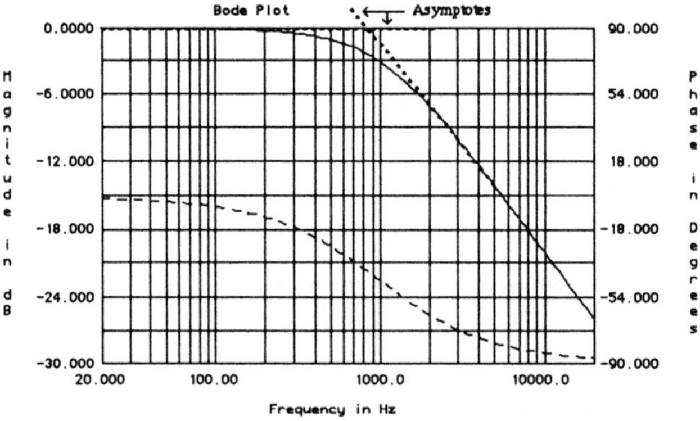

FIGURE 5.3 A typical Bode plot showing a logarithmic scale for magnitude and frequency. Note also the asymptotes meeting at the corner frequency.

5.4 TM SIMPLE RC AND RL LOW-PASS FILTERS

The elementary low-pass filters shown in Figure 5.1 include a pass band that provides for the normal transmission of signals up to a limit or cutoff frequency.

At or near this cutoff frequency, the signals start to be attenuated in the stop band. An ideal or brick-wall, low-pass filter would completely reject all frequencies above this cutoff frequency in the stop band. However, real-world analog brick-wall filters cannot be created and, even if they could, their phase response probably would be unacceptable. In simple low-pass filters, the response drops at a constant rate after the cutoff frequency, and the phase shift changes in a linear fashion.

MM SIMPLE RC AND RL LOW-PASS FILTERS

For the fundamental RC low-pass filter as in Figure 5.1 (A):

When $X_C = \frac{1}{2\pi fC}$ with f in Hz and C in farads, the cutoff

$$\text{frequency, } f_C = \frac{1}{2\pi RC} \qquad \text{(E 5.5)}$$

$$H_{oi}\,(j\omega) = \frac{V_o}{V_i} = \frac{X_C}{\sqrt{R^2 + X_C{}^2}} \text{ and the phase shift } \theta = -\tan^{-1}\frac{X_C}{R} \qquad \text{(E 5.6)}$$

For the fundamental RL low-pass filter as in Figure 5.1 (C):

Where $X_L = 2\pi fL$ with f in HZ and L in henrys, the cutoff

$$\text{frequency, } f_C = \frac{R}{2\pi L} \qquad \text{(E 5.7)}$$

$$H_{oi}\,(j\omega) = \frac{V_o}{V_i} = \frac{R}{\sqrt{R^2 + X_L{}^2}} \text{ and the phase shift } \theta = -\tan^{-1}\frac{X_L}{R} \qquad \text{(E 5.8)}$$

GM SIMPLE RC AND RL LOW-PASS FILTERS

FIGURE 5.4 Elementary low-pass filters.

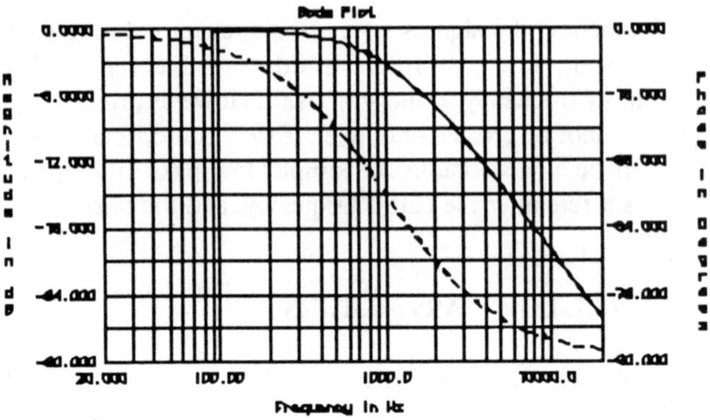

FIGURE 5.5 A dual plot of equations (E 5.6) with magnitude (in dB) and phase (in degrees) for the low-pass filter in Figure 5.4 (A). The values are R = 1000 ohms and C = 0.16 uf. Using equations (E 5.8) with L = 0.16 henrys would give a similar result.

5.5 SIMPLE RC AND RL HIGH-PASS FILTERS

TM ELEMENTARY HIGH-PASS FILTERS

The elementary high-pass filters shown in Figure 5.1 produce a stop band below cutoff and a pass band above cutoff.

MM ELEMENTARY HIGH-PASS FILTERS

For the fundamental RC high-pass filter as in Figure 5.1 (B):

Where $X_C = \dfrac{1}{2\pi fC}$ with f in Hz and C in farads, the cutoff

$$\text{frequency, } f_c = \frac{1}{2\pi RC} \qquad \text{(E 5.9)}$$

$$H_{oi}\,(j\omega) = \frac{V_o}{V_i} = \frac{R}{\sqrt{R^2 + X_c{}^2}} \text{ and the phase shift } \theta = \tan^{-1}\frac{X_c}{R} \qquad \text{(E 5.10)}$$

For the fundamental RL high-pass filter as in Figure 5.1 (D):

Where $X_L = 2\pi fL$ with f in Hz and L in henrys, the cutoff

$$\text{frequency, } f_c = 2\pi RL \qquad \text{(E 5.11)}$$

$$H_{oi}(j\omega) = \frac{V_o}{V_i} = \frac{X_L}{\sqrt{R^2 + X_L{}^2}} \text{ and the phase shift } \theta = \tan^{-1}\frac{R}{X_L} \qquad \text{(E 5.12)}$$

GM ELEMENTARY HIGH-PASS FILTERS

FIGURE 5.6 Elementary high-pass filters.

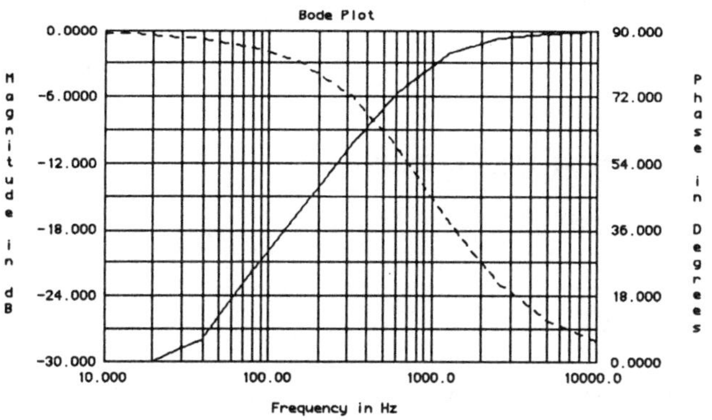

FIGURE 5.7 A dual plot of equations (E 5.10) with magnitude (in dB) and phase (in degrees) for the high-pass filter in Figure 5.6 (A). The values are R = 1000 ohms and C = 0.16 uf. Using equations (E 5.12) with L = 0.16 henrys would give a similar result.

5.6 TM SIMPLE RLC BAND-PASS FILTERS

Series and parallel RLC configurations are the keystone of analog communication circuits. There are many variations to the fundamental types shown in Figure 5.1. It should be remembered that these *tuned circuits* are, indeed, filters which are used to limit the transmitted energy to one (resonant) frequency or to a relatively restricted band or spectrum of frequencies.

For real-life circuits, some shunt or series resistance must be assumed and included in the design and analysis of these resonant circuits. Sometimes the major portion of the effective resistance is a separate, physical component.

However, the effects of the coil (inductor) resistance, skin effect, connection resistance and, at very high frequencies, the dissipation in the capacitor, must often be included. RLC circuits are covered in more detail in Chapter 6.

MM SIMPLE RLC BAND-PASS FILTERS

For the fundamental RLC series resonant circuit as in Figure 5.1 (E), where $X_C = \frac{1}{2\pi fC}$, $X_L = 2\pi fL$ and R is in ohms, the transfer function can give an output voltage across R:

$$H_{oi}(j\omega) = \frac{V_o}{V_i} = \frac{R}{\sqrt{R^2 + (X_L - X_C)^2}}$$ and the phase shift

$$\theta = \tan^{-1} \frac{R}{X_L - X_C} \qquad \text{(E 5.13)}$$

GM SIMPLE RLC BAND-PASS FILTERS

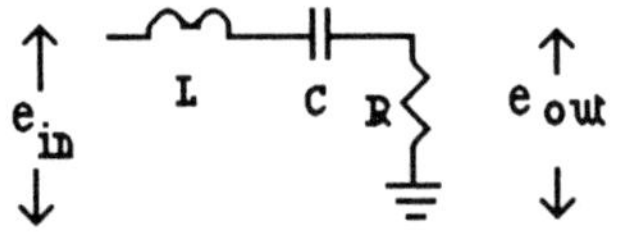

FIGURE 5.8 An elementary series resonant circuit band-pass filter.

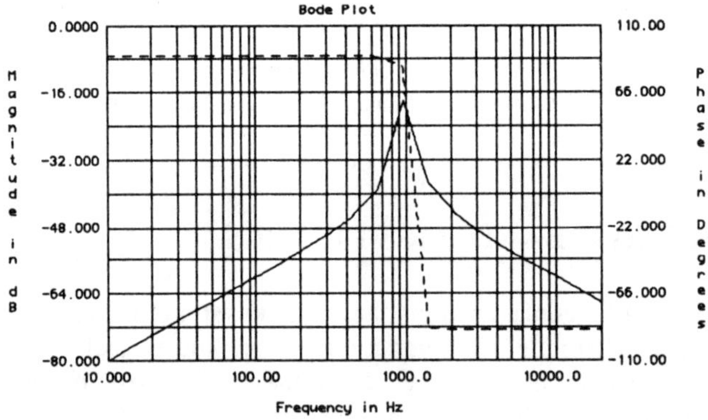

FIGURE 5.9 A dual plot of equations (E 5.13) with magnitude (in dB) and phase (in degrees) for the band-pass filter in Figure 5.8. The values are R =10 ohms, C = 0.16 uf and L = 0.16 henrys.

MM SIMPLE RLC BAND-PASS FILTERS

For the fundamental RLC parallel resonant circuit as in Figure 5.1 (F), when $X_C = \frac{1}{2\pi f C}$, $X_L = 2\pi f L$ and R is in ohms, the transfer function can be written in the form:

$$H_{oi}\,(j\omega) = \frac{V_o}{V_i} = \frac{L/C}{\sqrt{(L/C)^2 + R^2\,(X_L - X_C)^2}} \text{ is the magnitude and}$$

$$\theta = \tan^{-1} R\,\frac{X_L - X_C}{L/C} \text{ is the phase shift.}$$

Both are a function of frequency. (E 5.14)

GM SIMPLE RLC BAND-PASS FILTERS

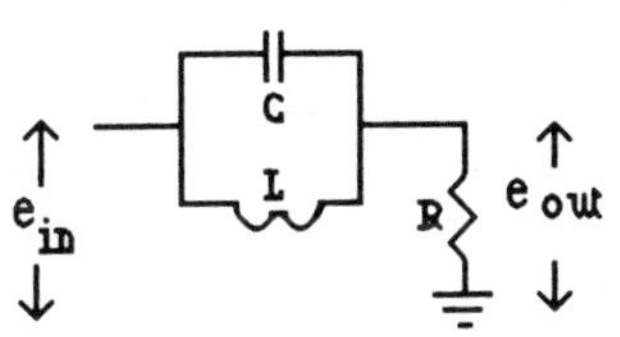

FIGURE 5.10 An elementary parallel resonant circuit band-pass (elimination) filter.

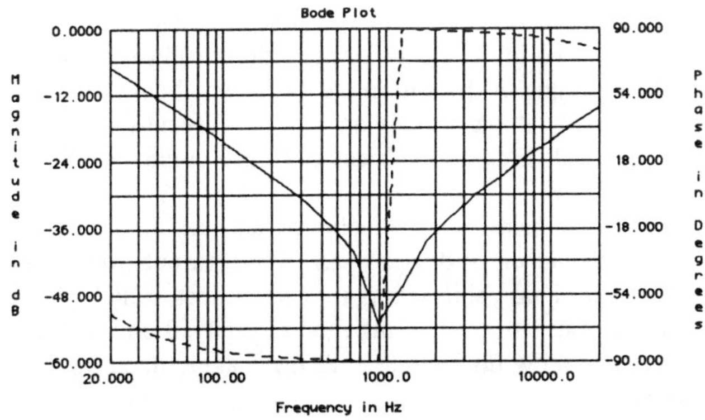

FIGURE 5.11 A dual plot of equations (E 5.14) with magnitude (in dB) and phase (in degrees) for the band-pass filter in Figure 5.10. The values are R = 10k ohms, C = 0.16 uf and L = 0.16 henrys.

5.7 TRANSMISSION-LINE ORIGINATED CONSTANT-K FILTERS

TM TRANSMISSION-LINE ORIGINATED CONSTANT-K FILTERS

Constant-k filters evolved from the theory of transmission lines. Their creation was based upon the concept that a transmission line, such as a coaxial cable, twin lead, twisted pair, etc., can be accurately modeled with a large number of individual R, L, and C components. These components—termed lumped constants—are usually configured in sections of a *T-network, π-network, H-network* or *ladder network*. The configurations can also be reduced to what are called *half sections*. Like transmission lines, constant-k and M-derived (to be discussed later) filters have a characteristic impedance, but their behavior varies with generator and load impedances. Also like transmission lines, they have attenuation in both the pass and stop bands if there are resistors (dissipating elements) in either of the series and shunt elements. Both constant-k and M-derived filters can be configured as low pass, high pass and band pass or band elimination.

BY DEFINITION, prototype constant-k filters contain only reactive elements. Mathematically, the product of their series and shunt (reactive) components always equals a constant or "k"—thus the name constant-k. Constant-k filters have three major problems. First, their input impedance equals the generator (characteristic) impedance at only one frequency. Second, the attenuation in the stop band, near the cutoff frequency, is relatively small. The third problem is more subtle. Although prototype constant-k filters, by definition, contain only reactive components, real-world inductors have series resistance and, at high frequencies, both capacitors and inductors have both series (skin effect) and shunt resistance. Thus, "real-world" constant-k filters do not always operate quite like their theoretical designs. Although, in the past, some books did give design graphs in terms of coil Q, it was usually an iterative cut-and-try process to finalize a working design. It is now much easier to use computer circuit design and plot programs.

Originally, these filters were based upon telephone transmission lines. The major parameters considered in early designs were their output-from-input voltage and current response. Phase shift or, as it is sometimes called, phase response, was NOT considered. However, using modern analysis and plotting practice, the examples that follow do include both the magnitude and phase response in their Bode plots.

GM TRANSMISSION-LINE ORIGINATED CONSTANT-K FILTERS

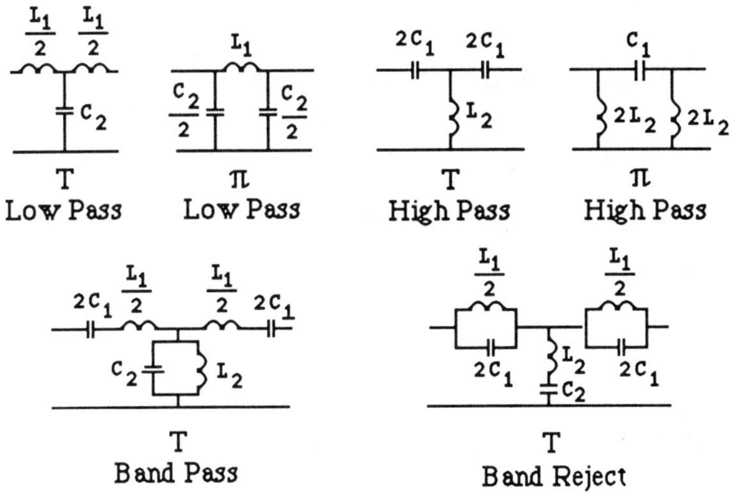

FIGURE 5.12 Prototype constant-k filters.

5.8 TM CONSTANT-K LOW-PASS FILTERS

Low-pass constant-k filters can be designed and analyzed using the equations based upon the constant-k, the characteristic impedance, and the cutoff frequency. Once the first-cut filter is designed and analyzed, resistors can be added to simulate coil Q and capacitor dissipation. Constant-k filters are usually configured as either three-element "T" or "π" circuits. T circuits have two series and one shunt element and π circuits have one series and two shunt elements.

MM REFERENCE TRANSMISSION LINE EQUATIONS

$$\gamma = \alpha + j\,\beta = \sqrt{(R + j\omega L)(G + j\omega C)}$$

where α is the line or network attenuation and β is the phase (delay) constant, both per unit length. R is the series resistance, L the series inductance, G the shunt conductance and C the shunt capacitance, all per unit length. (E 5.15)

$\gamma = j\omega\sqrt{LC}$ for an ideal network. γ is termed the *propagation constant*. (E 5.16)

$$Z_0 = \sqrt{\frac{R + j\omega L}{G + j\omega C}} = \sqrt{\frac{L}{C}}$$ for an ideal network. Z_0 is the characteristic impedance. (E 5.17)

MM CONSTANT-K LOW-PASS FILTERS

$$R_0 = \sqrt{\frac{L}{C}} \quad \text{(E 5.18)}$$

$$Z_{0T} = \sqrt{Z_1Z_2 + \frac{Z_1^2}{4}} = \sqrt{j\frac{-j\omega L}{\omega C} - \frac{\omega^2L^2}{4}} = \sqrt{\frac{L}{C}\left(1 - \frac{\omega^2LC}{4}\right)} \quad \text{(E 5.19)}$$

$$Z_{0\pi} = \frac{Z_1Z_2}{Z_{0T}} \quad \text{(E 5.20)}$$

$f_{0T} = \dfrac{1}{\pi\sqrt{LC}}$ where f_0 is the cutoff frequency (the limit of the pass band)(E 5.21)

$L = R_0^2C = \dfrac{1}{\pi^2 f_0^2 C}$ and, expanding, $R^2C^2 = \dfrac{1}{\pi^2 f_0^2}$ and $C^2 = \dfrac{1}{\pi^2 f_0^2 R_0^2}$ (E 5.22)

$C_2 = \dfrac{1}{\pi f_0 R_0}$ (farads) for a low-pass filter (E 5.23)

$L_1 = \dfrac{R_0}{\pi f_0}$ (henrys) for a low-pass filter (E 5.24)

$$Z_1Z_2 = j\omega L_1 * -j\frac{1}{\omega C_2} = \frac{L_1}{C_2} = k^2 \quad \text{(E 5.25)}$$

$Z_{0T} = \sqrt{Z_1Z_2 + \dfrac{Z_1^2}{4}}$ This formula is derived from transmission line sections. (E 5.26)

Note that both Z_1 and Z_2 are frequency dependent which makes Z_{0T} frequency dependent. The value of this equation goes to zero at f_0, the cutoff frequency, where $f = f_0$.

$$Z_{0T} = \sqrt{\frac{L}{C} * \left(1 - \frac{\omega^2 LC}{4}\right)} = \sqrt{\frac{L}{C} * \left(1 - \frac{\omega^2}{\omega_0^2}\right)}$$

$$= \sqrt{\frac{L}{C}} * \sqrt{1 - \frac{f^2}{f_0^2}} \qquad \text{(E 5.27)}$$

$$Z_{0\pi} = \frac{\sqrt{\frac{L}{C}}}{\sqrt{1 - \frac{f^2}{f_0^2}}} \qquad \text{(E 5.28)}$$

GM CONSTANT-K LOW-PASS FILTERS

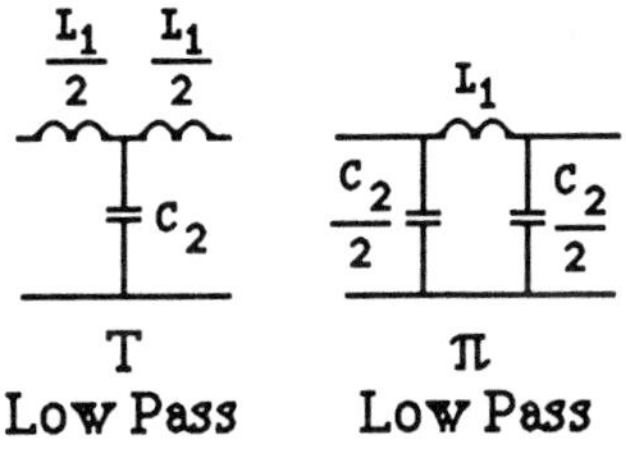

FIGURE 5.13 Prototype constant-k LP filters.

A CONSTANT-K LOW-PASS FILTER EXAMPLE

Design a T-type LP filter with $Z_0 = R_0 = 600$ ohms and a cutoff frequency (f_0) of 1000 Hz. Show the filter configuration, a Bode plot and a plot of the network impedance Z_{0T}.

$Z_0 = R_0 = 600$ ohms

$$C_2 = \frac{1}{\pi f_0 R_0} = \frac{1}{3.14*1000*600} = 0.53 \text{ uf} \qquad \text{(E 5.29)}$$

$$L_1 = \frac{R_0}{\pi f_0} = \frac{600}{\pi * 1000} = 0.19 \text{ h and } L_1/2 = 0.095 \text{ h.} \qquad \text{(E 5.30)}$$

$$Z_{0T} = \sqrt{\frac{L}{C}} * \sqrt{1 - \frac{f^2}{f_0^2}} = \sqrt{\frac{.19}{.53*10^{-6}}} * \sqrt{1 - \frac{f^2}{(1000)^2}}$$

$$= 598.7 * \sqrt{1 - \frac{f^2}{(1000)^2}} \qquad \text{(E 5.31)}$$

Note that, to be precise, $\sqrt{L/C}$ equals 598.7 rather than 600 ohms.

GM A CONSTANT-K LP FILTER EXAMPLE

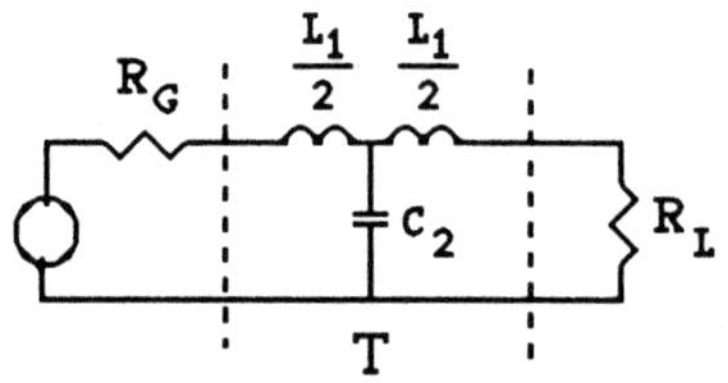

FIGURE 5.14 Circuit diagram of a constant-k LP filter example.

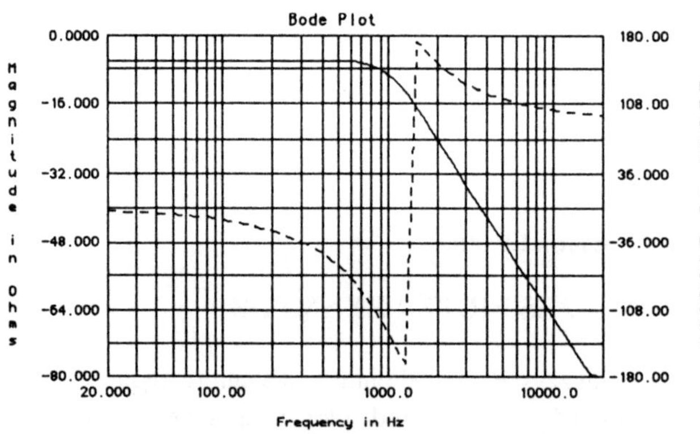

FIGURE 5.15 Bode plot of a constant-k LP filter example. (This plot includes the effect of R_G and R_L, thus the passband response is down 6 dB.)

A Special Note on Bode Plots for Constant-k Filters: Constant-k and M-derived filters were originally developed and used before phase shift, transient response and the idea of the transfer function (equations) were part of common filter and network thinking. For that reason, most of these original texts on transmission-line derived filters did not cover any of these subjects. The computer-generated Bode plots are included here for comparison and completeness.

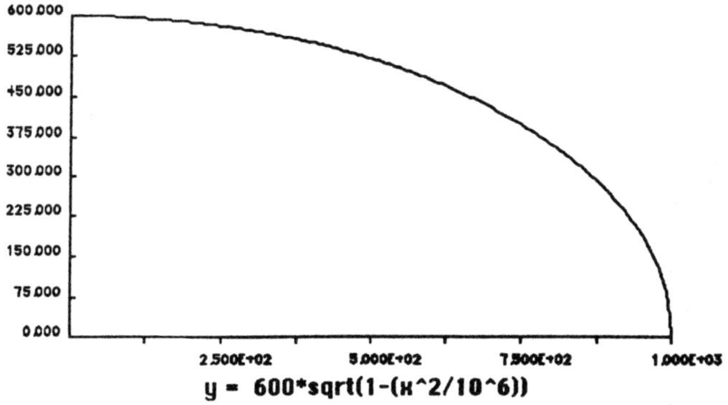

FIGURE 5.16 Filter Z_0 as a function of frequency; a constant-K LP filter example.

5.9 TM CONSTANT-K HIGH-PASS FILTERS

Constant-k high-pass filters can be designed and analyzed using the equations based upon the constant-k, the characteristic impedance, and the cutoff frequency. Once the first-cut filter is designed and analyzed, resistors can be added to simulate coil Q and capacitor dissipation.

MM CONSTANT-K HIGH-PASS FILTERS

$$R_0 = \sqrt{\frac{L}{C}} \qquad \text{(E 5.32)}$$

$$f_{0T} = \frac{1}{4\pi\sqrt{LC}} \qquad \text{(E 5.33)}$$

$$C_1 = \frac{1}{4\pi\, f_0\, R_0} \text{ (farads) for a high-pass filter} \qquad \text{(E 5.34)}$$

$$L_2 = \frac{R_0}{4\pi\, f_0} \text{ (henrys) for a high-pass filter} \qquad \text{(E 5.35)}$$

$$Z_{oT} = \sqrt{\frac{L}{C}} * \sqrt{1 - \frac{f^2}{f_0^2}} \qquad \text{(E 5.36)}$$

$$Z_{0\pi} = \frac{\sqrt{\frac{L}{C}}}{\sqrt{1 - \frac{f^2}{f_0^2}}} \qquad \text{(E 5.37)}$$

GM CONSTANT-K HIGH-PASS FILTERS

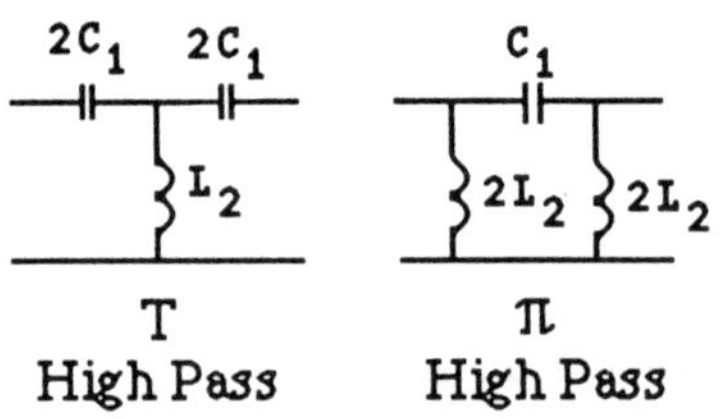

FIGURE 5.17 Prototype constant-k HP filters.

5.10 TM CONSTANT-K BAND-PASS FILTERS

Band-pass constant-k filters combine the components of a low-pass and a high-pass filter. There are two cutoff frequencies—one at the lower end of the band pass and one at the upper end.

MM CONSTANT-K BAND-PASS FILTERS

$$Z_o = \sqrt{Z_1 Z_2 + \frac{Z_1^2}{4}} = \sqrt{\frac{L_2}{C_1} + \frac{Z_1^2}{4}} = \sqrt{\frac{L_2}{C_1}} = \sqrt{\frac{L_1}{C_2}} \qquad \text{(E 5.38)}$$

$f_{mean} = \sqrt{f_{up}\, f_{lo}} = \frac{1}{\pi\sqrt{L_1 C_2}}$ where f_{up} is the upper cutoff and f_{lo}

is the lower cutoff frequency. (E 5.39)

$$L_1 = C_2 R_0^2 = \frac{1}{C_2(f_{up} - f_{lo})^2 \pi^2} = \frac{R_0}{\pi(f_{up} - f_{lo})} \qquad \text{(E 5.40)}$$

$$L_2 = C_1R_0^2 = \frac{R_0(f_{up}-f_{lo})}{4\pi f_{up}f_{lo}} \tag{E 5.41}$$

$$C_1 = \frac{f_{up}-f_{lo}}{4\pi f_{up}f_{lo}R_0} \tag{E 5.42}$$

$$C_2 = \frac{1}{\pi R_0(f_{up}-f_{lo})} \tag{E 5.43}$$

A CONSTANT-K BAND-PASS FILTER EXAMPLE

Design a constant-k band-pass filter with the following requirements: $Z_0 = R_0 =$ 600 ohms, $f_{lo} = 350$ Hz and $f_{up} = 4000$ Hz.

$$C_2 = \frac{10^6}{\pi*600*(4000-350)} = \frac{10^6}{2.19*10^6*\pi} = 0.145 \text{ uf}$$

$$C_2 = 0.145 \text{ uf} \tag{E 5.44}$$

$$L_1 = \frac{600}{\pi*(4000-350)} = \frac{600}{3650*\pi} = 0.0523 \text{ h}$$

$$L_1/2 = 0.0261 \text{ h} \tag{E 5.45}$$

$$C_1 = \frac{(4000-350)*10^6}{4\pi*4000*350*600} = \frac{3650*10^6}{4\pi*840*10^6} = 0.0345 \text{ uf}$$

$$2C_1 = 0.690 \text{ uf} \tag{E 5.46}$$

$$L_2 = \frac{600*(4000-350)}{4\pi*4000*350} = \frac{2.19*10^6}{4\pi*1.4*10^6} = 0.124$$

$$L_2 = 0.124 \text{ h} \tag{E 5.47}$$

GM CONSTANT-K BAND-PASS FILTERS

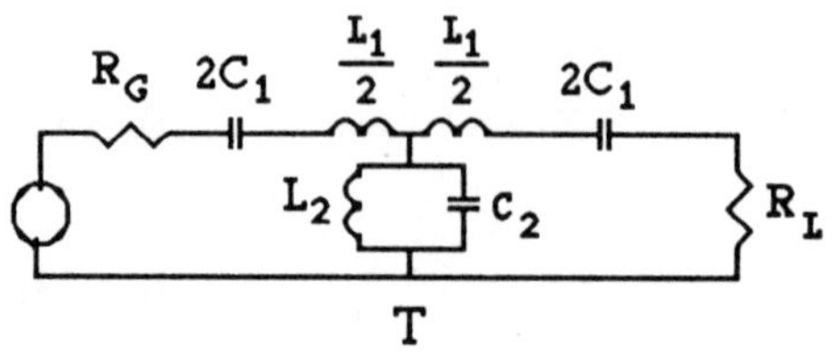

FIGURE 5.18 Circuit diagram of a constant-k band-pass filter example.

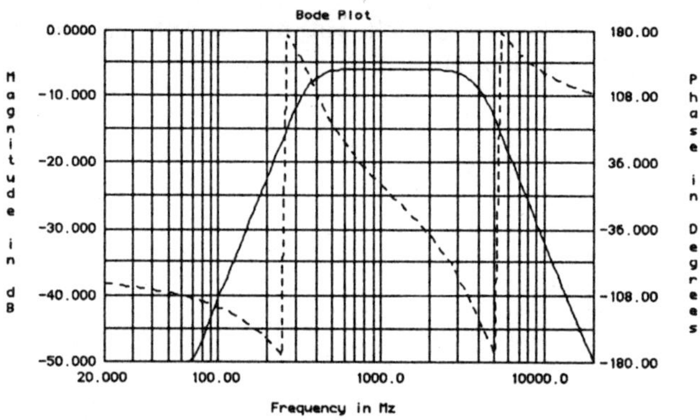

FIGURE 5.19 A constant-k BP filter example. (This plot includes the effect of R_G and R_L; thus the pass-band response is down 6 dB.)

5.11 TM CONSTANT-K BAND-REJECTION FILTERS

Band-rejection constant-k filters, like band-pass filters, uniquely combine low-pass and high-pass prototypes.

MM

$$L_1 = \frac{R_0(f_{up}-f_{lo})}{\pi f_{up} f_{lo}} \qquad \text{(E 5.48)}$$

$$L_2 = \frac{R_0}{4\pi(f_{up}-f_{lo})} \qquad \text{(E 5.49)}$$

$$C_1 = \frac{1}{4\pi R_0(f_{up}-f_{lo})} \qquad \text{(E 5.50)}$$

$$C_2 = \frac{f_{up} - f_{lo}}{\pi f_{up} f_{lo} R_0} \qquad \text{(E 5.51)}$$

GM CONSTANT-K BAND-REJECTION FILTERS

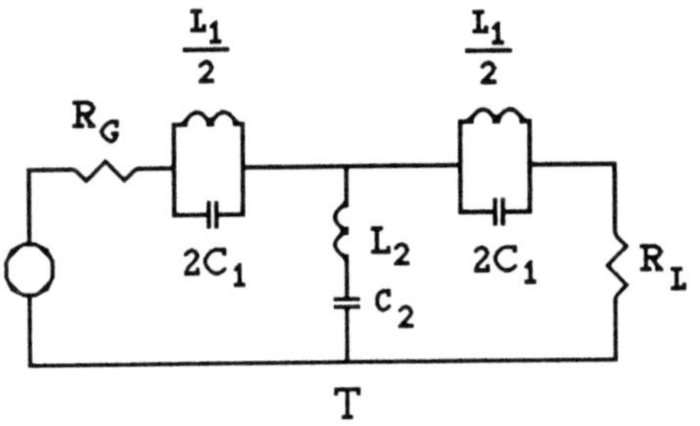

FIGURE 5.20 Circuit diagram of a constant-k band-rejection filter.

5.12 TRANSMISSION-LINE ORIGINATED M-DERIVED FILTERS

As has already been pointed out, constant-k filters have at least two major problems. First, the attenuation is small in the stop band near the cutoff frequency. Second, the characteristic impedance of the filter changes radically in both the pass and stop bands. A conceptual solution to the first problem would be to add a shunt (series) resonant circuit in the stop band near cutoff to improve the attenuation. The solution to the second problem is not so obvious. However, it can be shown that it is possible to derive special filter sections that are similar to the prototype constant-k models. When these new, special circuits are placed in tandem with the original constant-k sections, they greatly improve the near stop-band attenuation and give a much more constant impedance in the pass band.

TM THE M-DERIVED T-FILTER SECTION

The series elements in a T-section, m-derived filter are equal to the corresponding T sections in a constant-k filter multiplied by a factor m (which can vary from zero to one) and the characteristic impedance is equal to Z_0. The cutoff frequency of the m-derived section will be the same as the corresponding constant-k filter. It is assumed that, at some frequency in the near stop band, the shunt impedance will be very low or zero.

MM **THE M-DERIVED T-FILTER SECTION**

$$Z_0 = \sqrt{Z_1 Z_2 + \frac{Z_1^2}{4}} = \sqrt{mZ_1 Z_2' + \frac{m^2 Z_1^2}{4}} \text{ where} \quad \text{(E 5.52)}$$

$Z_{m\ series} = mZ_1$ (sometimes called Z_1')

$Z_{m\ shunt} = \frac{Z_2}{m} + \frac{(1+m^2)\,Z_1}{4m}$ (sometimes called Z_2') where m

can vary from 0 to 1 (E 5.53)

$$Z_{0Tm} = \sqrt{Z_{m\ series} Z_{m\ shunt} + \frac{Z_{m\ series}}{4}} \text{ and, likewise,} \quad \text{(E 5.54)}$$

$$= \sqrt{mZ_1\left(\frac{Z_2}{m} + \frac{1-m^2}{4m} Z_1 + \frac{m^2 Z_1^2}{4}\right)} \quad \text{(E 5.55)}$$

GM **THE M-DERIVED T-FILTER SECTION**

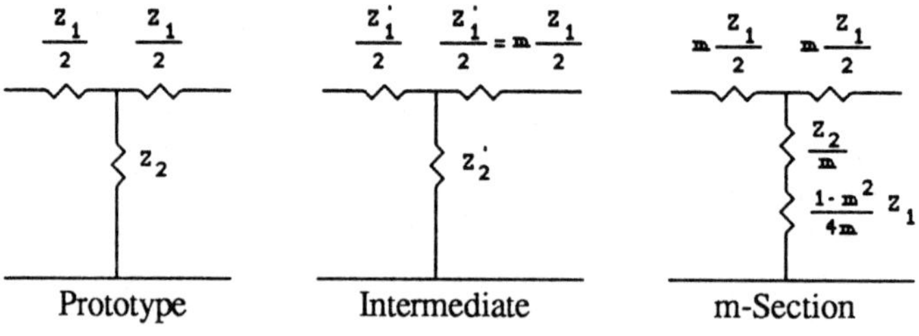

FIGURE 5.21 The development of m-derived filter models.

5.13 TM THE M-DERIVED Π-FILTER SECTION

As with most network models, the T configuration can be converted to the π configuration using well-established rules. This T to π transformation is especially useful because it provides an opportunity to investigate the effect of m on the characteristic impedance of these two prototype, m-derived networks. The equations show that the impedance of an m-derived T section is independent of m while the π section is dependent upon m. Further, it is possible to make the characteristic impedance of a π section almost constant in the pass band by using an m of 0.6.

MM THE M-DERIVED Π-FILTER SECTION

$$Z_{0\pi} = \frac{Z_{mseries} Z_{mshunt}}{Z_{0T}} \tag{E 5.56}$$

$$= \frac{mZ_1\left[\frac{Z_2}{m} + \frac{(1-m^2)Z_1}{4m}\right]}{\sqrt{Z_1Z_2 + \frac{Z_1^2}{4}}} \tag{E 5.57}$$

$$= \frac{Z_1Z_2 + (1-m^2)\frac{Z_1^2}{4}}{\sqrt{Z_1Z_2 + \frac{Z_1^2}{4}}} \tag{E 5.58}$$

GM THE M-DERIVED Π-FILTER SECTION

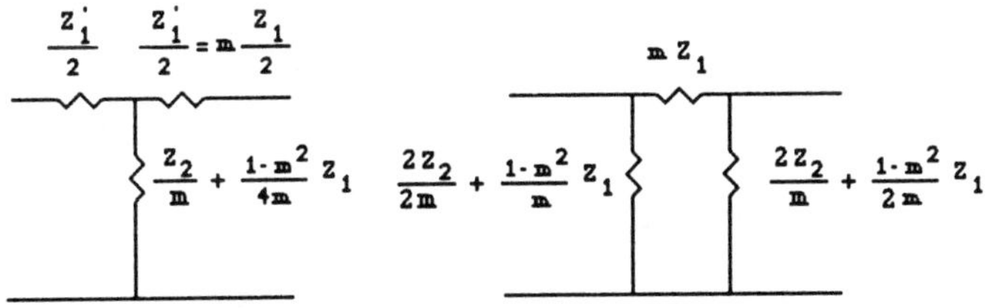

FIGURE 5.22 M-derived T and π sections.

GM THE M-DERIVED Π-FILTER SECTION IMPEDANCE

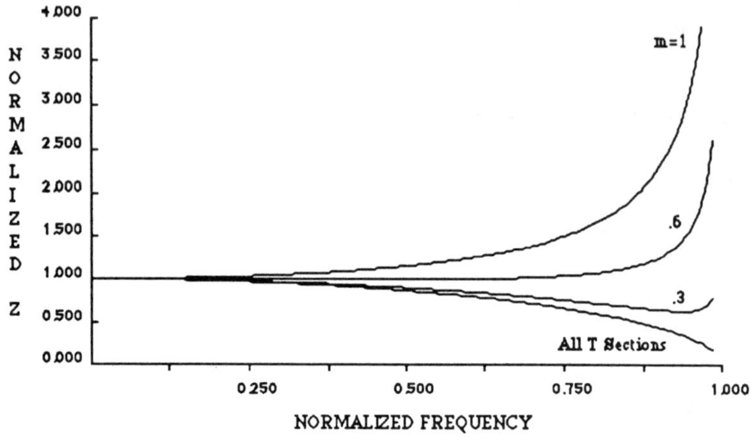

FIGURE 5.23 A plot of normalized impedance for a low-pass, m-derived π filter as a function of m up to cutoff. Note that the Z-plot of m = .6 is almost constant up to the normalized cutoff frequency of 1.0. For comparison, the lower plot shows the normalized impedance for T sections.

5.14 **MM** M-DERIVED LOW-PASS FILTER COMPONENT VALUES

$$mZ_{1\ \text{series}} = mj\omega L_k \quad \text{for a low-pass filter} \tag{E 5.59}$$

$$L_{1m} = mL_k \quad \text{for a low-pass filter} \tag{E 5.60}$$

$$L_{2m} = \frac{1-m^2}{4m} L_k \ \text{for a low-pass filter} \tag{E 5.61}$$

$$mZ_{2\ \text{shunt}} = \frac{-j}{m\omega C} + \frac{(1-m^2)j\omega L_k}{4m} \quad \text{for a low-pass filter} \tag{E 5.62}$$

$$C_{2m} = mC_k \ \text{for a low-pass filter} \tag{E 5.63}$$

$$m = \sqrt{1 - \frac{f_0^2}{f_r^2}} \quad \text{for a low-pass filter (see following derivation)} \tag{E 5.64}$$

$$f_r\,(\text{shunt}) = \frac{1}{2\pi\sqrt{\frac{1-m^2}{4m}\, mL_kC_k}} = \frac{1}{\pi\sqrt{(1-m^2)LC}} = \frac{f_0}{\sqrt{1-m^2}} \text{ where}$$

$f_0 = \frac{1}{\sqrt{\pi LC}}$ and where f_r (shunt) = $f_{resonance}$ (sometimes called f_∞).

$$\text{Since } \frac{f_0}{\sqrt{1-m^2}} \text{ and } 1-m^2 = \frac{1}{\pi^2 f_r^2 LC} = \frac{f_0^2}{f_r^2} \tag{E 5.65}$$

$$m = \sqrt{1-\frac{f_0^2}{f_r^2}} \text{ from E 5.64} \tag{E 5.66}$$

GM M-DERIVED LOW-PASS FILTER COMPONENT VALUES

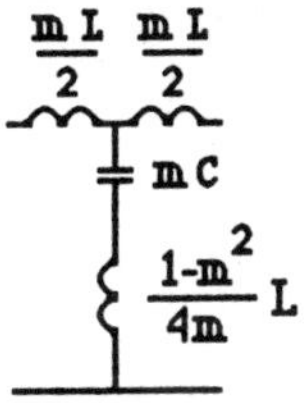

FIGURE 5.24 M-derived low-pass filter.

5.15 MM M-DERIVED HIGH-PASS FILTER COMPONENT VALUES

$$mZ_1 \text{ series} = Z_1' = \frac{-jm}{\omega C} \text{ for a high-pass filter} \tag{E 5.67}$$

$$mZ_2 \text{ shunt} = Z_2' = \frac{j\omega L}{m} - \frac{j(1-m^2)}{4m\omega C} \text{ for a high-pass filter} \tag{E 5.68}$$

$$C_{1m}' = \frac{C}{m} \text{ for a high-pass filter} \tag{E 5.69}$$

$$C'_{2m} = \frac{4mC}{1-m^2} \text{ for a high-pass filter} \quad \text{(E 5.70)}$$

$$L_{1m} = mL_k \text{ for a high-pass filter} \quad \text{(E 5.71)}$$

$$m = \sqrt{1 - \frac{f_r^2}{f_0^2}} \text{ for a high-pass filter} \quad \text{(E 5.72)}$$

$$\text{since } f_r = \frac{1}{2\pi\sqrt{\frac{4mC}{1-m^2} \cdot \frac{L}{m}}} = \frac{1}{4\pi}\sqrt{\frac{1-m^2}{LC}} = \sqrt{1-m^2}\, f_0 \text{ and}$$

$$1-m^2 = \frac{f_r^2}{f_0^2} \quad \text{(E 5.73)}$$

GM M-DERIVED HIGH-PASS FILTER COMPONENT VALUES

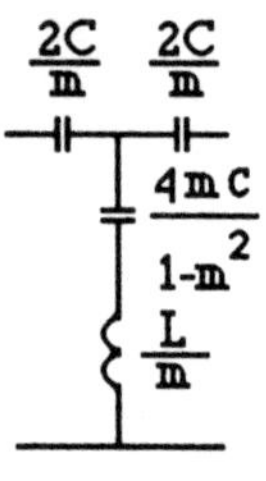

FIGURE 5.25 M-derived high-pass filter.

5.16 M-DERIVED FILTER HALF-SECTIONS

The generalized transmission line model can be constructed of T or π sections. Also, this general model can be extended with appropriate *half-sections* with no loss of generality or rigor. These special half-sections are used to match the impedance of one filter section to another or to match a section of the filter to the generator or load. It is important to match the impedances as closely as possible in order to maintain the desired filter characteristics and to provide the most possible power output in the pass band.

TM M-DERIVED FILTER HALF-SECTIONS

An m-derived π half-section can be added to the ends of a composite transmission line derived T filter to improve the match to the source and load characteristic impedance. Likewise, an m-derived T half-section can be added to the ends of a composite π filter to improve the match to the source and load characteristic impedance.

GM M-DERIVED FILTER HALF-SECTIONS

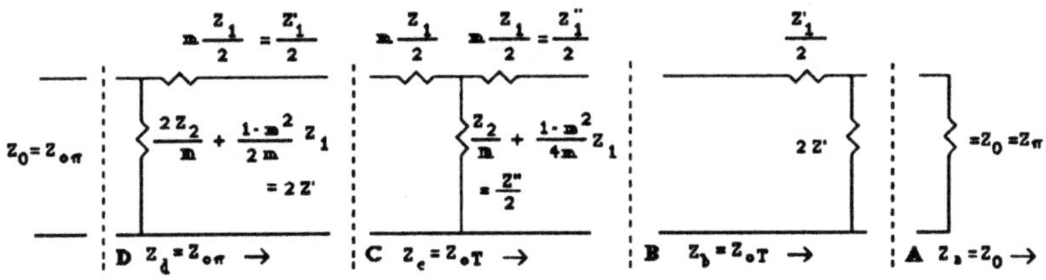

FIGURE 5.26 Matching with m-derived half-sections.

MM PROOF OF MATCHING WITH M-DERIVED HALF-SECTIONS—SEE FIGURE 5.26

At A, looking toward the load,

$$Z_a = Z_0 \tag{E 5.74}$$

At B, looking toward the load,

$$Z_b = \frac{Z_1'}{2} + \frac{2Z_2'Z_{0\pi}}{2Z_2'+Z_{0\pi}} = \frac{Z_1'Z_2' + \frac{Z_1'}{2}\frac{Z_1'Z_2'}{Z_{0T}} + \frac{2Z_2'Z_1'Z_2'}{Z_{0T}}}{2Z_2' + \frac{Z_1'Z_2'}{Z_{0T}}}$$

$$= \frac{Z_1'Z_2'Z_{0T} + 2Z_2'(Z_1'Z_2' + \frac{Z_1'^2}{4})}{2Z_2'Z_{0T} + Z_1'Z_2'} = \frac{(Z_1'Z_2'\ 2Z_2'Z_{0T})Z_{0T}}{2Z_2'Z_{0T} + Z_1'Z_2'} = Z_{0T} \quad \text{(E 5.75)}$$

At C, looking toward the load,

$$Z_c = \frac{Z_1''}{2} + \frac{Z_2''(\frac{Z_1''}{2} + Z_{0T})}{Z_2'' + Z_1'' + Z_{0T}} = \frac{Z_1''Z_2'' + \frac{Z_1''^2}{4} + Z_{0T}(\frac{Z_1''}{2} + Z_2'')}{Z_2'' + \frac{Z_1''}{2} + Z_{0T}}$$

$$= \frac{Z_{0T}(\frac{Z_1''}{2} + Z_2'' + Z_{0T})}{Z_2'' + \frac{'Z_1''}{2} + Z_{0T}} = Z_{0T} \quad \text{(E 5.76)}$$

At D, looking toward the load,

$$Z_d = \frac{2Z_2'(\frac{Z_1'}{2} + Z_{0T})}{2Z_2' + \frac{Z_1'}{2} + Z_{0T}} = \frac{Z_1'Z_2' + 2Z_2'\frac{Z_1'Z_2'}{Z_{0\pi}}}{2Z_2' + \frac{Z_1'}{2} + \frac{Z_1'Z_2'}{Z_{0\pi}}}$$

$$= \frac{Z_1' Z_2' \left(Z_{0\pi} \frac{Z_1'}{2} + Z_1 Z_2\right)}{\left(Z_1' Z_2' + \frac{Z_1'^2}{4}\right) Z_{0\pi} + Z_1' Z_2' \frac{Z_1'}{2}} = \frac{\frac{Z_{0\pi}}{2Z_2'} + 1}{\frac{1}{Z_{0\pi}} + \frac{1}{2Z_2'}} = \frac{(Z_{0\pi} + 2Z_2')\, Z_{0\pi}}{2Z_2' + Z_{0\pi}}$$

$$Z_{0\pi} = Z_0 \qquad \text{(E 5.77)}$$

Thus, $Z_a = Z_b = Z_c = Z_d = Z_0$ and the half-sections terminate the line!

TM AN M-DERIVED FILTER EXAMPLE

Design a composite low-pass constant-k filter with an additional m-derived section to improve the near cutoff attenuation in the stop band. Also use m-derived half-sections to make the filter's input and output impedance more constant in the pass band.

The pass band of the filter will be from zero Hz to 5000 Hz (f_0) with the m-derived shunt resonance (fr) at 6000 Hz.

GM AN M-DERIVED FILTER EXAMPLE

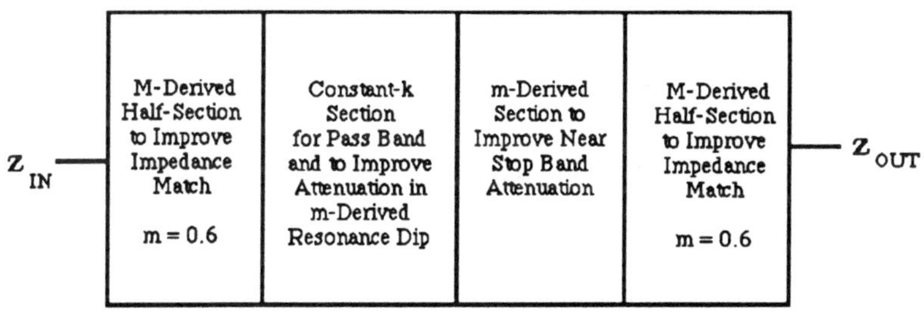

FIGURE 5.27 A block graphic of the example composite low-pass filter.

MM AN M-DERIVED FILTER EXAMPLE—FOR THE CONSTANT-K SECTION

$$C_k = \frac{1}{\pi f_0 R_0} = \frac{10^6}{\pi * 5000 * 600} = 0.106 \text{ uf} \qquad \text{(E 5.78)}$$

$$L_k = \frac{R_0}{\pi f_0} = \frac{600}{\pi * 5000} = 0.038 \text{ h}, \frac{L}{2} = 0.019 \text{ h} \qquad \text{(E 5.79)}$$

AN M-DERIVED FILTER EXAMPLE—FOR THE M-DERIVED SECTION

$$m = \sqrt{1 - \frac{f_0^2}{f_r^2}} \quad m = \sqrt{1 - \frac{5000^2}{6000^2}} = 0.306 \tag{E 5.80}$$

$$L_{1m\ series} = mL_k = 0.306 * 0.038 = 0.0116\ h \tag{E 5.81}$$

$$L_{2m\ shunt} = \frac{1-m^2}{4m} L_k = \frac{1-.306^2}{4 * .306} .038 = 0.677 * .036 = 0.028\ h$$

$$\frac{L}{2} = 0.014 \tag{E 5.82}$$

$$C_{2m} = mC_k = 0.306 * 0.016 * 10^{-6} = 0.032\ uf \tag{E 5.83}$$

AN M-DERIVED FILTER EXAMPLE—FOR EACH HALF-SECTION

First, design an m-derived T section with m = 0.6, and then divide it into two terminating half-sections. (Recall that the T section is chosen for the termination since its characteristic impedance is *independent* of the value of m.)

$$L_{1m} = mL_k = 0.6 * 0.038 = 0.0228\ h$$

$$\frac{1}{2} L_{1m} = 0.0114\ h = L_{series}\ \text{half-section}$$

$$L_{2m} = \frac{1-m^2}{4m} L_k = \frac{1-.6^2}{4 * .6} .038 = 0.01\ h$$

$$L_{2m} = 0.01\ h = L_{shunt}\ \text{half-section}$$

$$C_{2m} = mC_k = 0.6 * 0.106 * 10^{-6} = 0.0636\ uf$$

$$\frac{1}{2} C_{2m} = 0.0318\ uf = C_{shunt}\ \text{half-section}$$

GM AN M-DERIVED FILTER EXAMPLE—FOR THE HALF-SECTIONS

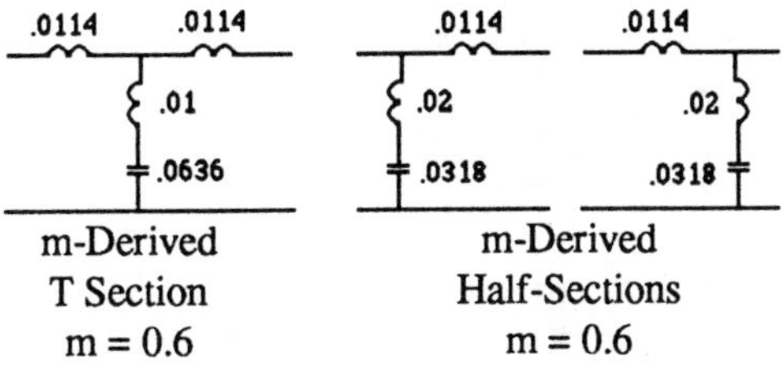

FIGURE 5.28 Derived half-sections.

GM AN M-DERIVED FILTER EXAMPLE—FOR ELEMENTARY COMPOSITE FILTER

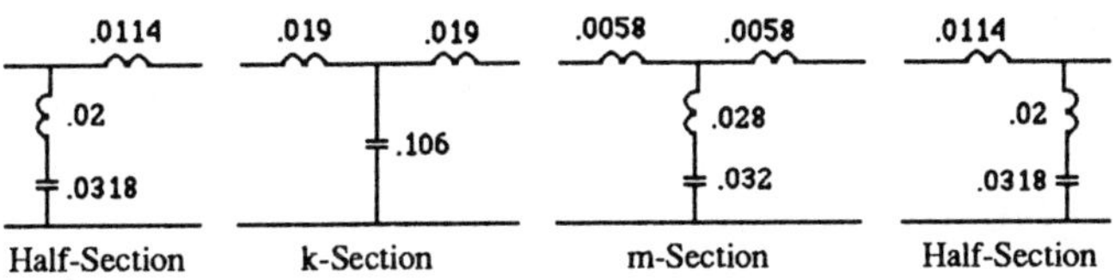

FIGURE 5.29 Elementary composite LP filter.

GM AN M-DERIVED FILTER EXAMPLE—FOR COMBINED COMPOSITE FILTER

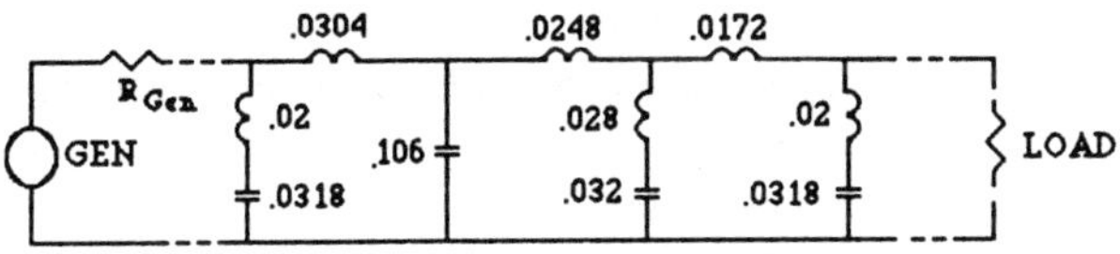

FIGURE 5.30 The example composite LP filter.

GM AN M-DERIVED FILTER EXAMPLE—A BODE PLOT OF THE COMPOSITE FILTER

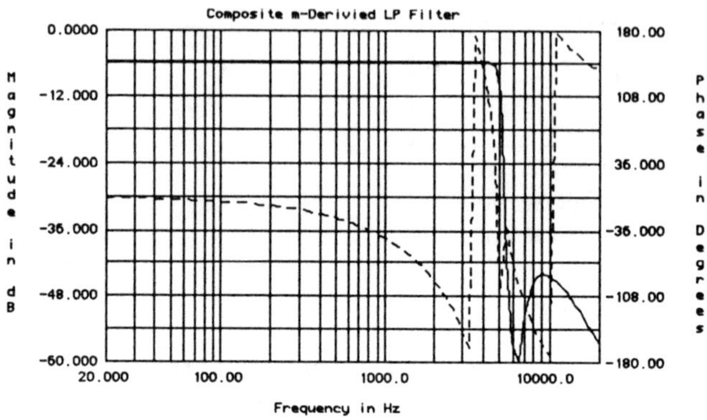

FIGURE 5.31 A Bode plot of amplitude and phase vs. frequency. (The pass band is down 6 dB because of the 600-ohm source generator and the 600-ohm load.)

5.17 POLYNOMIAL FILTERS

TM POLYNOMIAL FILTERS

Polynomial filter design has evolved from using both simple RL or RC networks and networks derived from transmission-line models advancing into a technique of using the mathematics of the transfer function. Polynomial filters include a wide range of circuits, including RF or IF tuned circuits; passive, lumped-constant networks; active, analog filters; and a variety of the appropriate circuit duals in digital configurations.

Polynomial filter design is usually based upon such parameters as overshoot and damping, ringing or ripple, and phase response and phase (group) delay as well as frequency response and cutoff. They are considered *approximate*, or compromise, filters since it is usually impossible to exactly match such requirements as overshoot *and* rapid cutoff or rapid cutoff *and* linear phase—thus, a compromise is required. An insight into the theory as well as the practical design and analysis of polynomial filters can be obtained by a review of the elementary concepts of the transfer function.

An elementary transfer function is a mathematical statement of division. For an ideal low-pass filter, the transfer function ratio $H = V_o/V_i$ would be unity in the pass band and 1/N in the stop band where N would be a large number approaching infinity. Practical considerations make it impossible to have a unity value for all of the pass band or to have any fixed attenuation number in all of the stop band. However, it is usually possible to find a combination of numbers (a polynomial) that will give a transfer function (response) of *about* unity in the pass band and *about* some attenuation value in the stop band. For example, using the simple RC low-pass filter,

$$H_{oi}\,(j\omega) = \frac{V_o}{V_i} = \frac{X_c}{\sqrt{R^2 + X_c^2}} \text{ and} \tag{E 5.84}$$

$$V_o = V_i * \frac{X_c}{\sqrt{R^2 + X_c^2}} \tag{E 5.85}$$

Using E 5.85, V_o is about one times (about equal to) V_i at very low frequencies, but changes to a smaller fraction of V_i as the frequency increases. Some of the polynomial equations are not continuous in both the pass band and the stop band and, when they are plotted together, require graphic manipulation to get a continuous curve.

Polynomial filters are usually specified by order—the order relating to the highest power of radian frequency that appears in the transfer function. The fundamental transfer function equation is usually considered a polynomial in "s" where s is equal to the angular frequency jω. The factored s terms in the numerator and denominator can be plotted in the complex s plane, which will give a good first approximation for both the frequency and the transient response.

MM POLYNOMIAL FILTERS—THE GENERAL TRANSFER FUNCTION

$$\frac{e_{in}}{e_{out}} = f(S) = K \frac{1}{S^n + aS^{n-1} + bS^{n-2} + \cdots\cdots dS + 1} \tag{E 5.86}$$

where K is the value of the function at ω=0 (DC).

In factored form,

$$f(S) = K \frac{1}{(s-d_n)(s-d_{n-1}) \cdots (s-d)} \tag{E 5.87}$$

GENERAL POLYNOMIAL EQUATION EXAMPLES

A general fifth order:

$$f(S) = K\frac{1}{S^5 + aS^4 + bS^3 + cS^2 + dS + 1} \quad \text{(E 5.88)}$$

A general third order:

$$f(S) = K\frac{1}{S^3 + cS^2 + dS + 1} \quad \text{(E 5.89)}$$

A general second order:

$$f(S) = K\frac{1}{S^2 + dS + 1} \quad \text{(E 5.90)}$$

A general first order:

$$f(S) = K\frac{1}{S + 1} \quad \text{(E 5.91)}$$

For any order of filter above the first, the appropriate coefficients a, b, c, ···· can be modified to change the shape of the voltage transfer function response both in the pass band and the stop band. These factors are derived from several polynomials known as Butterworth, Chebyshev, Bessel, Elliptic, Cauer, etc. Some of these polynomials, including Chebyshev, also have subsets that will produce certain modifications on the general response of the main polynomial.

5.18 POLYNOMIAL CHARACTERISTICS

TM BUTTERWORTH

The simplest polynomial. The filter response somewhat resembles that of simple cascaded RC or RL network. The response is smooth in both the pass and stop bands, but it does not have a linear phase response. The attenuation approaches 6n dB per octave in the far stop band when n equals the number of polynomial coefficients, which is also equal to the number of RC or RL sections.

MM BUTTERWORTH

$$|H_{lp}(j\omega)|^2 = \frac{1}{1+\left(\frac{\omega}{\omega_{co}}\right)^{2n}} = \frac{1}{1+\left(\frac{\omega}{1}\right)^{2n}} = \frac{1}{1+\omega^{2n}} \text{ for } \omega_{co} \text{ normalized}$$

to unity. (E 5.92)

GM BUTTERWORTH

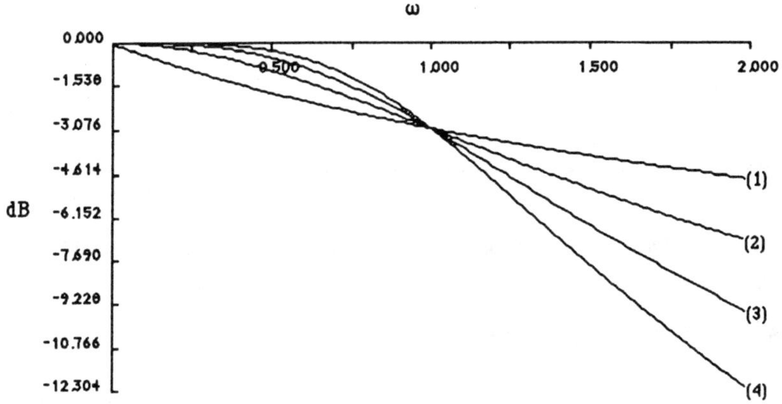

FIGURE 5.32 The response of Butterworth filters having one–four poles.

TM CHEBYSHEV

A family of filters that can produce rapid roll-off in the stop band at the expense of ripple (smooth up and down changes) in the response in the pass band. There are also subsets of the main polynomial that control the amplitude of this ripple as a tradeoff against rapid roll-off. This polynomial can give a rapid roll-off, but it does not have linear phase in the pass band or good transient response.

MM CHEBYSHEV

$$|H_{lp}(j\omega)|^2 = \frac{1}{1+\varepsilon C_n^2(\omega_n)}$$ where ε is the ripple amplitude constant ($0 \lesssim \varepsilon \gtrsim 1$) and

C stands for Chebyshev. (E 5.93)

GM CHEBYSHEV

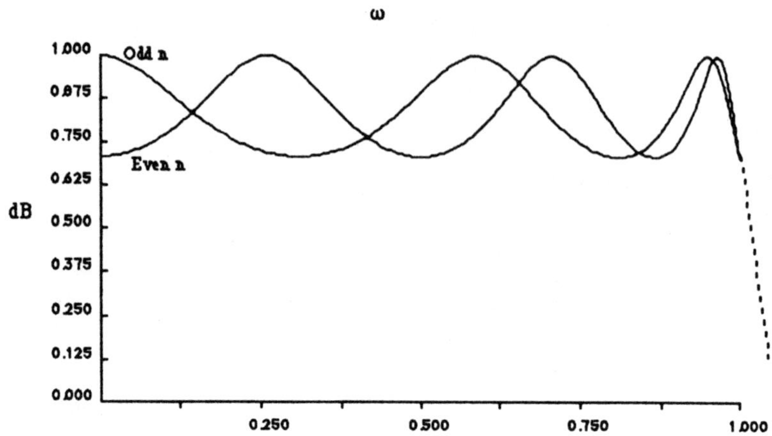

FIGURE 5.33 The general response of a Chebyshev filter with n even or n odd number of poles.

TM BESSEL OR MAXIMALLY FLAT

The filter design that produces linear phase (constant delay) in the pass band. These filters are used in video or radar where the fidelity of the phase and wave shape transient response is especially important. It does not cut off as fast as the Butterworth. The design of Bessel filters is based upon a series of equations in higher mathematics called Bessel polynomials. See the Weinberg, Rabnier and Gold references.

MM BESSEL OR MAXIMALLY FLAT

$$H(j\omega) = \frac{(2n)! / 2^n n!}{B_n}$$ where n is the filter order and B_n is the Bessel polynomial (E 5.94)

GM BESSEL OR MAXIMALLY FLAT

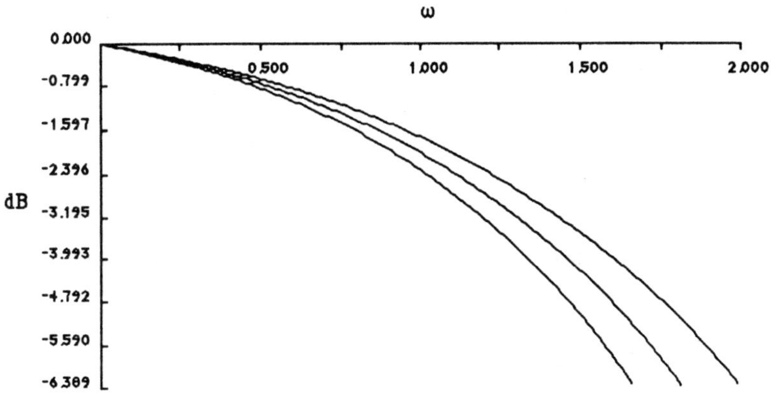

FIGURE 5.34 The general response of Bessel filters.

TM ELLIPTICAL OR CAUER

Filters that produce steep roll-off and infinite rejection of some frequencies in the stop band at the expense of rather severe ripples (ellipsis) in the stop band. Elliptical filters have very poor phase and transient response.

Elliptical filters really fall in a special class of polynomial filters since, unlike Butterworth, Chebyshev and Bessel, they have zeros in the numerator as well as poles in the denominator; thus they are not all-pole filters. The numerator zeros improve the roll-off rate, but produce ripples in the stop band. Although their transfer function somewhat resembles the Chebyshev filter, there are now three parameters that must be considered in their design. These include ε and C (now sometimes called CC for Cauer-Chebyshev) as well as a selectivity factor (commonly called R) that controls the sharpness of the cutoff. R is usually defined as ω_{2s}/ω_{1p}, where ω_{1p} is the upper edge frequency of the pass band ripple and ω_{2s} is the lower edge frequency of the ripple in the stop band—both defined for a low-pass filter.

MM ELLIPTICAL OR CAUER

There is no simple transfer function.

GM ELLIPTICAL OR CAUER

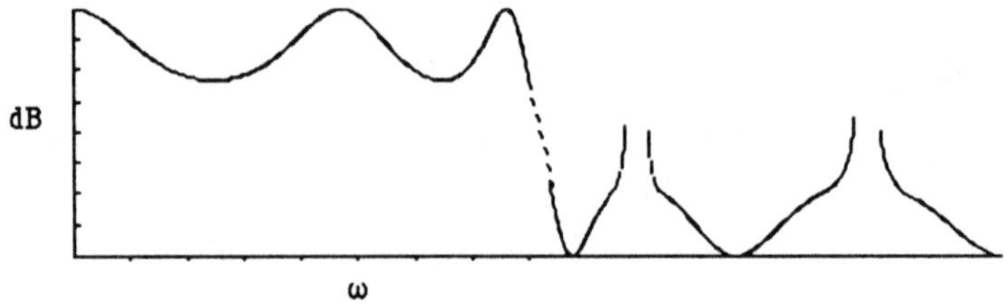

FIGURE 5.35 A generalized sketch to illustrate elliptical filter characteristics.

5.19 RULES FOR THE DESIGN OF POLYNOMIAL FILTERS

Because the various polynomials provide so many variations on the filter characteristics, the actual design procedure requires many, sometimes confusing and frustrating, choices of parameters and techniques. The following list is a guide to some of the questions that must be answered before an actual design should begin.

1. What is the rate of cutoff required? A Chebyshev or elliptic design may be the first choice but there may be other requirements that will limit their use.
2. Can ripple in the pass band and/or the stop band be tolerated? If so, how much?
3. Is good (linear) phase and transient response required? In most cases, this limits the design to the use of Bessel polynomials.
4. Is a given design procedure based upon *normalized* impedance? If it is, the preliminary design must be scaled to the characteristic impedance of the original design specifications.
5. Is a given design procedure based upon normalized frequency? Again, if it is, the preliminary design must be scaled.
6. Is it a low-pass, high-pass, band-pass or band-rejection filter? Most design techniques are based upon a prototype low-pass design and require a scaling transformation; i.e., design a low-pass filter and use a set of rules to scale (transform) it into a high-pass, etc.
7. If it is a band-pass or band-rejection filter, does it require a narrow-band or wide-band design. A wide-band design creates the pass or rejects band-pass with the combination of an appropriate low-pass and high-pass filter. A narrow-band design requires an entirely different design approach based upon a tuned circuit.

8. Will the filter use an odd or even number of poles? There are certain limitations in the polynomials that require either even or odd poles for a workable filter.
9. Will the input (the first element) be capacitive or inductive; i.e., is this a basic T or π configuration? Each one may have separate design rules.

In the past, polynomial filters have been designed using very well-executed tables and charts. Even with several of these references at hand, the design procedure could be mysterious and certainly required several iterations—it was as much an art as a science. A more contemporary approach to these problems, especially the ones involving repeated cut and try, would be to use an appropriate computer-aided design program. There are a number of these programs available for either personal or mainframe computers. As an illustration, the following is an abbreviated, edited menu of a program called LCFIL, published by BV Engineering in Riverside, California.

Filter Operations:

1. Enter new filter (enter heading, number of poles, family type)
2. Load (old) filter
3. Save filter
4. Compute response
5. Compute noise equivalent bandwidth
6. Compute spectral data
7. Print schematic
8. Print values
9. Edit components, frequencies, order, impedance/Q, topology, family

Topology:

1. Low pass—capacitor or inductor input
2. Band pass—Mesh or Nodal capacitor coupled including *all pole* or Cauer which includes zeros
3. High pass—capacitor or inductor input

For any of the above, enter source Z (no normalization or scaling is required), inductor Q, etc.

The filter response can now be calculated. Any or all of the component values can be modified for another response calculation. When the response listing looks promising, the program will print a simplified schematic and the response data can easily be used in other programs to obtain a Bode plot or transient response plot. Computer programs of this type make polynomial filter design much faster as well as giving designers added versatility by allowing them to change the value of inductor Q.

PSpice (MicroSim Corporation) also has an excellent "Advanced Filter Designer" which makes polynomial filter design fast and accurate. One can choose the filter type, pass and stop band frequency limits and attenuation, and the pass-band ripple, etc. The menu makes it easy to compare different designs.

5.20 ELEMENTARY ACTIVE FILTERS

TM ELEMENTARY ACTIVE FILTERS

By using wide band operational amplifiers or, in some cases, simple emitter or source followers, filters can be created that do not need physical inductors. This reduces the size, weight and cost of these filters and also has inspired some very clever variations on the basic classic filters. Not only is an op amp needed to isolate very simple cascaded filter circuits, but the fact that it is indeed ACTIVE and can feed back energy makes it possible to simulate very high-Q inductors and resonant circuits. Some elementary passive filter circuits (see Figure 5.1), along with their equivalent operational amplifier configurations, are shown below. The major limitation is the bandwidth of the operational amplifiers.

GM ELEMENTARY ACTIVE FILTERS

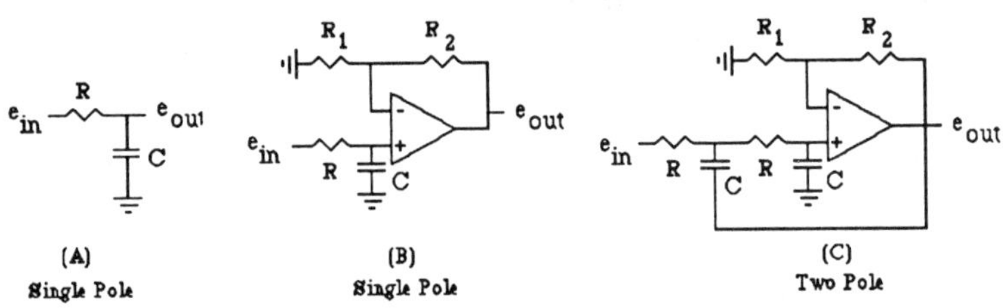

FIGURE 5.36 An elementary low-pass filter in component and operational amplifier form.

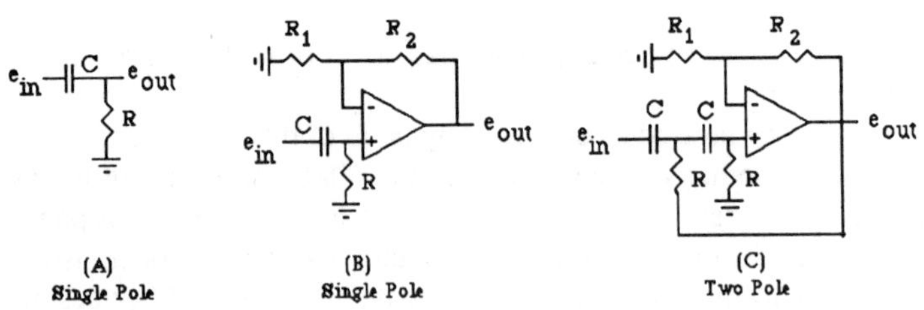

FIGURE 5.37 An elementary high-pass filter in component and operational amplifier form.

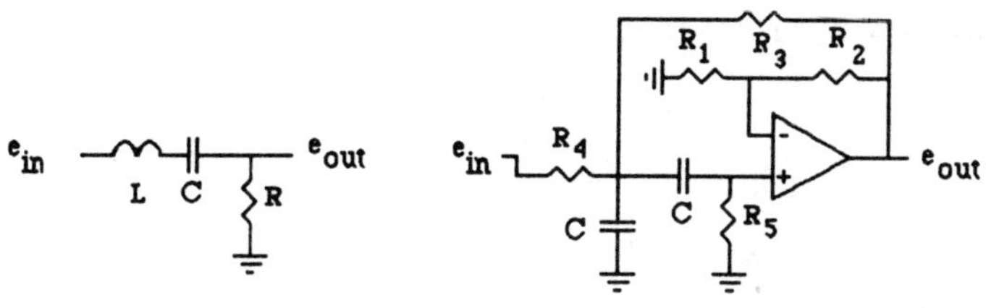

FIGURE 5.38 An elementary second order band-pass filter in component and operational amplifier form (the band-pass filter must have 2n poles).

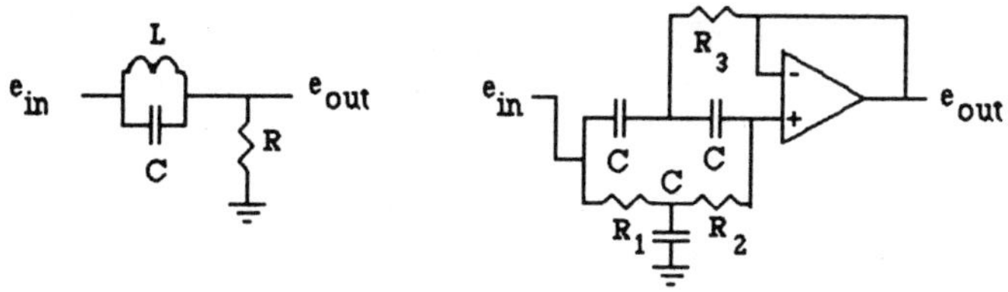

FIGURE 5.39 An elementary second order band-reject filter in component and operational amplifier form (the band-reject filter must have 2n poles).

5.21 ADVANCED ACTIVE FILTERS

TM ADVANCED ACTIVE FILTERS

The models for operational amplifier active filters can take several generic forms. The most common are the Sallen-Key, which are really adaptations of the above elementary filters, and the multiple feedback. The very clever biquad and the state variable forms will also be included.

The Sallen-Key circuits work well and can easily be configured in n-pole Butterworth, Bessel, Chebyshev, and, in some cases, elliptical families. The Q of Sallen-Key Filters is somewhat limited in a single amplifier band-pass implementation. Multiple feedback active filters usually use at least one less component per filter section than the Sallen-Key. Although the biquad filters require more components than the Sallen-Key or the multiple feedback, this circuit has the advantage of not only simulating very high-Q tuned circuits but the Q of these circuits is easily controlled. Although the biquad can be configured in both high and low passes, it is usually used as a variable-Q, band-pass or band-reject filter. The state-variable configuration is very similar to the biquad, but has the advantage that one filter can produce a variable circuit gain as well as low-pass, high-pass and band-pass outputs.

GM ADVANCED ACTIVE FILTERS

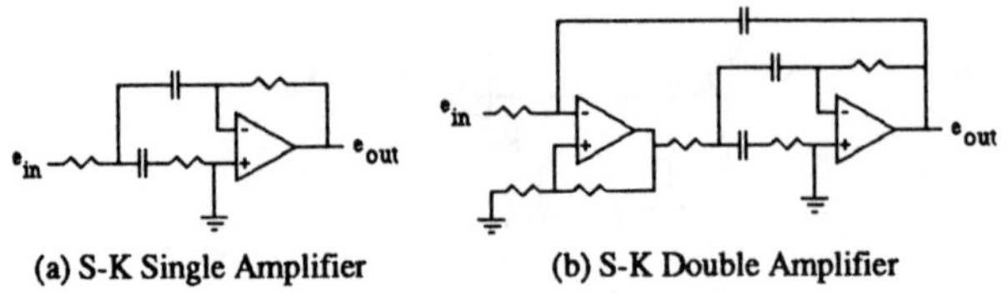

FIGURE 5.40 Two Sallen-Key band pss amplifier configurations.

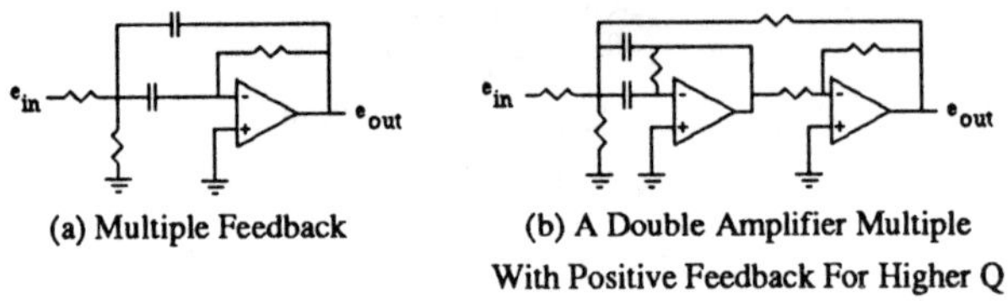

FIGURE 5.41 Multiple feedback band pass filters.

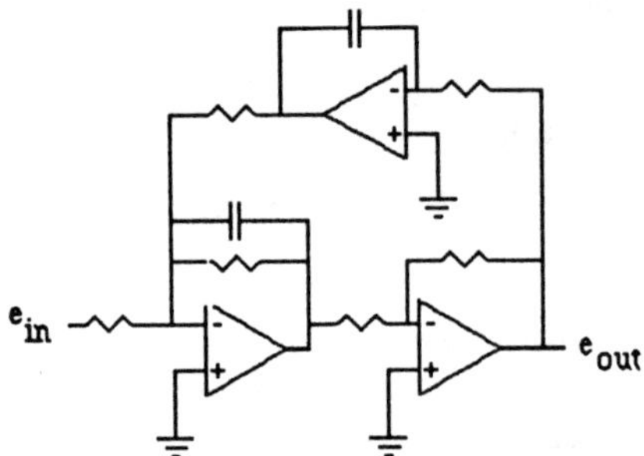

FIGURE 5.42 A second order biquad band pass filter.

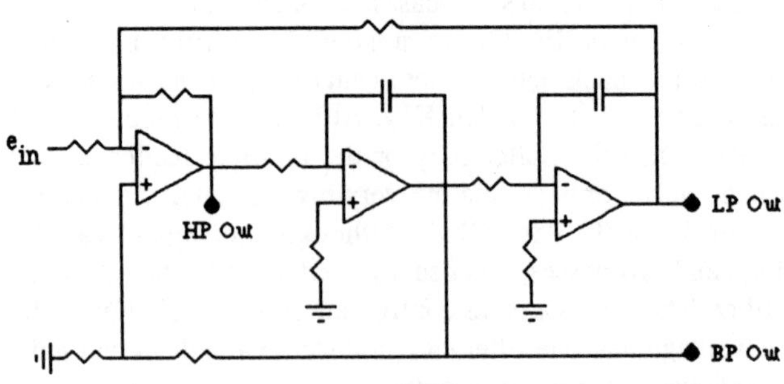

FIGURE 5.43 The state-variable filter.

5.22 DESIGNING ACTIVE FILTERS

The use of active filters was popularized circa 1960–70 in analog music synthesizers. Many books, especially those of Lancaster, have detailed the analysis and use of operational amplifiers in polynomial filter circuits. However, these references, in turn, use design tables. The problems of tables were outlined in previous sections.

There are two approaches to computer active filter design. General circuit analysis programs, such as PSpice, do an excellent job once the form of the filter, the first attempt number of poles, etc., has been established. Circuit modifications are convenient, as are steady-state and transient plots. The disadvantage is the quite considerable knowledge of polynomial filters AND operational amplifier filter circuits required before the computer design can begin.

A second, less demanding, approach is to use a specific active filter-design program. As an example, ACTFIL, produced by BV Engineering, or "Advanced Filter Design," a specialized addition to the general PSpice program, require less detailed knowledge. To start, they provide the user with several filter models and an easy-to-use design menu. ACTFIL is limited to Butterworth and Chebyshev filters in a Sallen-Key, multiple feedback, or biquad configuration. ACTFIL and "Advanced Filter Design" not only compute the design parameters, but they can compute LaPlace transfer function coefficients and the filter's magnitude, phase, and delay characteristics. These specific programs are excellent learning and design tools.

6

RLC Circuits: Forced and Natural Response

6.1 ABOUT THIS CHAPTER

This chapter is about energy—the storage of energy and the release of energy. Inductors and capacitors store energy—resistors dissipate energy. However, resistors are extremely important in circuit energy exchanges because they govern the rate at which the energy is stored or released. If the circuit is driven (forced) with a sine wave, the relative values of the resistors in the RC's, RL's, or RLC's determine how rapidly they can store or release the sinusoidal current or voltage and thus determine the frequency response of the circuit. Likewise, if the circuit is forced with a step or square wave, the component relationships, i.e., the time constants, determine the rate of rise or fall of the output wave relative to the input. For LINEAR circuits, the driven or forced response to a continuing sine wave is a sine wave. For circuits forced with a step or pulse, the response is an oscillation (a special case) or an increasing or decreasing exponential.

THE NATURAL RESPONSE is a second, more subtle, relationship between the RLC components and their voltages and currents. This *natural* response is sometimes overlooked but it may be more important to the proper operation of the circuit than the forced response. The release or exchange of component energy that is not caused directly by the input forcing function is termed the natural response of the circuit. In practical terms, the natural response can occur when the forcing function is radically changed or removed (1) in a low-impedance, constant-voltage generator for a series circuit or (2) in the high-impedance constant-current generator for a parallel circuit. When these forcing functions are removed, the stored energy will dissipate in its own circuit. It is the natural response that, many times, produces the dangerous or damaging transient in high-powered TV or radar transmitters. It is also the natural response, from power supply turn-off, that can produce the high reverse-voltage spike, which may destroy critical integrated circuits in a system.

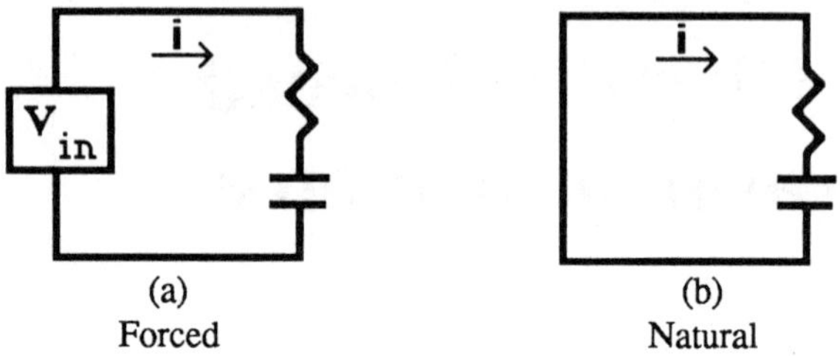

FIGURE 6.1 Elementary crcuits to illustrate circuit response.

In this chapter, it is assumed that all complex networks have been simplified and reduced to an equivalent RC, RL, or RLC circuit. The networks are usually one pole, which are termed first order, or two pole, which are termed second order. The phrases *first order* and *second order* refer to the highest value of the exponents, one or two, in the equations that describe these networks. By using these simplified networks, it is easier to understand the underlying principles of natural response. More complex, real-world networks can be analyzed by such computer programs as BV Engineering's SPP or MicroSim's PSpice. The models that are produced include the Kirchhoff's equation relationships for the network, the circuit impedance and transfer function, and the (Bode-type) steady state and (s-plane) transient information.

THE MATHEMATICAL MODELS for finding the voltage and current response in RLC circuits include the following general steps:

1. For the *frequency* (usually sine-wave) response, the impedances (Z) and the transfer function (H) must be defined in terms of (jω). Using the voltage division rule, the voltage output across any *series* component a, b, $\cdots$ n is equal to the input voltage times the ratio of that one complex impedance $Z_n(\omega)$ divided by the total series impedance $Z_{Tot}(\omega)$ or $v_n = v_{in} * Z_n/Z_{Tot}$. Likewise, using the current divider rule, the current in any *parallel* component is $i_n = i_{in} * Z_{Tot}/Z_{other}$ (remember, when using the current divider rule, use the *other* circuit element in this formula). Since the impedances are functions of jω, there are usually only algebraic equations to solve.
2. For the *time* (usually transient) response, such as from an input step, all impedances and the transfer function H must be defined in terms of (s).

3. For the time (transient) response, the Kirchhoff equations are written in the form $v_{in} = v_a + v_b + \cdots v_n$ for series circuits and $i_n = i_a + i_b + \cdots i_n$ for parallel circuits. These equations usually can be simplified by differentiation—see Chapter 14.
4. The final differential equations are then solved for the common series transient current i_{tr} or the common parallel voltage v_{tr}. As a first guess, the FORM of the solution is usually an increasing or decreasing exponential (e^x).
5. In a series current, any individual component voltage response is $v_n = i_{ser}Z_n$ and, in a parallel circuit, any individual component current response is $i_n = e_{par}/Z_n$.

THE GRAPHIC MODELS are presented to show the FORM of the output response. In general, the outputs will be increasing or decreasing exponentials. The Graphic Models usually will also show an s-plane plot of the circuit transfer function. This pole-zero plot indicates the FORM of the transient response (oscillation, overshoot, or decaying exponential) of the circuit.

THE FOLLOWING DEFINITIONS of terms and procedures will be used in this chapter:

1. The fundamental Kirchhoff equations $v_1 + v_2 + \cdots = 0$ or $i_1 + i_2 + \cdots = 0$ are presented in the form $v_{in} = v_1 + v_2 + \cdots$ or $i_{in} = i_1 + i_2 + \cdots$
2. The steady state (sine wave) equations are presented as functions of radian frequency (ω). The corresponding amplitude and phase response is described in Chapter 3.
3. The transfer function is assumed to be e_{out}/e_{in}, i_{out}/i_{in}, etc., where the e or i will be indicated.
4. The transient equations are presented as functions of the complex plane operator (s).
5. The input step forcing function can be visualized as coming from an ideal battery and a switch, an ideal step in voltage or current, or a *long* square wave of voltage or current.
6. The following equations relate to the steady-state (sine wave) and transient voltages and currents for RLC components.

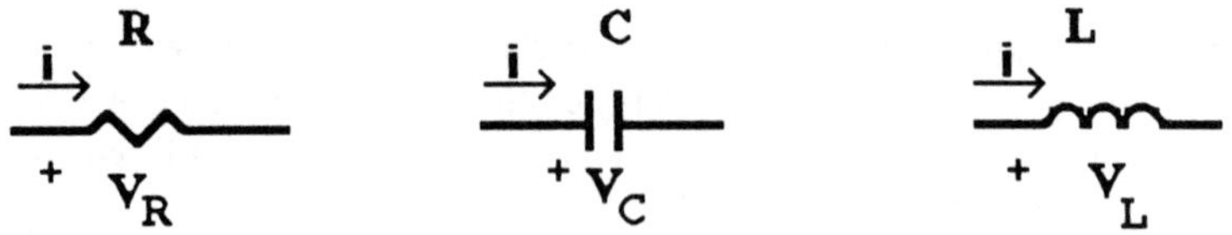

$$v_R = R\,I \qquad v_C = \frac{1}{j\omega C}\,i \qquad v_L = j\omega L\,i$$

$$v_C = \frac{1}{C}\int_{t0}^{t} i\,dt + v_0 \qquad v_L = L\frac{di}{dt}$$

$$I_R = \frac{1}{R}\,V \qquad i_C = j\omega C\,v \qquad i_L = \frac{1}{j\omega L}\,v$$

$$i_C = C\frac{dv}{dt} \qquad i_L = \frac{1}{L}\int_{t0}^{t} v\,dt + i_0$$

$$Z_R = \frac{V}{I} = \frac{Ri}{i} \qquad Z_C = \frac{v}{i} = \frac{v}{sCv} = \frac{1}{sC} \qquad Z_L = \frac{v}{i} = \frac{sLi}{i} = sL$$

Note: $s = j\omega$ for sine/cosine waves only!

FIGURE 6.2 The voltage, current and impedance for RCL components.

6.2 THE SERIES RC OR RL CIRCUIT

TM THE RESPONSE OF AN RC OR RL SERIES CIRCUIT

In any SERIES circuit, one common current, produced by the input forcing voltage and/or the stored energy in the circuit, flows in all the components. The instantaneous voltage across any component is this common current times the impedance of that component. Thus, the analysis problem is to find the common current and then to find the voltage across one or more of the components. The problem can be simplified if only one E_{out} is required. By using the series circuit voltage devider rule, $v_n = v_{out} = v_{in} * Z_n / Z_{Total}$.

MM THE FREQUENCY (SINE-WAVE) RESPONSE OF AN RC OR RL SERIES CIRCUIT

$Z_{(\omega)} = \sqrt{R^2 + X_c^2}$ is the sinusoidal impedance of a series RC circuit.

$= \sqrt{R^2 + X_L^2}$ is the sinusoidal impedance of a series RL circuit.

(E 6.1)

$H_{(\omega)} = \dfrac{E_{out}}{E_{in}} = \dfrac{iX_C}{\sqrt{iR^2 + iX^2}} = \dfrac{X_C}{\sqrt{R^2 + X^2}}$ and

$= \dfrac{X_L}{\sqrt{R^2 + X^2}}$ where $H_{(\omega)}$ is the steady-state transfer function for

a series RC or RL circuit. (E 6.2)

$i_{series} = v_{in} / Z_{(\omega)}$ (E 6.3)

$v_{(\omega in)} = R\,i + jX_C\,i$ or

$= R\,i + jX_L\,i$ is the steady-state Kirchhoff voltage equation (E 6.4)

$v_{(\omega out)} = H_{(\omega)}\, v_{(\omega in)}$ and is the steady-state response—

see Chapter 3. (E 6.5)

MM THE TIME (STEP) RESPONSE OF AN RC OR RL SERIES CIRCUIT

$Z_{(s)} = R + \dfrac{1}{sC}$ or $R + sL$ is the series impedance in (s). (E 6.6)

$H_{(s)} = \dfrac{1}{s + \dfrac{1}{RC}}$ or $\dfrac{s}{s + \dfrac{R}{L}}$ is the series RC or RL transfer function

in (s) where RC or L/R can be reduced to a time constant T. (E 6.7)

$i_{ser} = v_{in} / Z_{(s)}$ (E 6.8)

$v_{in} = v_1 + v_2 = v_R + v_C$ or $v_R + v_L$ for a series RC or RL circuit. (E 6.9)

AN EXAMPLE: AN RC SERIES CIRCUIT EXAMPLE—FINDING THE COMMON CURRENT:

$v_{step\ in} = vR + vC$

$v_{step\ in} = Ri + \frac{1}{C}\int_0^t i\,dt + vc0$ is the transient (in time) Kirchhoff equation, where $v_{step\ in}$ is the input, and vc0 is any voltage on C at time zero—assumed to be zero. (E 6.10)

$R\frac{di}{dt} + \frac{1}{C}i = 0$ or $\frac{di}{dt} + \frac{1}{RC}i = 0$ is an equation in i and di that can be obtained by differentiating the above E 6.10. (E 6.11)

$i_{series} = i_0\,e^{-t/RC}$ is the solution of the above differential equation when i_0 is the current at t_0. It shows the familiar exponential output form—see Chapter 14.

Since the capacitor voltage cannot change instantaneously,

$i_0 = \frac{V_{step} - V_{c0}}{R}$ or $i = \frac{V_{step} - V_{c0}}{R}e^{-t/RC}$ or $i = \frac{V_{step}}{R}e^{-t/RC}$ if $V_{c0} = 0$ (E 6.12)

$v_C = V_{step} - iR$ is Kirchhoff's equation for the series RC circuit or, from above (E 6.13)

$v_C = V_{step} - V_{step}\,e^{-t/RC}$ which is the step-response output. (E 6.14)

AN EXAMPLE: AN RL CIRCUIT EXAMPLE—FINDING THE COMMON CURRENT:

$v_{step\ in} = v_R + v_L$ (E 6.15)

$v_{step\ in} = Ri + L\frac{di}{dt}$ and, solving the differential equation (see Chapter 14), (E 6.16)

$i_{series} = \frac{E}{R}(1 - e^{-\frac{R}{L}t}) = \frac{E}{R} - \frac{E}{R}e^{-\frac{R}{L}t}$ where $\frac{E}{R}$ is the steady-state part and $-\frac{E}{R}e^{-\frac{R}{L}t}$ is the transient part (E 6.17)

MM THE NATURAL RESPONSE OF AN RC OR RL SERIES CIRCUIT

AN EXAMPLE: A SERIES RC CIRCUIT NATURAL RESPONSE EXAMPLE:

$\Sigma_v = 0 = v_C - v_R = V_0 - \frac{1}{C}\int_0^t i\,dt - Ri$ is the Kirchhoff equation

for the series natural response. (E 6.18)

$C\frac{dv}{dt} + \frac{v}{R} = 0$ is the natural response loop current for a series CR

circuit, or differentiating, (E 6.19)

$R\frac{di}{dt} + \frac{1}{C}i = 0$ is a (Kirchhoff) differential equation, in terms of

i and di, that gives the instantaneous voltage drops around the closed loop. (E 6.20)

To find the actual current i, this differential equation must be solved. A good guess about the FORM of the solution is to remember that most electric circuits change (grow or decay) in an exponential (e^x or e^{-x}) manner. Using the form e^x, the solutions for i and di become $i = A\,e^{st}$ and $di/dt = sAe^{st}$ where A is an amplitude constant and s is in terms of time constants which can, in turn, be reduced to dimensions of frequency. (E 6.21)

$RsAe^{st} + \frac{1}{C}Ae^{st} = 0$ is the solution of E 6.20 where s = -1/RC

(first order time constant). (E 6.22)

To find A, using physical reasoning, the instant the circuit goes into its natural response the capacitor is a short circuit. Thus, the current at that t_0 is V_0/R where V_0 was the voltage across the capacitor just before the circuit went to natural response. Therefore,

$A = \frac{V_0}{R}$ and, using this and E 6.21, (E 6.23)

$i_{series} = A\,e^{st}$ or $\frac{V_0}{R}e^{st} = \frac{V_0}{R}e^{(-1/RC)\,t}$ for all time after t_0 (E 6.24)

AN EXAMPLE: A SERIES RL CIRCUIT NATURAL RESPONSE EXAMPLE:

$$\Sigma_v = 0 = v_C + v_L = L\frac{di}{dt} + Ri \text{ or } \int_{t_0}^{t} \frac{di}{i} = -\frac{R}{L}\int_{t_0}^{t} \frac{dt}{1} \quad \text{(E 6.25)}$$

$i_{nat} = I_0\, e^{t/\tau} = I_0\, e^{-t(R/L)}$ where I_0 was the current in the inductor at the instant the circuit went into its natural response. (E 6.26)

$$v_R = iR = I_0 R\, e^{-t(R/L)} \quad \text{(E 6.27)}$$

GM THE SINE-WAVE RESPONSE OF AN RC AND RL CIRCUIT

Refer to Figures 5.4 and 5.5.

GM THE STEP RESPONSE OF A SERIES RC CIRCUIT

See Figure 3.15.

GM THE STEP RESPONSE OF A SERIES RL CIRCUIT

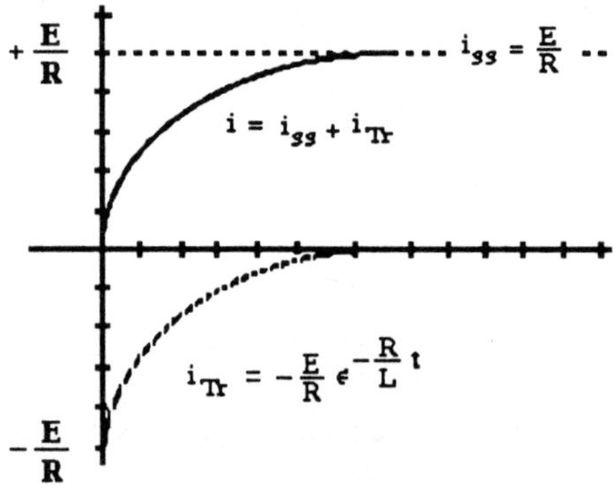

FIGURE 6.3 The step response of a series RL circuit showing that the series current is a combination of a transient component (i_{Tr}) and a steady-state component (i_{ss}).

GM THE NATURAL RESPONSE OF AN RC AND RL CIRCUIT

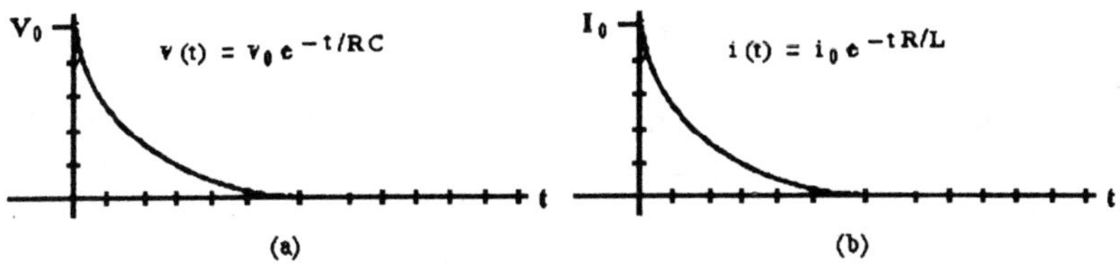

FIGURE 6.4 The natural response of a series RC circuit (a) and an RL circuit (b).

GM THE POLE PLOT OF AN RC AND RL CIRCUIT

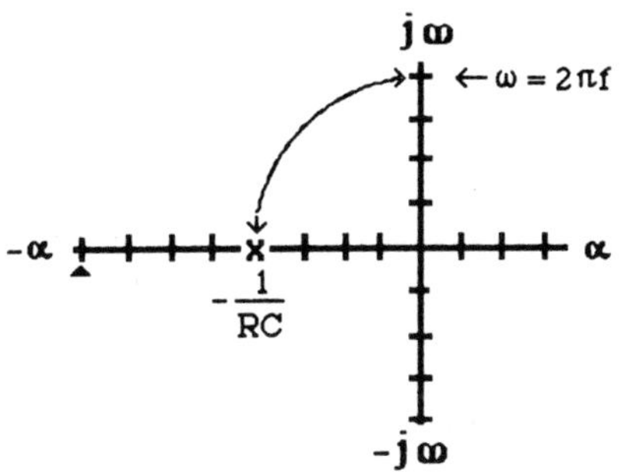

FIGURE 6.5 An s-plane pole plot of a series RC circuit. (For a series RL circuit, its -R/L pole would be a the same location.)

Special Note: A few more words about the s-plane and poles and zeros (refer to the last part of Chapter 4).

Many times the grid marks on s-plane plots are not numbered. This is because the s-plane is primarily used to show the FORM of the transfer function equation and thus gives an indication of the circuit's tendency to overshoot or oscillate. However, numbers certainly may be assigned to the grid on the axes to indicate the relationship between the poles (and zeros). For instance, in Figure 6.5, the one pole was arbitrarily positioned four marks to the left of 0 (the $j\omega$ axes) on the $-\alpha$ axis. This position also prescribes an ω, or its cutoff frequency, on the $j\omega$ axis as indicated by the arrowed part of a semicircle. If the circuit had contained a second pole R_2C_2, where that R_2C_2 might have a cutoff frequency that was twice the value of the first RC circuit, the second pole would have been placed at minus eight marks, indicated here with a small triangle, and the associated second-pole ω frequency would be at eight marks up on the $j\omega$ axis (not shown). Thus, if desired, the axis grid marks can be used to show precise numerical relationships as well as showing the relative positions of the circuit poles.

6.3 THE PARALLEL RC OR RL CIRCUIT

TM THE RESPONSE OF AN RC OR RL PARALLEL CIRCUIT

In any parallel circuit, one common voltage, produced by the input forcing current and/or the stored energy, appears across all the parallel components. The instantaneous current in any one component is this common voltage divided by the impedance of that component. Although there are several approaches to the analysis of parallel circuits, the use of a current input and the actual, physical impedance of the components makes the most "real-world" sense and is used by most computer analysis programs. As has been stated in earlier chapters, a collector or drain (or a pentode plate) output is close enough to a perfect constant-current source so that they may be used as both theoretical and practical models in circuit analysis. If there are only two parallel impedances, the parallel circuit current divider rule can be used to find the current in one component by $i_1 = i_{in} * (Z_2 / (Z_1 + Z_2))$ or $i_2 = i_{in} * (Z_1 / Z_1 + Z_2))$. Note that when using the current-divider rule, the "other" impedance is used in the left side of the equation. A complete (complex) voltage and current analysis of a parallel circuit will usually involve differential equations that can be solved using the steps listed at the start of this chapter.

MM THE FREQUENCY (SINE-WAVE) RESPONSE OF AN RC OR RL PARALLEL CIRCUIT

$$Z_{(\omega)} = \frac{RX_C}{\sqrt{R^2 + X_C^2}}$$ is the sinusoidal impedance of a parallel RC circuit

$$= \frac{RX_L}{\sqrt{R^2 + X_L^2}}$$ is the sinusoidal impedance of a parallel RL circuit

(E 6.28)

$$H_{(\omega)C} = \frac{I_{out}}{I_{in}} = \frac{Z_{TOT}}{Z_{other}} = \frac{\frac{RX_C}{\sqrt{R^2 + X_C^2}}}{R}$$ where $H_{(\omega)}$ is the

steady-state current transfer function of a parallel RC circuit with current input (E 6.29)

$$v_{par} = i_{in} * Z_{(\omega)} \qquad \text{(E 6.30)}$$

$i_{(\omega in)} = i_R + i_{XC} = \frac{v}{R} + \frac{v}{X_C}$ is the steady-state Kirchhoff current equation (E 6.31)

MM THE TIME (STEP) RESPONSE OF AN RC OR RL PARALLEL CIRCUIT

$$Z_{(s)} = \frac{R\frac{1}{sC}}{R + \frac{1}{sC}}$$ and, after the appropriate algebra,

$$= \frac{1}{C}\frac{1}{s + \frac{1}{RC}}$$ is the impedance in (s) of a parallel RC circuit,

where 1/C and 1/RC are constants for a given circuit. (E 6.32)

$$Z_{(s)} = \frac{R\,sL}{R + sL} = R\frac{s}{s + \frac{R}{L}}$$ is the impedance in (s) of a parallel RL circuit. (E 6.33)

$$H_{(s)c} = \frac{I_C}{I_{in}} = \frac{\frac{v}{X_C(s)}}{\frac{v}{R} + \frac{v}{X_C(s)}} = \frac{\frac{v}{\frac{1}{sC}}}{\frac{v}{R} + \frac{v}{\frac{1}{sC}}} = \frac{s}{s + \frac{1}{RC}}$$ is the transfer function

in (s) for the capacitor current in a parallel RC circuit and, (E 6.34)

$$= \frac{I_L}{I_{in}} = \frac{\frac{v}{X_L(s)}}{\frac{v}{R} + \frac{v}{X_L(s)}} = \frac{\frac{v}{sL}}{\frac{v}{R} + \frac{v}{sL}} = \frac{1}{1 + s\frac{L}{R}}$$ is the transfer function for the

inductor current in a parallel RL circuit. (E 6.35)

$v_{par} = i_{in} * Z_{(s)}$ (E 6.36)

$i_{in} = i_1 + i_2 = i_R + i_C$ or $i_R + i_L$ for a parallel RC or RL circuit (E 6.37)

AN EXAMPLE: AN RC CIRCUIT—FINDING THE COMMON PARALLEL VOLTAGE:

$i_{step\ in} = iR + iC$

$i_{step\ in} = I_S = C\frac{dv_C}{dt} + \frac{v_C}{R}$ or, dividing by C and rearranging, (E 6.38)

$\frac{dv_C}{dt} + \frac{v_C}{RC} = \frac{I_S}{C}$ or, solving the differential equation, (E 6.39)

$v_C = I_SR + (V_0 - I_SR)\ e^{-t/RC}$ where V_0 is any charge on C before the step—assumed zero (E 6.40)

$i_C = C\frac{dv_C}{dt} = (I_{step} - \frac{V_0}{R})\ e^{-t/RC}$ (E 6.41)

AN EXAMPLE: AN RL CIRCUIT EXAMPLE—FINDING THE COMMON VOLTAGE:

$i_{step\ in} = iR + iL$

$i_{step\ in} = I_S = \frac{v}{R} + \frac{1}{L}\int_0^t v\ dt + I_0$ and differentiating and rearranging,

(E 6.42)

$\frac{1}{R}\frac{dv}{dt} + \frac{v}{L} = 0$ or $\frac{dv}{dt} = \frac{-R}{L}v$ or $\int_0^t \frac{di}{i} = \frac{-R}{L}\int_0^t dt$ (see Chapter 14) (E 6.43)

$iL = \frac{1}{L}\int_{t0}^{t} v\ dt + I0$ (E 6.44)

$= I_{step} - \frac{1}{R}(V_{step} - I_0R)\ e^{-(R/L)\,t} = \frac{V_{step}}{R} - (\frac{V_{step}}{R} - I_0)\ e^{-(R/L)\,t}$

(E 6.45)

$vL = R(I_S - I_0)\ e^{-(R/L)\,t}$ where I_0 is this initial current in the inductor

(E 6.46)

MM THE NATURAL RESPONSE OF AN RC OR RL PARALLEL CIRCUIT

AN EXAMPLE: A PARALLEL RC CIRCUIT—FINDING THE NATURAL RESPONSE:

$$\Sigma_i = 0 = i_C - i_R = 0$$

$$\frac{v}{R} = -C\frac{dv}{dt} \text{ or } \int_0^t \frac{dv}{v} = -\frac{1}{RC}\int_0^t dt \text{ for a parallel RC Circuit} \qquad \text{(E 6.47)}$$

$v_C = V_C\, e^{-t/RC}$ which is the same result but a slightly different form than E 6.20 to E 6.23 (E 6.48)

AN EXAMPLE: A PARALLEL RL CIRCUIT EXAMPLE—FINDING THE NATURAL RESPONSE:

$$\Sigma_i = 0 = i_L - i_R = 0$$

$$L\frac{di}{dt} = Ri \text{ or } \int_{t_0}^{t} \frac{di}{i} = -\frac{R}{L}\int_{t_0}^{t} \frac{dt}{1} \qquad \text{(E 6.49)}$$

$$i_n = I_0\, e^{-(R/L)\,t} \text{ and } v_R = RI_0\, e^{-(R/L)\,t} \qquad \text{(E 6.50)}$$

GM

Special Note: The FORM of the frequency, time and pole response of the parallel RC and RL circuits are the same as the corresponding series circuits.

6.4 AN INTRODUCTION TO THE RESPONSE OF RLC CIRCUITS

Capacitors and inductors are combined to produce the most useful phenomenon of electrical resonance. Any resistance, whether it is separate or part of the L or C components, will not only affect the rate of energy storage or release but, to a lesser degree, will also affect the value of the resonance frequency. The first step in analyzing RLC circuits is, as with two-component circuits, to write the Kirchhoff equations. However, the solution of these second order equations is more difficult than for first order circuits and requires new techniques. Also, for the general case of series or parallel RLC circuits, there is not one but *three* possible solutions to the equations, depending on the effect of the resistor. One of these three solutions will show that a circuit is *overdamped* (OD), which means the effect of the resistance is quite dominant. For this overdamped circuit, the step response rise time is slow and the resonant frequency output is not much different than the circuit's output above and below resonance. A second solution will show a circuit is *critically damped* (CD) which indicates that, among other things, the value of the resistance makes the rise time for an input step the fastest possible without a transient overshoot. A third *underdamped* (UD) solution will show that the resistance is much less dominant and, although the rise time is fast, the circuit may show both significant overshoots and a pronounced tendency to oscillate. The techniques used to solve the three types of the equation usually require the finding of the roots of the equations—see Chapter 14. The use of these roots in the pole plots shows the *form* of these three types of responses.

The characteristics of the energy storage and release are also interesting. Whereas in first order circuits, the voltage or current would increase or decrease with time, in resonant (second order) circuits, a complete oscillating wave form or transient *envelope* may increase or decrease with time. An s-plane pole plot of the transfer function is especially helpful in both quickly showing the form of the resonant-circuit character and also indicating what must be done to increase or decrease the damping to tailor the transient response.

TO FIND THE RESPONSE OF AN RLC CIRCUIT

1. Write the voltage (series circuit) or current (parallel circuit) Kirchhoff differential equation.
2. Using the numerical value of the components given in the problem, find the damping constant α and the natural frequency ω_0. Find the type of damping (OD,UD or CD) using the values in Table 6.1.
3. Determine the initial conditions. The equation for a series or parallel resonant circuit has the general form of $f(t) = A_1e^{s_1t} + A_2e^{s_2t}$. This means that to solve the resonant equation, the two constants, A_1 and A_2, must be determined. The evaluation of constants, by using initial conditions, is the most tricky part of finding the response of both first order and second order circuits. In general, the steps include (1) finding the voltage across the capacitor just before and just after the transient change (the switch is thrown) and, likewise, (2) finding the current in the inductor just before and after the transient change. Remember that the voltage across the capacitor or the current in the inductor is the same just before and just after an instantaneous change. If there is any initial voltage across a capacitor in a parallel resonance circuit, that voltage is also across the inductor and the resistor.

4. Differentiate the Kirchhoff differential equations from step 1. Constants such as E or I will thus become zero. The solution, equal to zero, is termed the *homogeneous* or transient equation. A second or *particular* differential equation can also be written based upon the forcing function and the initial conditions. The particular equation is the steady-state equation.
5. Solve the Kirchhoff differential equations or, using the circuit type, its component values, and the damping, choose the appropriate equation form from table 6.2. Assume the solutions will take the general form of $i = Ae^{st}$ or $v = Ae^{st}$. Since there are two time constants (in s) in second order equations, the general FORM will expand to $A_1e^{s_1t} + A_2e^{s_2t}$. This general form can be separated into underdamped, critically damped, and overdamped.
6. Using the operators $s^2 = d^2/dt^2$ and $s = d/dt$, write the simple characteristic equation in the form $s^2 + s\,\tau_1 + \tau_2$.
7. Find the roots of the characteristic equation.
8. Solve for the A, B, or D constants using the initial conditions for i and di or v and dv or refer to Figure 6.2 and Table 6.3.
9. Solve for the appropriate i or v.

The following tables help to organize the solutions:

TABLE 6.1 EQUATIONS FOR ALL RLC RESONANT CIRCUITS

$s_1 = -\alpha + \sqrt{\alpha^2 - \omega_0^2}$ s_1 and s_2 are the roots of

$s_2 = -\alpha - \sqrt{\alpha^2 - \omega_0^2}$ the characteristic equation.

$\alpha = \frac{1}{2RC}$ for parallel RLC circuits

$= \frac{R}{2L}$ for series RLC circuits

$\omega_0^2 = \frac{1}{LC}$ for parallel and series RLC circuits

$\omega_d = \sqrt{\omega_0^2 - \alpha^2}$ is a special case for overdamped circuits

$\omega_0^2 < \alpha^2$ is the criterion for an overdamped response. The roots s_1 and s_2 are simple but unequal.

$\omega_0^2 > \alpha^2$ is the criterion for an underdamped response. The roots s_1 and s_2 are complex.

$\omega_0^2 = \alpha^2$ is the criterion for a critically damped response. The roots s_1 and s_2 are simple and equal.

TABLE 6.2 THE GENERAL FORM OF RLC EQUATIONS

The Form of Equations for RLC OVERDAMPED Circuits:

$i_L(t) = I_f + A_1e^{s_1t} + A_2e^{s_2t}$ parallel circuit step response

$0 = A_1e^{s_1t} + A_2e^{s_2t}$ parallel circuit natural response

$v_c(t) = V_f + A_1'e^{s_1t} + A_2'e^{s_2t}$ series circuit step response

$0 = A_1'e^{s_1t} + A_2'e^{s_2t}$ series circuit natural response

The Form of Equations for RLC UNDERDAMPED Circuits:

$i_L(t) = I_f + B_1e^{-\alpha t} \cos \omega_d t + B_2e^{-\alpha t} \sin \omega_d t$ parallel circuit step response

$0 = B_1e^{-\alpha t} \cos \omega_d t + B_2e^{-\alpha t} \sin \omega_d t$ parallel circuit natural response

$v_c(t) = V_f + B_1'e^{-\alpha t} \cos \omega_d t + B_2'e^{-\alpha t} \sin \omega_d t$ series circuit step response

$0 = B_1'e^{-\alpha t} \cos \omega_d t + B_2'e^{-\alpha t} \sin \omega_d t$ series circuit natural response

The Form of Equations for RLC CRITICALLY DAMPED Circuits:

$i_L(t) = I_f + D_1 t\, e^{-\alpha t} + D_2\, e^{-\alpha t}$ parallel circuit step response

$0 = D_1 t\, e^{-\alpha t} + D_2\, e^{-\alpha t}$ parallel circuit natural response

$v_c(t) = V_f + D_1't\, e^{-\alpha t} + D_2't\, e^{-\alpha t}$ series circuit step response

$0 \quad = D_1't\, e^{-\alpha t} + D_2't\, e^{-\alpha t}$ series circuit step response

I_f and V_f = the final value

TABLE 6.3 GENERAL RLC CIRCUIT EQUATIONS FOR THE ARBITRARY CONSTANTS A, B, AND D

FOR PARALLEL AND SERIES CIRCUITS

For OVERDAMPED Circuits:

$$A_1\ \text{par} = \frac{V_0 s_2 + (1/C)[(V_0/R) + I_0]}{s_2 - s_1} \qquad A_1\ \text{ser} = \frac{-s_2 I}{s_2 - s_1}$$

$$A_2\ \text{par} = \frac{V_0 s_1 + (1/C)[(V_0/R) + I_0]}{s_2 - s_1} \qquad A_2\ \text{ser} = \frac{-s_1 I}{s_2 - s_1}$$

For UNDERDAMPED Circuits:

$$B1\ \text{par} = V0 \qquad B1\ \text{ser} = vc - Vin$$

$$B2\ \text{par} = \frac{\alpha}{\omega d} V0 - \frac{1}{\omega d C}\left(\frac{V0}{R} + I0\right) = -\frac{\alpha}{\omega d}(V0 + 2I0R) \qquad B2\ \text{ser} = B1\,\frac{\alpha}{\omega 0}$$

For CRITICALLY DAMPED Circuits:

$$D_1\ \text{par} = -\alpha(V_0 + 2I_0R) \qquad D_1\ \text{ser} = -\alpha I$$

$$D_2\ \text{par} = V_0 \qquad D_2\ \text{ser} = -I$$

V_0 = The stored capacitor voltage at time t_0.

I_0 = The inductor current at time t_0.

6.5 **TM** THE RESPONSE OF A SERIES RLC CIRCUIT

In a series resonant circuit, the current is maximum at resonance and the impedance is minimum. Also, the voltage across either the inductor or the capacitor can exceed the input voltage by a factor $Q = \omega L/R$.

MM THE FREQUENCY (SINE-WAVE) RESPONSE OF A SERIES RLC CIRCUIT

$$Z_{(\omega)} = R + jX_L - jX_C = R + j(X_L - X_C) \quad \text{(E 6.51)}$$

$$H_{(\omega)} = \frac{E_{out(c)}}{E_{in}} = \frac{X_C}{R + j(X_L - X_C)} \quad \text{(E 6.52)}$$

$\omega_0 = \dfrac{1}{\sqrt{LC}}$ is the natural resonant frequency without any resistance effects (E 6.53)

$Q_{series} = \dfrac{\omega L}{R} = \dfrac{1}{\omega_0 CR} = \dfrac{1}{R}\sqrt{\dfrac{L}{C}}$ is the ratio of the resonant frequency to bandwidth (E 6.54)

$BW = Z_{BW} = \sqrt{2}\,R$ = the bandwidth of the circuit from the lower half-power frequency to the upper half-power frequency. (E 6.55)

$$Z_{BW} = \sqrt{R^2 + (\omega L - 1/\omega C)^2} = \sqrt{2}\,R \quad \text{(E 6.56)}$$

$\omega_{LO} = -\dfrac{R}{2L} + \sqrt{(\dfrac{R}{2L}) + \dfrac{1}{LC}}$ is the lower half-power frequency

$\omega_{HI} = +\dfrac{R}{2L} + \sqrt{(\dfrac{R}{2L}) + \dfrac{1}{LC}}$ is the upper half-power frequency (E 6.57)

$$i_{series} = \frac{V_{in}}{Z_{(\omega)}} \quad \text{(E 6.58)}$$

$v_{(\omega)in} = Ri + jX_L + jX_C$ is the Kirchhoff series equation (E 6.59)

$$v_{outC} = H_{(\omega)C} * v_{(\omega)in} \quad \text{(E 6.60)}$$

AN EXAMPLE: THE SINE-WAVE RESPONSE OF A SERIES RLC RESONANT CIRCUIT:

Find the series current and the capacitor (output) voltage for a series RLC circuit. Show the appropriate phase angle for each term. R = 12500 ohms, L = .001 henrys, and $C = 25.4 * 10^{-12}$ farads, given a resonant frequency of about 1 MHz.

$v_{in} = V \cos(\omega t + \phi) = V \cos(2\pi\, 10^6\, t + 0^o)$ for a 1 MHz cosine wave input with no phase shift.

$Z_{in} = R + jX_L - jX_C$ (The phase angle of the generator is assumed to be 0°.)

$= 12500 + j6280 - j6269$ (The capacitor is not the exact value for 1 MHz resonance.)

$= 12500 + j11 \approx 12500$

$$I_{ser} = \frac{1 \angle 0^o}{12500 \angle 0^o} = 80 * 10^{-6} \angle 0^o \text{ amps.}$$

$$V_C = I_{ser} * X_C = (80 * 10^{-6} \angle 0^o) * (6269 \angle -90^o) = .5 \angle -90^o$$

GM THE FREQUENCY (SINE-WAVE) RESPONSE OF A SERIES RLC CIRCUIT

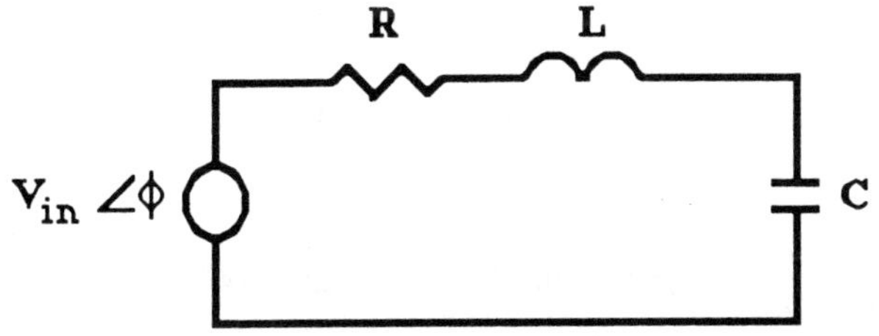

FIGURE 6.6 A series RLC circuit forced by a sine-wave generator.

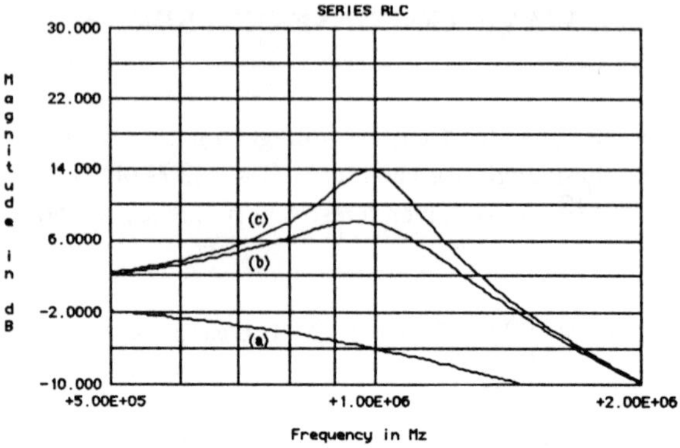

FIGURE 6.7 The sine-wave response of a series resonant circuit: The output voltage across the capacitor (a) with a resistor value to give critical damping; (b) with a resistor = one-fifth the CD value; (c) with a resistor = one-tenth the CD value.

MM THE TIME (STEP) RESPONSE OF A SERIES RLC CIRCUIT

$$Z_{(s)} = R + sL + \frac{1}{sC} \quad \text{(E 6.61)}$$

$$H(s) = \frac{E_{out(c)}}{E_{in}} = \frac{\frac{1}{sC}}{R + sL + \frac{1}{sC}} = \frac{1}{s^2LC + sRC + 1} \quad \text{(E 6.62)}$$

$v_{step\ in} = v_R + v_L + v_C$ for the series resonant circuit, or (E 6.63)

$Ri + L\frac{di}{dt} + \frac{1}{C}\int_0^t i\,dt + V_0$ and, differentiating with respect to t, (E 6.64)

$R\frac{di}{dt} + L\frac{d^2i}{dt^2} + \frac{i}{C}$ or $\frac{d^2i}{dt^2} + \frac{R}{L}\frac{di}{dt} + \frac{i}{LC}$ puts the differential equation in terms of s and the time constants $\frac{R}{L}$ and LC (E 6.65)

E 6.65 can be solved using the well-honored calculus technique of assuming a solution and seeing if it works. From the previous work on RC and RL circuits, it is reasonable to assume a FORM of $i = Ae^{st}$, so

$s^2Ae^{st} + s\frac{R}{L}Ae^{st} + \frac{1}{LC}Ae^{st}$ or $Ae^{st}(s^2 + s\frac{R}{L} + \frac{1}{LC})$ is the form of a general solution, and (E 6.66)

$(s^2 + s\frac{R}{L} + \frac{1}{LC}) = 0$ is termed the *characteristic equation* since, in this simplified form, it shows the character of what the response will be—see Chapter 14. The solution (the roots) of this equation will lead to the final response of E 6.66. (E 6.67)

$s_1 = -\frac{R}{2L} + \sqrt{(\frac{R}{2L})^2 - \frac{1}{LC}}$ or $-\alpha + \sqrt{\alpha^2 - \omega_0^2}$ is one root with

$\alpha = \frac{R}{2L}$ and $\omega_0 = \frac{1}{LC}$ and $s_1 = -\frac{R}{2L} - \sqrt{(\frac{R}{2L})^2 - \frac{1}{LC}}$ or

$-\alpha - \sqrt{\alpha^2 - \omega_0^2}$ is the second root (solution) (E 6.68)

$\alpha = \frac{R}{2L}$ for series RLC circuits—from Table 6.1 (E 6.69)

$\omega_0^2 = \frac{1}{LC}$ for series RLC circuits—from Table 6.1 (E 6.70)

To find i(t):

1. Find α and ω_0^2 using Table 6.1 and determine the type of damping.
2. Solve the characteristic equation for the roots s_1 and s_2.
3. Find the appropriate equation form in Table 6.2.
4. Use Table 6.3 to find the value of the constants A, B, and D.
5. Find the numerical solution for v_C or i_L.

AN EXAMPLE: THE TRANSIENT (STEP) RESPONSE OF A SERIES RLC RESONANT CIRCUIT:

Find the capacitor (output) voltage for a series RLC circuit. R = 2500 ohms, L = .001 henrys, and C = 25.4 $*$ 10^{12} farads and the resonant frequency is about 1 MHz. V_{in} is 10 $\angle 0^{o}$.

$\alpha \quad = \frac{R}{2L} = \frac{2.5 * 10^3}{2 * 10^{-3}} = 1.25 * 10^6$

$\alpha^2 \quad = 1.56 * 10^{12}$

$\omega_0 \quad = \frac{1}{\sqrt{LC}} = \frac{1}{10^{-3} * 25.4 * 10^{-12}} = 6.274 * 10^6$

$\omega_0^2 \quad = 39.37 * 10^{12}$

$\omega_0^2 \quad > \alpha^2$ so this circuit is underdamped.

$s_1 \quad = -\alpha + \sqrt{\alpha^2 - \omega^2}$

$= -1.25 * 10^6 + \sqrt{1.56 * 10^{12} - 39.37 * 10^{12}} = -1.25 * 10^6 +$

$\sqrt{-37.81 * 10^{12}}$

$= -1.25 * 10^6 + j6.148 * 10^6$

$s_2 \quad = -1.25 * 10^6 - j6.148 * 10^6$

$v_c(t) \quad = V_{in} + B_1 e^{-\alpha t} \cos \omega_d t + B_2 e^{-\alpha t} \sin \omega_d t$ (from Table 6.2)

$B_{1\,ser} = v_c - V_{in} = 0 - 10 = -10$ (from Table 6.3). v_c is assumed zero

at $t = 0$.

$B_{2\,ser} = B_1 \frac{\alpha}{\omega_0} = -10 \frac{1.25*10^6}{6.148*10^6} = 2.03$

$v_c(t) \quad = 10 - 10e^{-1.25 * 10^6 t} \cos 6.148 * 10^6\, t + 2.03\, e^{-1.25 * 10^6 t}$

$\sin 6.148 * 10^6\, t$

GM SERIES RLC CIRCUIT TRANSIENT RESPONSE

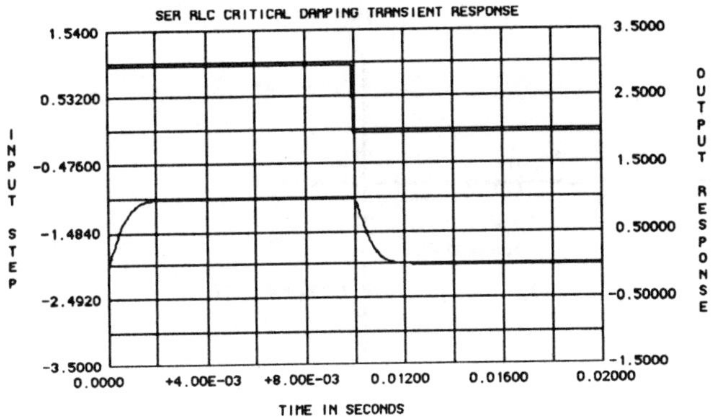

FIGURE 6.8 The critically damped step response of example series RLC circuit.

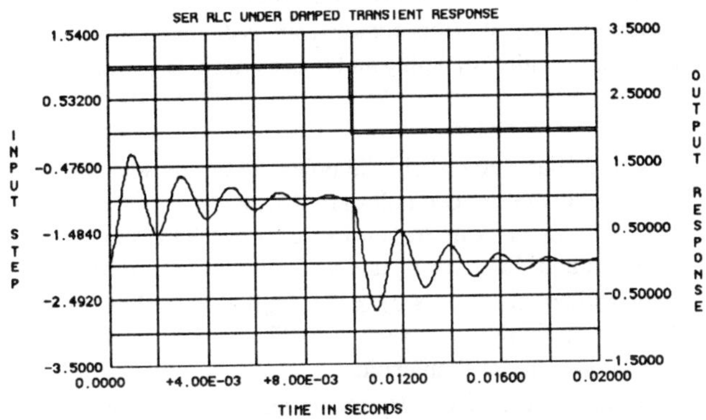

FIGURE 6.9 The underdamped step response of example series RLC circuit.

GM SERIES RLC CIRCUIT S–PLANE ANALYSIS

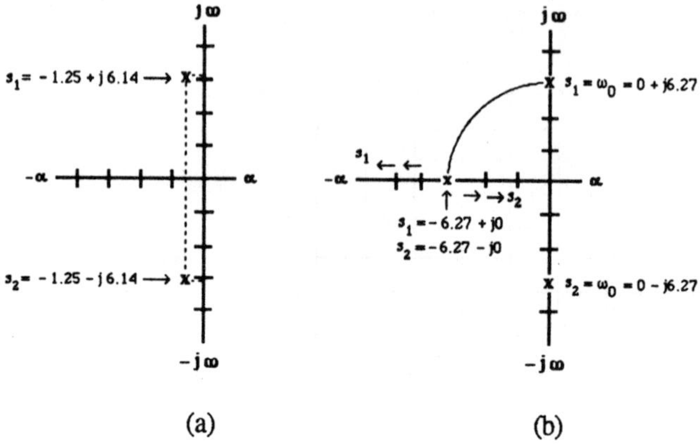

FIGURE 6.10 (a) The s-plane plot of the poles in the series RLC example. (All values are divided by 10^6.) Note that this circuit is quite underdamped. (b) A general s-plane plot showing the path of the poles from complete underdamping, which would mean oscillation, down to critical damping and then to the pole path of an overdamped circuit.

1. For a completely underdamped circuit, the poles are on the +jω and –jω axis. In this figure the underdamped poles are $0 \pm j6.27$.
2. As the value of the resistor is increased in the series RLC circuit, the damping is increased and the poles s_1 and s_2 follow the semicircle (the lower part is not shown) to the –α axis for critical damping. In this figure, critical damping would be $s_1 = s_2 = -6.27$.
3. As damping is increased more than critical, the s_1 pole will stay on the –α axis but move to the left. Likewise, the s_2 pole will stay on the –α axis but move to the right. Values for overdamped poles are not shown in the figure.

Thus, from this figure, the FORM of the use of pole plots on the s-plane can be better understood. If more or less damping with the resulting more or less overshoot is desired, the general magnitude of the required ciruit values can thus be quite easily approximated.

TM THE NATURAL RESPONSE OF A SERIES RLC CIRCUIT

$Ri + L\frac{di}{dt} + \frac{1}{C}\int_0^t i\,dt + V_0 = 0$ is the Kirchhoff voltage equation for the natural response. Note that this equation is the same as E 6.64 and, therefore, the solution for the natural response will follow that solution for the step response.

6.6 TM THE RESPONSE OF A PARALLEL RLC CIRCUIT

In a parallel resonant circuit, the voltage and impedance are maximum at resonance and the signal current is minimum. The analysis uses current division.

MM THE FREQUENCY (SINE-WAVE) RESPONSE OF A PARALLEL RLC CIRCUIT

$$\frac{1}{Z}(\omega) = \frac{1}{R} + \frac{1}{j\omega L} + j\omega C = \frac{R + j\omega L\ \omega^2 RLC}{j\omega RL} \quad \text{(E 6.71)}$$

$$Z(\omega) = \frac{j\omega RL}{R + j\omega L\ \omega^2 RLC}$$ is the (complex) parallel circuit impedance (E 6.72)

$$H(\omega) = \frac{E_{out}}{I_{in}} = Z(\omega) = \text{the circuit transfer impedance} \quad \text{(E 6.73)}$$

$v_{par} = I_{in}\,Z(\omega)$ is the common (complex) voltage across the parallel circuit (E 6.74)

$i_n = \frac{v_{par}}{Z_n}$ is the complex current in any parallel component (or series of components) (E 6.75)

GM THE FREQUENCY (SINE-WAVE) RESPONSE OF A PARALLEL RLC CIRCUIT

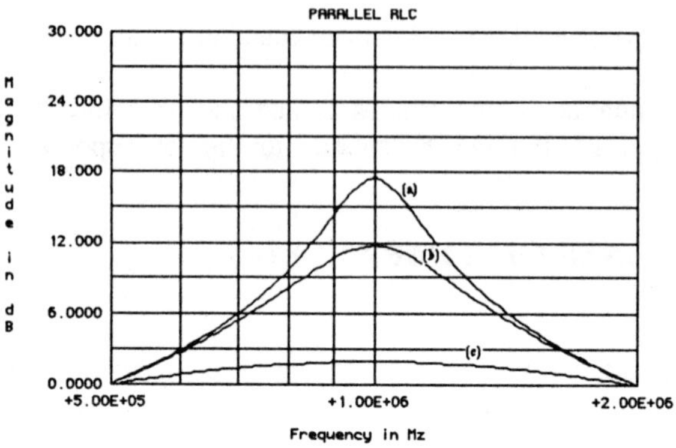

FIGURE 6.11 The sine-wave response of a parallel resonant circuit: The output voltage (c) with a resistor value to give critical damping; (b) with a resistor = five times the CD value; (a) wth a resistor = ten times the CD value.

MM THE TIME (STEP) RESPONSE OF A PARALLEL RLC CIRCUIT

$$Z(s) = \frac{1}{\frac{1}{R} + \frac{1}{sL} + sC} = \frac{sRL}{s^2RLC + sL + R}$$

$$= \frac{1}{C}\,\frac{s}{s^2 + s\frac{1}{RC} + \frac{1}{LC}} \qquad \text{(E 6.76)}$$

$$H(s) = \frac{E_{out}}{I_{in}} = Z(s) \qquad \text{(E 6.77)}$$

$$I_{step} = i_R + i_L + i_C$$

$$= i_L + \frac{v}{R} + C\frac{dv}{dt} \text{ also, } v = L\frac{di}{dt} \text{ and } \frac{dv}{dt} = L\frac{d_2i}{dt^2} \quad \text{therefore,}$$

$$I_{step} = i_L + \frac{L}{R}\frac{di}{dt} + LC\frac{d_2i}{dt^2} \qquad \text{(E 6.78)}$$

As with the series RLC circuit,

1. Find α^2 and ω_0^2 using Table 6.1 and determine the type of damping.
2. Solve the characteristic equation for the roots s_1 and s_2.
3. Find the appropriate equation form in Table 6.2.
4. Use Table 6.3 to find the value of the constants A, B, and D.
5. Find the numerical solution for v_C or i_L.

AN EXAMPLE: THE IMPEDANCE AND THE TRANSIENT (STEP) RESPONSE OF A PARALLEL RLC RESONANT CIRCUIT:

Find circuit impedance and the transient output voltage for a parallel RLC circuit. The circuit will be forced (driven) by a voltage-controlled constant current generator with a normalized output of 1.0 ampere. For completeness, the resonant circuit will contain both a parallel resistor and a series resistor R_{ser} that simulates the series resistance of the inductor.

R_{par} = 3140 ohms for critical damping and 15700 ohms for the underdamped case, R_{ser} = 10 ohms, L = .001 henrys, and C = 25.4 $*$ 10^{-12} farads and the resonant frequency is about 1 MHz.

FIND THE PARALLEL RLC IMPEDANCE:

$$\frac{1}{Z_{tot}} = \frac{1}{Z_1} + \frac{1}{Z_2} + \frac{1}{Z_3} = \frac{1}{Z_R} + \frac{1}{Z_L} + \frac{1}{Z_C}$$

$$= \frac{1}{3140} + \frac{1}{10+j6280} + \frac{1}{-j6269}$$

$$= \frac{1}{3140} + \frac{1}{10+j6280}\left(\frac{10-j6280}{10-j6280}\right) + \frac{1}{-j6269}$$

$$= \frac{1}{3140} + \frac{10-j6280}{100+39438400} + \frac{1}{-j6269}$$

$$\approx \frac{1}{3140} + \frac{1}{+j6280} + \frac{1}{-j6269}$$

$$\approx .000318 + j.0001592 - j0001595$$

$$\approx .000318$$

Z_{tot} $= 3144 \approx 3140$, which is the value of the parallel resistor.

The number 3144 is caused by rounding errors and using the slightly inaccurate value of 1 MHz.

Special Note: The above model illustrates the technique of using 1/Z when working with parallel circuits. The ten-ohm inductor resistance was not significant at one MHz but, if the frequency had been much lower, for example ten KHz, the inductor impedance would be 10 + j62.8 and the resistance would have to be included in all calculations.

v_{par} $= IZ = 1 * 3140 = 3140$ volts normalized form the one-ampere input.

$$i_C = \frac{v_{par}}{X_C} = \frac{3140}{-j6269} \approx -j0.5 \text{ amps}$$

$$i_L = \frac{v_{par}}{X_C} = \frac{3140}{j6269} \approx j0.5 \text{ amps}$$

FIND THE PARALLEL RLC TRANSIENT (STEP) RESPONSE:

$$\alpha = \frac{1}{2RC} = \frac{1}{2 * 15.7 * 10^3 * 25.4 * 10^{-12}} = 1.254 * 10^6$$ (using a resistor for underdamping)

$$\alpha^2 = 1.572 * 10^{12}$$

$$\omega_0 = 6.274 * 10^6$$

$$\omega_0^2 = 39.37 * 10^{12}$$

$\omega_0^2 > \alpha^2$ shows the circuit is underdamped

$$s_1 = -\alpha + \sqrt{\alpha^2 - \omega^2}$$

$$= -1.254 * 10^6 + \sqrt{1.572 * 10^{12} - 39.37 * 10^{12}}$$

$$= -1.254 * 10^6 + \sqrt{-39.37 * 10^{12}}$$

$$= -1.254 * 10^6 + j6.148 * 10^6$$

$s_2 \quad = -1.254*10^6 - j6.148*10^6$

$i_L(0) \quad = 0 = I_{in} + B_1 = 10 + B_1$ where, as an initial condition, $i_L(0) = 0$

$B_1 \quad = -10$

$\frac{di_L}{dt}(0) \quad = 0 = \omega_d B_2 - \alpha B_1$, so $B_2 = (-I)\frac{\alpha}{\omega_d} = (-10)\frac{1.2254*10^6}{6.148*10^6}$

where $I_0 = 0$ (see Table 6.3)

$B_2 \quad = -2.03$

$i_L(t) \quad = I_f + B_1 e^{-\alpha t}\cos\omega_d t + B_2 e^{-\alpha t}$ from Table 6.2

$= 10 - 10e^{-1.254*10^6 t}\cos 6.148*10^6 t - 2.03e^{-1.254*10^6 t}$
$\sin 6.148*10^6 t$

MM THE NATURAL RESPONSE OF A PARALLEL RLC CIRCUIT

$i_R + i_L + i_C = 0$ (E 6.79)

$\frac{v}{R} + \frac{1}{L}\int_0^t v\,dt + I_0 + C\frac{dv}{dt} = 0$ or, differentiating, (E 6.80)

$\frac{1}{R}\frac{dv}{dt} + \frac{v}{L} + C\frac{d^2v}{dt^2} = 0$ and dividing through by C and rearranging, (E 6.81)

$\frac{d^2v}{dt^2} + \frac{1}{RC}\frac{dv}{dt} + \frac{v}{LC}$ assuming the *form* of the solution for $v = Ae^{st}$ (E 6.82)

$As^2e^{st} + A\frac{s}{RC}e^{st} + A\frac{1}{LC}e^{st} = 0$ or (E 6.83)

$Ae^{st}(s^2 + s\frac{1}{RC} + \frac{1}{LC}) = 0$ is the *form* of the solution, and (E 6.84)

$(s^2 + s\frac{1}{RC} + \frac{1}{LC}) = 0$ is termed the *characteristic equation* because the *roots* of this quadratic equation will determine the mathematical character of v(t). (E 6.85)

$$s_1 = -\frac{1}{2RC} + \sqrt{(\frac{1}{2RC})^2 - \frac{1}{LC}}$$ and

$$s_2 = -\frac{1}{2RC} - \sqrt{(\frac{1}{2RC})^2 - \frac{1}{LC}}$$ are the two roots that form the characteristic equation. Remember, either of these roots should work as a substitute for the FORM $v = Ae^{st}$ in (E 6.85) above. (E 6.86)

For the currents, follow the five steps used before.

GM PARALLEL RLC

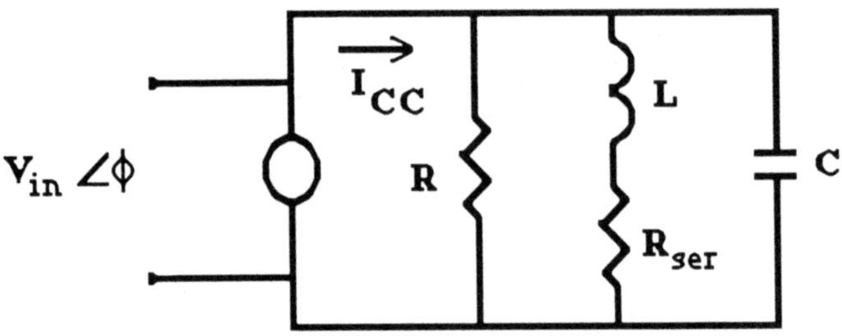

FIGURE 6.12 A parallel RLC resonant circuit forced with a voltage-controller constant-current generator.

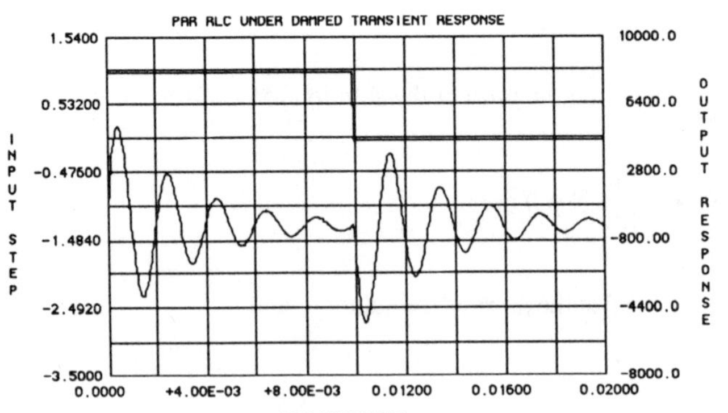

FIGURE 6.13 The step response of an underdamped parallel RLC circuit.

6.7 THE USE OF COMPUTERS IN CALCULATING AND PLOTTING RESONANCE RESPONSE

A complete understanding of the steady-state and transient currents and voltages in resonance circuits requires voluminous calculations and repeated plots. The steady-state (sine-wave frequency) calculations and plots are at home in the frequency domain, while the transient (step) operations usually are carried out in the time domain. The use of computers gives an ability to analyze in much more detail and depth than hand calculations. However, magnificent as they are, computers only do what they are told. It is the responsibility of the user to understand the initial problem as well as the techniques and methods of the computer program being used. Just as the problems in this chapter require both the frequency and time domains, so do the programs that help to solve these problems and to graph their solutions. A somewhat detailed explanation of how the circuits in this chapter were analyzed and how the plots were obtained may be helpful. The reader must remember that there are many circuit analysis programs and many ways of using these programs. Many circuit analysis programs are more sophisticated than the one illustrated here and have several of the following steps built in. However, this elementary and detailed procedure can be used as a general conceptual model. Thus, the following explanation will also be used as a foundation for later chapters using frequency and time analysis.

As a starting point, the series RLC values were first entered into a steady-state circuit analysis program and the voltages or currents were calculated for the node of interest as a function of frequency. The program gave the user the ability to choose the frequency range. The output of this type of analysis program is a table giving the voltage or current magnitude and phase, relative to the input, for a number of frequency points throughout the specified frequency (sweep) range. The user can choose the desired frequencies or the computer will do it automatically. The user can also choose a linear or logarithmic sweep and display. The complex (magnitude and phase) output calculations were saved and that file was used in a plotting program to plot the steady-state output as in Figure 6.7. Because this particular figure was designed to show only relative magnitude, the phase information was not used. Also, the figures that show three magnitude plots required the use of a paint program to produce the graphic overlays.

Parallel resonant circuits require special manipulation. Most analysis programs assume a constant voltage input. If the user requires a constant-current input, a special circuit must be added. One choice would be a transistor amplifier with the collector driving the (parallel) resonant circuit. The problem with this approach is that the transistor must be an extremely simple prototype with no frequency compensation built in. We used an alternative choice, the voltage-controlled constant-current generator. This is an active circuit in the program's file, just as are transistors, operational amplifiers, etc. They can be entered just as another component in the original circuit. The constant-current generator can be controlled by any selected node voltage. The constant-voltage input appears on the first (input) node in the program and the constant-current generator was programmed to provide a simple one-ampere output per volt input.

The transit analysis and plotting are more complicated than the steady state. They involve both the steady-state (frequency domain) analysis program and a time/frequency domain processing program. However, the transient analysis and plots *start* in the steady-state program. Once the circuit is entered or recalled, a *spectral* analysis is taken. This is a set of complex numbers that define the response of the circuit to a sinusoidal input with 255 harmonics of the fundamental frequency. A "Spectral.2" file is created and saved for use in the time/frequency domain processing program.

Using the time/frequency domain processing program, an input step or square wave was first produced (in the time domain) and a special sub-program called a Fast Fourier Transform (covered in detail in Chapter 12) was used to produce its frequency spectrum. This was saved as Spectral.1. The spectra of this input step and the original circuit—Spectral.1 and Spectral.2—are now multiplied to form a combined Spectral.3. It should be remembered that Spectral.1, Spectral.2, and now Spectral.3 are all frequency-domain data. The final transient output will be in the time domain. Therefore, the combined result of Spectral.3 must be *retransformed* (called *inverse* Fast Fourier Transformed) to the time domain. This final time-domain data is used by the plotting program to give a graph of the input square wave and the output transient response.

One final detail about plots and plotting programs. Although some analysis programs include plotting capability, special plotting programs have been created to give the highest possible graphing resolution. They include the ability to add headings, scales and labels to all sides and also label the curves in some cases. Careful examination of the scales that have been used may indicate a rather peculiar choice of left and right scale numbers. For each graph, the scale numbers were chosen to best display the relationship of the curves rather than to indicate the actual "real-world" numbers that might have resulted from precise laboratory data.

7

Diode and Transistor Models

7.1 AN INTRODUCTION

The 1948 invention of the transistor by Bardeen, Brattain, and Shockley and the later inventions of the semiconductor chip by both Noyce and Kilby created the foundation for the information age. From the concept of moving electrical charge across the barrier or junction of two similar, but not identical, pieces of a specially prepared germanium or silicon has come inventions that have again changed the world the way Gutenberg's printing press did.

This chapter will use the semiconductor junction as the fundamental building block. It is used to develop models for one-junction diodes, an almost endless variety of two-junction transistors, and three-junction devices that rectify, latch and control in many clever and very useful ways.

7.2 THE CONCEPTUAL DIODE

THE ELEMENTARY RECTIFYING DIODE

The use of the semiconductor rectifier can be generalized into two classes—large and small (signal) voltage. For large ac voltages at power-line frequencies, the rectifier diode properties do not materially affect the output voltage and the diode can be modeled as a simple polarity-sensing switch. When the voltage across the diode is in the *forward* polarity, which makes the diode conduct, the switch is on and current will flow. When the voltage across the diode is *reverse*, the switch is open and no current will flow. Of course, this very simple model neglects such things as leakage current, forward drop and problems relating to the reverse voltage such as breakdown.

When the ac voltage is relatively small and/or the frequency is higher, the diode properties do affect the output and the switch model must be expanded. The first modification is to include a dc battery to simulate the forward voltage drop across the diode. As a starting point, this battery can have a dc voltage of 0.3 for germanium and 0.7 for silicon diodes. Likewise, as one works with small voltages (small signals), it soon becomes apparent that besides the forward-voltage battery there is also a forward-voltage series resistance. Some experimentation will show that this resistance is usually 10 ohms or less. Again, this value is only a first attempt model. There is also a diode capacitance that must be considered at high frequencies. A still more accurate model will include details about the reverse-voltage, non-conducting character of the diode including its reverse-voltage leakage, its reverse-voltage breakdown and other interesting high-frequency problems. The following paragraphs will investigate the characteristics mentioned above and show how practical, working diode models can be derived.

SEMICONDUCTORS

Materials that are used in electronics are often classified by their ability to conduct electric current. Every material has an inherent physical property called resistivity, which is dependent upon its atomic structure. The resistance of a piece of material is proportional to this characteristic resistivity, its length, and inversely proportional to its cross-sectional area. Materials such as porcelain, grass, rubber, and most plastics are very poor conductors or insulators. They possess very high resistivity, with typical values being in the order of 10^{10} to 10^{15} ohms-centimeters. The atoms in these insulators possess very few free electrons and, therefore, it is very difficult to force an electric current through them. Conductors, such as copper, aluminum, and gold, have a much lower resistivity, about 10^{-5} ohms-centimeter. The atoms in conductors possess many free electrons and thus cannot resist electric current nearly as much.

Semiconductors, such as silicon, germanium and gallium arsenide, have a resistivity of about 10 to 10^4 ohms-cm and have a few free electrons. Pure (intrinsic) semiconductors have a negative temperature coefficient since the resistivity decreases with rising temperature, making them conduct more currents as the temperature is increased. Another important property of semiconductors is the ability to have their conductivity changed when a small amount of a special foreign material, called an impurity, is combined into the pure material. This combining, using various processes, is called doping. Two types of impurities are used—donors and acceptors. Donors increase the available free negative charges or electrons (N-doped) and acceptors increase the available free positive charges or holes (P-doped).

TM THE PHYSICAL, OPERATING P-N JUNCTION

A single, isolated piece of P-doped or N-doped material would behave something like a high-value resistor with no particular use. However, when a section of P-doped material is fabricated with, and adjacent to, a section of N-doped material, the two materials form a P-N junction, a device with new, unique properties. The action of the charges (carriers) near the junction makes possible the rectifying ability of this two-section device.

Even when the P-N junction is unbiased, i.e., when it is not connected to any external voltage or current source, thermal energy still causes atomic action. Near the junction, the motion of the free carries produces two actions with a rather unexpected result. Because of the thermal energy, the free electrons move or migrate (diffuse) in all directions. At and near the junction, some of these free electrons will leave the n-side and move across the junction into the p-side. Here some of these free electrons soon drop into the p-side holes and change from free electrons back to combined or valence electrons. In a like manner, some of the holes diffuse from the p-side and recombine on the n-side. This diffusion thus removes both free electrons and holes in the area near the junction, and forms what is termed a *depletion layer* in the two sides near the junction. The amount of recombination in the depletion layer is limited because the added electrons on the p-side soon tend to repel more electron diffusion. Likewise, on the n-side, the added holes soon tend to repel more hole diffusion. Thus, in the near-junction region, a charge equilibrium is reached that is termed the *barrier potential*. When the P-N junction is forward biased, i.e., when an external voltage is connected with its positive potential on the p-side and the negative potential on the n-side, if the external voltage exceeds this barrier potential (about 0.7 volts for a silicon diode), a current will flow in the diode.

GM THE P-N SEMICONDUCTOR DIODE

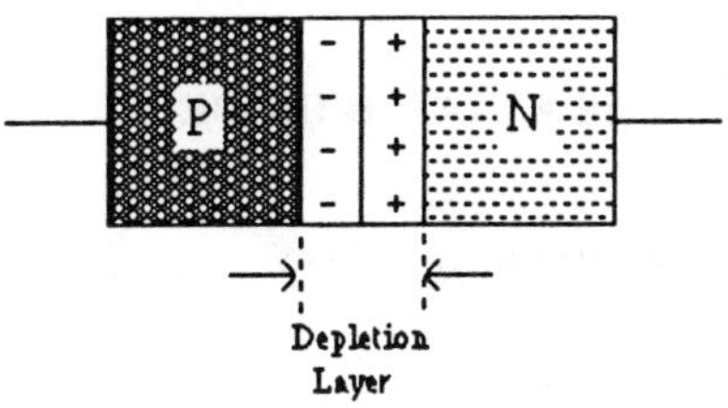

FIGURE 7.1 A semiconductor P-N diode model.

GM THE DIODE CURVE

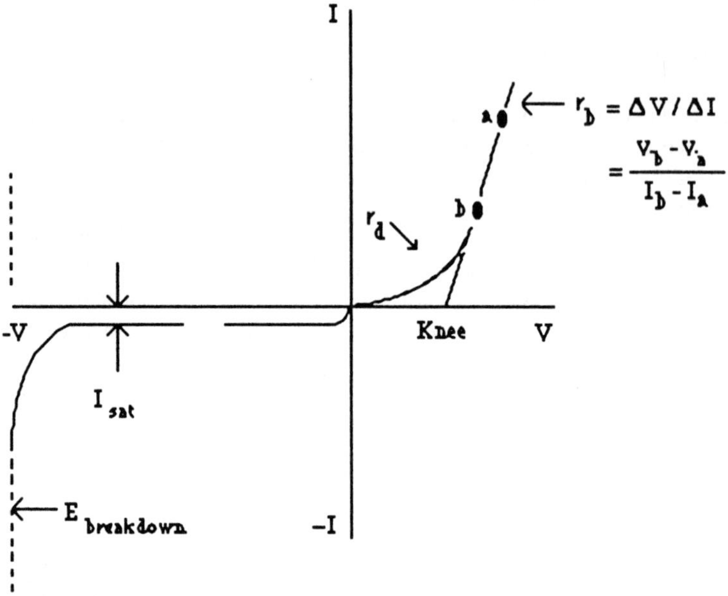

FIGURE 7.2 The forward and reverse characteristic of a semiconductor diode (values are not to scale and are exaggerated for illustration).

THE PARTS OF THE DIODE CURVE

The equations for the forward and reverse currents in a diode are very complicated since they involve several separate but related concepts in theoretical semiconductor physics. The following simplified equations, based upon Shockley's original papers (see the Shockley reference), are given only as a brief background for some of the programmed device models that will be presented later.

MM THE FORWARD VOLTAGE ACROSS A BIASED DIODE

$$V \approx \frac{kT}{q} \ln \left(\frac{N_A N_D}{n_i^2}\right) \approx V_T \ln \left(\frac{N_A N_D}{n_i^2}\right) \text{ is the diode voltage where} \qquad \text{(E 7.1)}$$

k is the Boltzman constant = 1.38 * 10^{-23} joules/°K, T is °K, q is the electron charge = 1.6 * 10^{-19} coulombs and, combining, V_T = 0.026 volts at 300°K. N_A

and N_D are acceptor and donor doping densities in the N and P material and n_i is the intrinsic impurity density. (V is about 0.7 volts for a silicon diode at room temperature (see the Bogar reference.)

$$I_D \approx I_{sat}\,(e^{qV/kT} - 1) \text{ is the forward biased diode current, where} \qquad \text{(E 7.2)}$$

e is 2.7183, q is the electronic charge, v is the diode voltage, k is the Boltzman constant having to do with the thermal energy in the junction and T is the absolute temperature of the junction. (Note that this equation follows the general e^x curve.)

$$I_{sat} \approx \frac{qn_i^2 D_p A}{n_N L_N} \text{ is the saturation current in the junction where} \qquad \text{(E 7.3)}$$

n_i is the density of ionized charges (see the Hurley reference).

7.3 THE OPERATING PARAMETERS AND LIMITATIONS OF DIODES

THE FORWARD RESISTANCE

The diode resistance R_D is composed of a very non-linear portion r_d below the knee or transition region and r_b above the knee. The r_b is a linear resistor ($\Delta v/\Delta i$) and is primarily the *bulk resistance* of the p and n sides of the diode. When the forward current is increased, so is the I^2R_{bulk} power in the diode. This I^2R_{bulk} power will heat the diode and, in some cases, go high enough to melt the junction materials. Each diode will have a maximum operating current above which the diode will eventually destroy itself.

MAXIMUM FORWARD CURRENT

Forward current is limited by the diode's ability to get rid of the I^2R heat generated in the junction. The maximum current may be specified in different ways, such as peak, RMS, instantaneous or dc.

DIODE CAPACITANCE

The charges in the p and n areas of the depletion areas act like the plates of a real capacitor. The capacitance of a fast, high-frequency diode may be only a few picofarads while the capacitance of a large power rectifier may be several microfarads.

REVERSE SATURATION CURRENT

When the polarity of the P-N diode is reverse biased or off, there is a small reverse current. This small current, termed the saturation current or I_S, is caused by a small number of thermally generated holes and electrons around the junction. This saturation current is not influenced by the magnitude of the reverse voltage, but is influenced by temperature. This saturation current approximately doubles for every increase of 10°C. Silicon diodes, which dominate the electronics industry, have much smaller reverse-saturation currents than germanium. There is also some reverse current due to the diode surface leakage I_{SL}, which is both temperature and voltage sensitive. Most data sheets lump I_S and I_{SL} together and call it I_R or reverse leakage.

MAXIMUM REVERSE VOLTAGE AND REVERSE VOLTAGE (AVALANCHE) BREAKDOWN

As the value of the reverse voltage increases, it starts to affect the thermally generated free electrons (and holes) in the depletion area. As the voltage field increases, so does the speed of the free electrons. Some of these free electrons will gain enough speed (energy) to dislodge some of the valence electrons. Some of the freed valence electrons will, in turn, then gain enough energy to dislodge other valence electrons and so on until there is an ever-increasing or avalanche current produced. If the reverse voltage exceeds the diode's design limits, the diode may be destroyed or have its life shortened.

7.4 DIODE MODELS

TM ELEMENTARY DIODE MODELS

The three elementary diode models include a polarity sensing switch, a switch and a battery, and a switch, a battery and a diode resistor.

GM ELEMENTARY DIODE MODELS

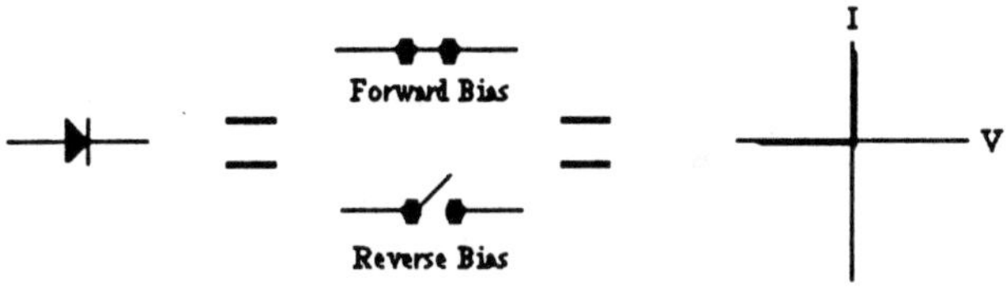

FIGURE 7.3 The ideal diode as a switch.

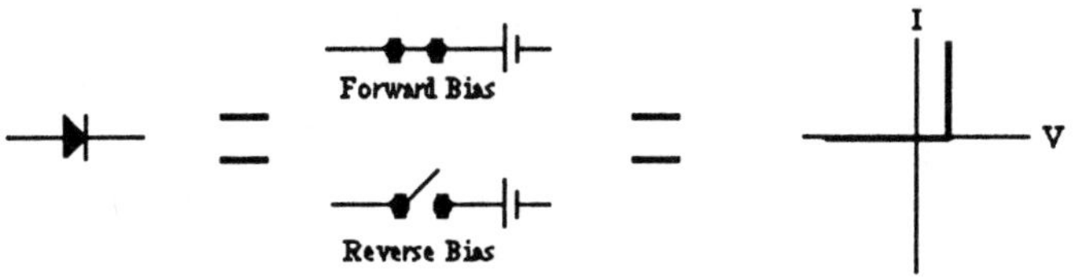

FIGURE 7.4 The diode as a switch and battery.

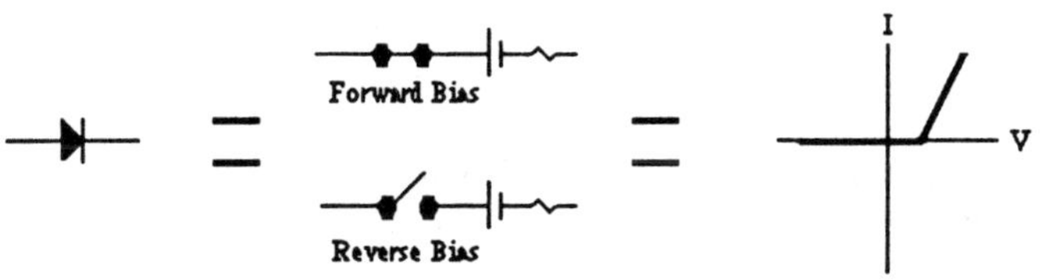

FIGURE 7.5 The diode as a switch, battery and series resistor.

THE WORKABLE EQUIVALENT CIRCUIT FOR A SEMICONDUCTOR DIODE

Some models ignore the switch and assume only forward bias. Likewise, some circuit analysis programs give only a brief description of their models. Some of the better models not only give the details, but also give the user the option to change or not use some parameters. The three models that follow are examples of diode models used in computer analysis.

GM COMPUTER DIODE MODELS

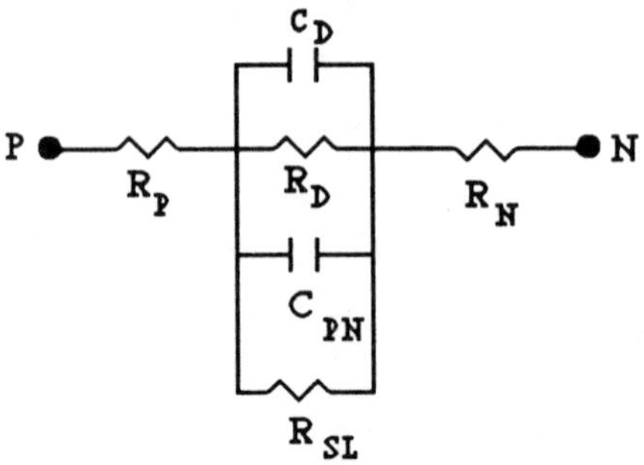

FIGURE 7.6 An elementary P-N diode model. (see the Hurley reference.). R_P and R_N are the ohmic resistances of the P and N regions. R_{SL} is the equivalent surface-leakage resistance. C_{PN} is the diode capacity as defined above. C_D is a non-linear diffusion capacitance.

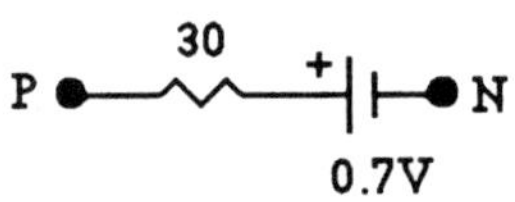

FIGURE 7.7 The diode dc equivalent circuit used in BV Engineering's circuit analysis program DCNAP.

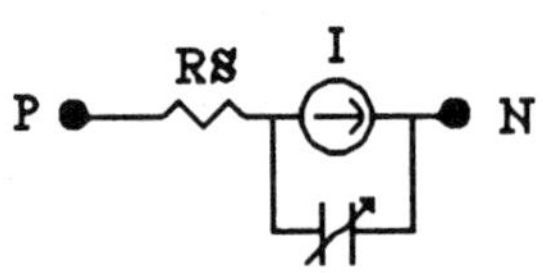

FIGURE 7.8 The diode equivalent circuit used by MicroSim's PSpice. This graphic is deceptively simple. PSpice includes 20 parameters that can be set for dc, ac, transient, temperature and noise analysis.

7.5 THE CONCEPTUAL TWO-JUNCTION TRANSISTOR

The simple, non-operating transistor model can be visualized by connecting two N-P diodes back-to-back, forming an N-P-P-N configuration. One of the N sections is called the emitter, the other N section the collector, and the lead connecting the two P sections is called the base. This elementary model can be made a bit more realistic if the two P sections are lumped together as an N-P-N configuration. It is this combined middle P element that makes a transistor work.

Junction transistors are fabricated as N-P-N or P-N-P where the middle P or N element is the base. This size and *doping* of the base will affect the gain, the frequency response and power output from the transistor.

TM THE ELEMENTARY TRANSISTOR

The two junctions in an elementary transistor can behave somewhat as junctions in a P-N diode. In transistor circuits, one of these junctions, the emitter-base diode is forward biased and the collector-base diode is reverse biased. However, because the regions are not doped alike, the depletion layers differ in the two junctions. The emitter is heavily doped, which results in a very narrow depletion region, but the base is thin and lightly doped, which produces a wide depletion region. In fact, the base depletion area extends across the base and somewhat into the emitter. Thus, the emitter depletion area is small but, because of doping, the collector depletion area is large.

GM THE ELEMENTARY TRANSISTOR

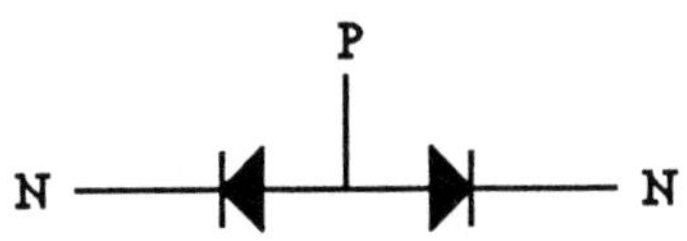

FIGURE 7.9 A conceptual, but unworkable, NP-PN two-diode transistor model.

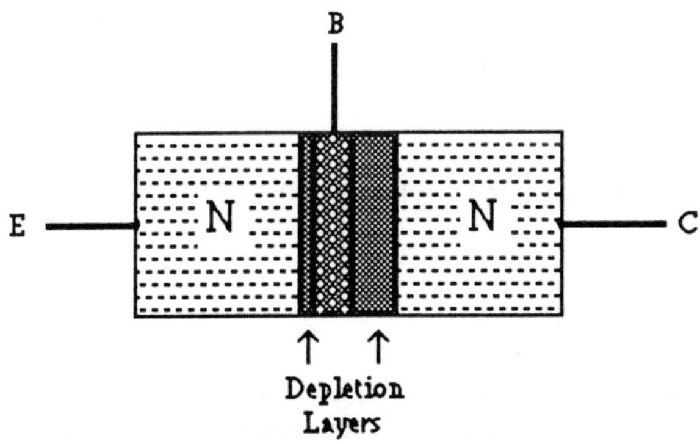

FIGURE 7.10 The N-P-N junction transistor model. Note the thin base region and the uneven depletion regions.

THE PHYSICAL, OPERATING TRANSISTOR

The forward bias of the emitter-base junction produces an expected forward current across the junction. However, most of the electrons (in an N-P junction) do not go out the base lead to the bias power supply but, because of the thin, lightly doped base, travel through the base into the collector, where they are attracted by the collector battery potential. In most transistors, more than 95 percent of the emitter-injected electrons flow to the collector and less than 5 percent flow directly out of the base. The ratio of the collector current to the emitter current is termed dc alpha (α) and is one of the indicators of transistor quality. Many transistors can have an alpha of 0.98 or more—going asymptotically to one. A similar transistor quality indicator termed beta (β or h_{FE}), is the ratio of dc collector current to dc base current. Betas of 100 or more are common in transistors.

TRANSISTOR DC VOLTAGE AND CURRENT LIMITATIONS

Transistor manufacturers give detailed information about the limitations on such parameters as voltage, current, power, rise and fall time and temperature for their products. The following is an abbreviated list of some specifications:

V_{EBO}	maximum breakdown voltage, emitter to base, collector open
V_{CEO}	maximum breakdown voltage, collector to emitter, base open
V_{CBO}	maximum breakdown voltage, collector to base, emitter open
I_C	maximum collector current
P_D	maximum power dissipation at some specified temperature
T_J	maximum junction operating temperature

7.6 THE COMMON-EMITTER (CE) TRANSISTOR AMPLIFIER

TM THE COMMON-EMITTER (CE) TRANSISTOR AMPLIFIER

A transistor is a three-lead or three-terminal amplifier in which one of the terminals is the input, one the output and one the common (ground) return for both the input and output. Transistor amplifiers can use either the emitter or the base as the input section. The most universally used amplifier configuration has the base as the input, the collector as the output, and the emitter as the common terminal.

GM THE COMMON-EMITTER (CE) TRANSISTOR AMPLIFIER

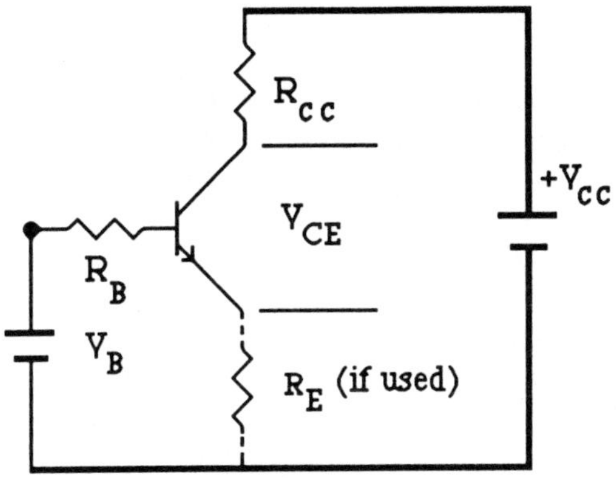

FIGURE 7.11 The elementary CE dc transistor amplifier. This circuit can be used to obtain the data shown in the curves of Figure 7.12.

TM COMMON-EMITTER DC GAIN

A common emitter transistor is fundamentally a current amplifier. Voltage amplification, derived from this current amplification, involves additional circuit components and is covered in a later section. The dc current gain, beta (β), for a common-emitter amplifier is the ratio of dc collector current to dc base current.

MM COMMON-EMITTER DC CURRENT GAIN

$$\beta = h_{FE} = I_C/I_B \text{ or } I_C = \beta I_B \qquad \text{(E 7.4)}$$

COMMON-EMITTER DC VOLTAGE GAIN

Not a readily predictable quantity for a specific circuit.

TM COMMON-EMITTER DC INPUT RESISTANCE

The (base) input to a common-emitter amplifier might seem to be the resistance of the base-emitter diode. However, this is not the case. The base current is much less than the emitter current since most of the emitter current goes through the base to form the collector current. Since the base current is a fraction (1/β) of the emitter current, it follows that the base input impedance is approximately the emitter-base resistance times β. Thus, if the normal base-emitter resistance is 5 ohms and the β is 100, the base dc input resistance is about 500 ohms.

MM COMMON-EMITTER DC INPUT RESISTANCE

$R_B \approx R_E * \beta$ where R_E is a physical resistor.

If R_E is not used (a short circuit), the dc internal emitter resistance is very non-linear and difficult to define. (R_E is the ac internal resistance that will be described later.)

TM COMMON-EMITTER DC OUTPUT RESISTANCE R_C

The output resistance is complicated but, to a first approximation, it resembles the resistance of the reverse biased collector diode. It will be remembered that there is a reverse-leakage current in a reverse biased P-N diode. Since this reverse current is ALMOST constant, the current vs. voltage line is ALMOST flat. If this reverse biased curve were completely flat, it would represent zero current change for any increasing voltage, which would mean that $R_d = \Delta V/\Delta I$ would represent $\Delta V/0$ or an infinite value of resistance. However, in both a reverse biased P-N diode and a reverse collector diode, the current does, in fact, change a wee bit as the voltage is increased. Even so, the $\Delta V/\Delta I$ is very large, which is another way of stating the resistance is very large.

Starting with zero base current, which means a collector current of I_{co} (see below) for each incremental increase in base current, there is an incremental increase in collector current and a new collector $\Delta V/\Delta I$. Thus, if the base current is increased ten unit steps, it produces ten values of collector current, which is really a family of ten collector output curves of $\Delta V_{collector}/\Delta I_{collector}$. As stated above, the output resistance is complicated. This is due to the fact that as the emitter, base and collector currents increase, other factors (such as increasing element resistance) increase and become non-linear, which results in non-linearities in the output. These non-linearities are evident when a family of collector curves is displayed. This family of curves will show that, for stepped increases in base input current, the steps of output current do not increase in an equal linear fashion. Likewise, the slope of the collect curves, which represents the graph of the output resistance, will increase with increased collector current.

MM COMMON-EMITTER DC OUTPUT RESISTANCE $\approx R_C$

TM COMMON-EMITTER SATURATION VOLTAGE

Starting with a very small base current, the collector current is increased as the base current is increased by the dc current gain β_{dc}. This increased collector current will, in turn, increase the $I_C R_{CC}$ voltage drop and reduce the available transistor voltage V_{CE}—see Figure 7.11. At some point, the increased base and collector current will force the $I_C R_{CC}$ voltage drop to almost equal the supply voltage V_{CC}. Even then, however, there is still a small voltage V_{sat} (usually less than one volt) across the transistor. If the base current is even further increased, V_{sat} is still across the transistor.

TM COMMON-EMITTER CUTOFF CURRENT

When there is zero base current, there is still some leakage current in the collector. This leakage, noted as I_{co}, is similar to the leakage or I_{sat} current in a P-N diode. I_{co} is very low and can usually be ignored in silicon transistors, but is large enough to affect designs using germanium transistors.

TM COMMON-EMITTER COLLECTOR BREAKDOWN

Since the collector-base diode is reverse biased, it has a reverse breakdown voltage just like a P-N diode.

GM COMMON-EMITTER COLLECTOR CURVES SHOWING OUTPUT RESISTANCE

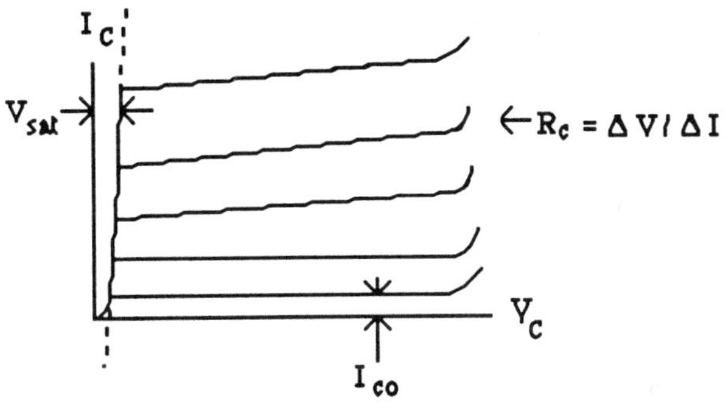

FIGURE 7.12 The collector curves for a common-emitter transistor amplifier showing the R_c resistance slope (the value of I_{co} is exaggerated).

The curves in Figure 7.12 represent only the currents in the transistor and ignore any load that the transistor might drive.

TM COMMON-EMITTER DC LOAD LINES

For the simple amplifier circuit in Figure 7.11, the battery voltage V_{CC} is across the transistor and the resistor. When there is a collector current in the transistor, this current flows through R_C and produces an IR drop. This IR drop will change with changing collector current. For example, assume three conditions.

First, if the transistor is cut off (no base current), to a first approximation, there is no collector current and thus no $I_C R_C$ drop, which means that all of V_{CC} is across the transistor as V_{CE}. Second, when base and thus collector current flow, part of the V_{CC} voltage is across the transistor and part of it is the resistor IR_C drop. Third, if the base current increases enough to saturate the transistor, there is very little voltage (a few tenths of a volt) across the transistor and most of the battery voltage is now across the resistor. Thus, for the two extremes, the current in the resistor will be either zero or V_{CC}/R_C.

MM COMMON-EMITTER DC LOAD LINES

Using Kirchhoff's voltage law for the transistor output current loop:

$$V_{CC} = I_C R_C + V_{CE} \text{ or} \qquad \text{(E 7.5)}$$

$$I_C = \frac{-1}{R_C} V_{CE} + \frac{V_{CC}}{R_C} \qquad \text{(E 7.6)}$$

Note the minus sign in the $-1/R_C$ term. This indicates that the slope of the $1/R_C$ line is negative.

GM

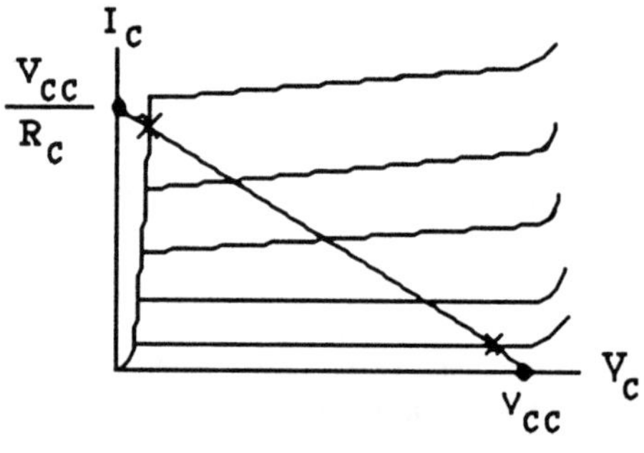

FIGURE 7.13 Transistor collector curves with a dc load line.

TM DC LOAD LINES AND THE TRANSISTOR SWITCH

The elementary transistor switch, which is used in control and digital circuits, is a good example of the use of the dc load line. This circuit uses a two level base-input to obtain an on/off or high/low output. When there is no base input voltage, and thus no base current, the collector is cut off and thus the collector (output) voltage is very near the supply voltage V_{CC}—it equals V_{CC} minus any slight IR_C drop caused by I_{C0}. When the base input is high or on, the collector is driven into saturation with a voltage very near to zero—it will equal V_{sat}. Thus, referring to Figure 7.13, the output voltage of the transistor switch is either high at V_{CC} minus any $I_{Co}R_C$ or low at zero (ground) plus V_{sat}. Note that the position of the two x's is very exaggerated.

GM THE ELEMENTARY CE TRANSISTOR SWITCH

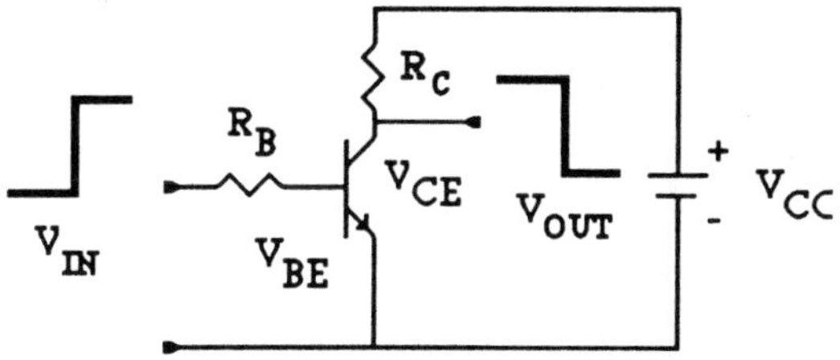

FIGURE 7.14 An elementary transistor switch.

7.7 TRANSISTOR CE BIASING FOR SMALL-SIGNAL OPERATION

Figure 7.11 shows a separate battery to bias an N-P-N common-collector transistor so that the EB junction is forward biased and the CB junction is reverse biased. Although this circuit illustrates the dc operation of a transistor and can be used to obtain the data which represents the dc operating curves, it is rarely used in practical designs. The following three transistor biasing circuits are presented to show the development of transistor biasing techniques.

TM SINGLE POWER SUPPLY, SINGLE RESISTOR BASE BIASING

Since the object of the base bias is to provide a voltage source for R_B that is more positive than the potential of the emitter, R_B can be connected to the positive side of V_{CC}. However, this is a very unsatisfactory biasing method because of variations in β_{dc} (two to one or more) from one transistor to another. This method of biasing SHOULD NEVER BE USED IN LINEAR TRANSISTOR AMPLIFIERS, although it is a common circuit in transistor switching circuits.

MM SINGLE POWER SUPPLY, SINGLE RESISTOR BASE BIASING

$$R_B \approx \frac{V_{CC} - 0.7}{I_B} \tag{E 7.7}$$

GM SINGLE POWER SUPPLY, SINGLE RESISTOR BASE BIASING

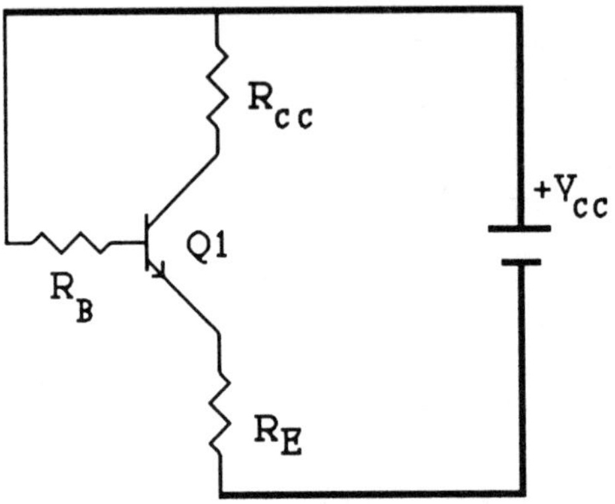

FIGURE 7.15 Base bias using one power supply and one biasing resistor (not recommended for linear circuits).

TM COLLECTOR-FEEDBACK BASE BIASING

If the bias resistor R_B is connected directly to the collector rather than V_{CC}, some improvement in bias stability vs. a change in β_{dc} is obtained. With this method, if a change in β_{dc} makes the collector current increase, this increased current will increase the $I_C R_{CC}$ drop which, in turn, will lower the collector voltage and reduce the base current. With less base current, the collector voltage will decrease or stabilize close to its original, intended value. Likewise, if the β_{dc} tends to lower the collector current, a process opposite to the above will take place and the base current will again be somewhat stabilized. Although collector-feedback bias gives some improvement in bias stability, it is still a very poor scheme and is not recommended for any design.

MM COLLECTOR-FEEDBACK BASE BIASING

$$V_{BE} - V_{CC} + (I_C + I_B)R_C + I_B R_B = 0 \text{ or} \tag{E 7.8}$$

$$V_{BE} - V_{CC} + I_C R_C + I_B R_B \approx 0 \text{ and since } I_B = I_C / \beta_{dc} \tag{E 7.9}$$

$$I_C = \frac{V_{CC} - V_{BE}}{R_C + R_B/\beta_{DC}} \quad \text{(E 7.10)}$$

GM COLLECTOR-FEEDBACK BASE BIASING

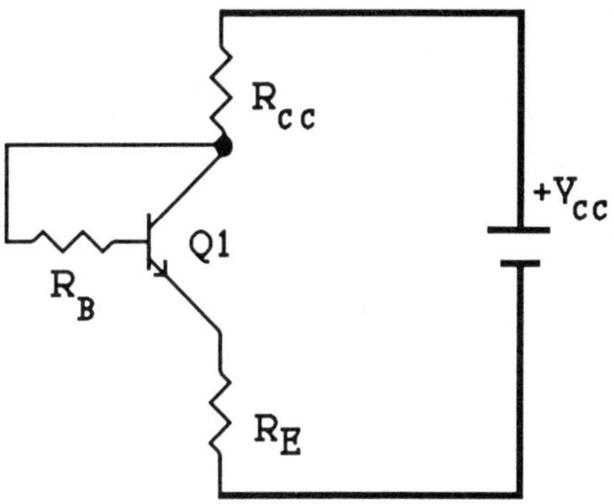

FIGURE 7.16 Collector-feedback Bias.

TM VOLTAGE-DIVIDER BIAS

The ideal base biasing circuit should operate from V_{CC} and provide a fixed (sometimes termed stiff) bias that is not changed by temperature or β_{dc}. A two-resistor voltage divider bias, while not completely perfect, works well for both germanium and silicon transistors. If the series current in these resistors is made five to ten times the desired dc base current, this voltage divider will provide a near constant voltage base bias that is independent of β_{dc} and almost independent of temperature. If a series emitter resister R_E is added to the circuit, the dc bias becomes almost immune to practical changes in transistor parameters. The price that is sometimes paid for this improvement in bias stability is overall circuit gain, since the Thevenin equivalent of the two bias resistors may load the previous amplifier.

MM VOLTAGE-DIVIDER BIAS

$$V_{Base} = \frac{R_2}{R_1 + R_2} V_{CC} \quad \text{(E 7.11)}$$

$$I_E = \frac{V_{Base} - 0.7}{R_E} = \frac{V_E}{R_E} \text{ when the emitter resistor } R_E \text{ is used} \quad \text{(E 7.12)}$$

$$V_C = V_{CC} - I_C R_C \quad \text{(E 7.13)}$$

GM VOLTAGE-DIVIDER BIAS

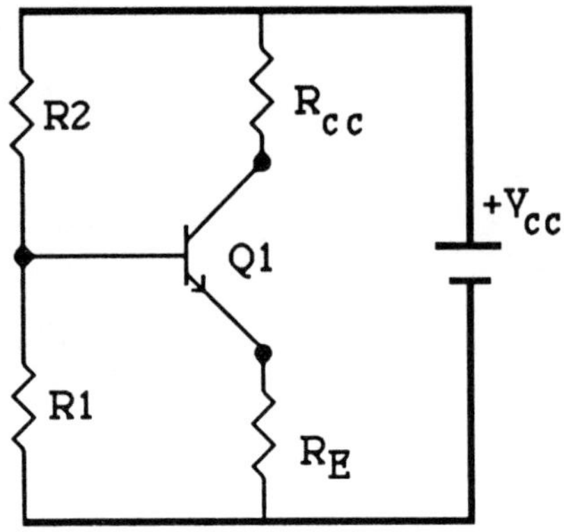

FIGURE 7.17 Common emitter voltage-divider bias, including the emitter resistor R_E.

TM CE DC BIASING AND THE OPERATING OR Q-POINT

The bias is used to establish dc quiescent operating Q-points. It is assumed that an input signal will increase or decrease the base current and thus move base voltage up and down from this quiescent point. This changing base current will, in turn, move the collector output current (and the collector voltage) up and down from its established operating Q-point. It should be noted that an increase or decrease in signal base current will produce a corresponding increase or decrease in the collector current. However, for the collector output voltage, there is a phase reversal and, therefore, an increase in base VOLTAGE (and current) causes a DECREASE in collector voltage. A single-transistor CE amplifier inverts the input voltage.

GM CE DC BIASING AND Q-POINT

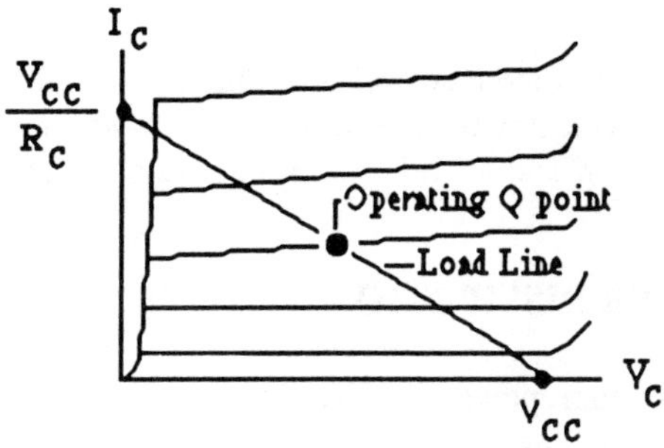

FIGURE 7.18 The collector quiescent or Q-point. (The effects of V_{sat} and I_{co} are not shown.)

7.8 COMMON EMITTER SMALL-SIGNAL AC AMPLIFIERS

Transistor amplifiers for alternating current are more common than dc amplifiers. Once the dc operating Q-points are established, several individual transistor stages can be capacitively coupled for increased gain.

GM A CAPACITIVELY COUPLED AC TRANSISTOR AMPLIFIER

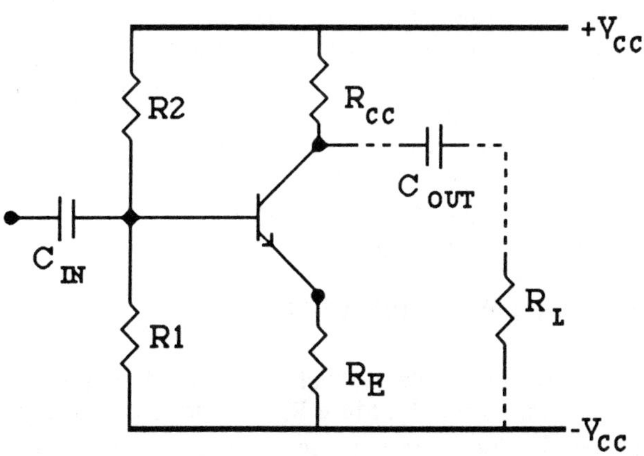

FIGURE 7.19 A single-stage CE AC transistor amplifier.

TM CAPACITIVELY COUPLED AC TRANSISTOR AMPLIFIER ANALYSIS

The analysis of an ac transistor amplifier is more complicated than the analysis of a dc amplifying stage. It is usually necessary to consider a new *ac load line*, calculate the current and voltage swing and signal distortion, and consider transistor dissipation.

TOTAL OR MAXIMUM POWER DISSIPATION

Each transistor will have a maximum power dissipation listed by the manufacturer. This power equals the voltage across the transistor times the current through the transistor which is, to a good approximation, $V_{CE} * I_C$. This maximum power can be plotted as a parabolic curve on the transistor V-I graph. This gives a maximum power locus that can be used with the load line to assure proper operation.

COMMON EMITTER AC INPUT IMPEDANCE

The input RESISTANCE of the amplifier in Figure 7.19 is approximately the value of three resistors in parallel—R1 in parallel with R2 (R1||R2) plus the parallel equivalent of the base input resistance. The equivalent base input resistance is β times the ac emitter-base diode resistance plus any external emitter resistance R_E. This ac emitter-base diode resistance, r'e, is derived from an emitter-base diode V-I curve and is the dynamic resistance $\Delta V_{BE}/\Delta I_E$.

The input IMPEDANCE is the series impedance of the reactance of the input capacitor and the ac base input resistance described above. At some frequencies, often termed mid-band, the input capacitor can be considered an ac short circuit and thus the input impedance is equivalent to the input resistance. However, at some lower frequencies, the capacitive reactance is significant and the equivalent input impedance is $R - jX_C$ (see Chapter 3).

There is also an input capacitance, C_{ib}, listed in most transistor specifications, which affects the input impedance at high frequencies.

COMMON EMITTER AC OUTPUT IMPEDANCE

The output or looking-back resistance is the parallel equivalent of the resistor R_C and the transistor output resistance R_c. Since R_c is quite large, usually several hundred thousand ohms or more, the output resistance can be considered R_{CC}. At high frequencies, there is an output capacitor, C_{ob}, and a *Miller feedback capacitor* C'_c.

COMMON EMITTER AC VOLTAGE GAIN

The CE ac voltage can be made almost transistor *independent* by the use of the emitter resistor R_E. This "trick," along with voltage-divider bias, makes ac transistor amplifiers very reliable even with changing temperature and wide variations in β_{dc}.

MM CAPACITY COUPLED AC TRANSISTOR AMPLIFIER ANALYSIS

MAXIMUM POWER DISSIPATION

$$P_{Max} = I_C V_{CE} \qquad \text{(E 7.14)}$$

COMMON EMITTER AC INPUT IMPEDANCE

$$R_{in} = R_1 \parallel R_2 \parallel \beta(R_E + r'e) \text{ where } r'e \approx 0.025/I_E \qquad \text{(E 7.15)}$$

$$Z_{in} = R_{in} - jX_{C\ in} \qquad \text{(E 7.16)}$$

COMMON EMITTER AC OUTPUT IMPEDANCE

$$Z_{out} \approx R_{CC} \text{ which is the “looking back” } Z_{out} \qquad \text{(E 7.17)}$$

COMMON EMITTER AC VOLTAGE GAIN

$$A = \frac{V_{out}}{V_{in}} \qquad \text{(E 7.18)}$$

$$v_{in} = i_e r'_e \approx i_c r'_e \text{ where } R_E = 0 \qquad \text{(E 7.19)}$$

$$v_{out} = -i_e R_{CC} \qquad \text{(E 7.20)}$$

$$A \approx \frac{-i_c R_{CC}}{i_c r'_e} = \frac{-R_{CC}}{r'_e} \text{ where } R_E \text{ is 0} \qquad \text{(E 7.21)}$$

$$A = \frac{-i_c R_{CC}}{i_c r'_e + R_E} = \frac{-R_{CC}}{r'_e + R_E} \text{ if } R_E \text{ is included, and} \qquad \text{(E 7.22)}$$

$$A = \frac{-R_{CC}}{R_E} \text{ where } R_E >> r'_e \qquad \text{(E 7.23)}$$

TM **THE COMMON EMITTER AC LOAD LINE**

THE COMMON EMITTER AC LOAD LINE graph differs from the dc load line graph since the ac load must include R_L as shown in Figure 7.14.

MM **THE COMMON EMITTER AC LOAD**

$$R_{col\ ac} = R_{CC} \parallel R_L \qquad \text{(E 7.24)}$$

$$= R_{CC} \parallel R_L - jX_{Cout} \text{ if } C_{out} \text{ is not considered an ac short circuit} \qquad \text{(E 7.25)}$$

GM **THE COMMON EMITTER AC LOAD**

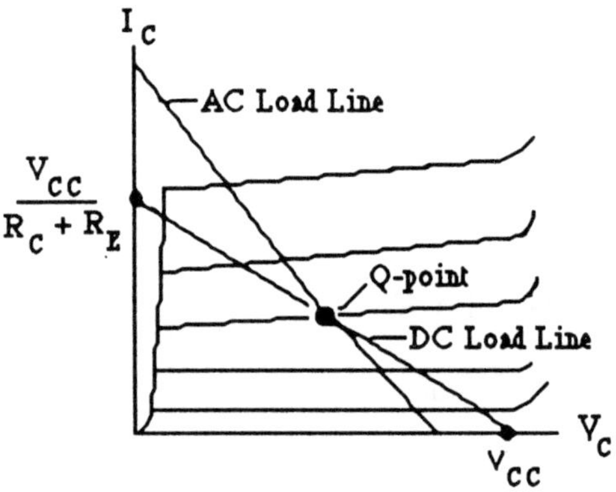

FIGURE 7.20 The ac load line added to the collector curve graph.

7.9 THE COMMON BASE AMPLIFIER

TM **THE COMMON BASE AMPLIFIER**

The common base amplifier has a low input impedance, but does not have a phase reversal between input and output voltage or current. This circuit is not widely used, but does have appeal in some instrumentation and communications circuits with low impedance signal source generators. The common base is sometimes at ac ground only. Often this configuration requires an additional emitter power supply V_{EE} as well as the collector supply V_{CC}.

MM THE COMMON BASE AMPLIFIER

$$I_E = \frac{V_{EE} - V_{BE}}{R_E} \approx I_C \qquad \text{(E 7.26)}$$

$$R_{in\ ac} \approx r'_e \qquad \text{(E 7.27)}$$

$$\approx r'_e + R_E \text{ if a series emitter resistor is used in the input} \qquad \text{(E 7.28)}$$

$$R_{out\ ac} = R_C \parallel R_{CC} \approx R_{CC} \qquad \text{(E 7.29)}$$

$$A = \frac{R_{out\ ac}}{R_{in\ ac}} \qquad \text{(E 7.30)}$$

$$V_C = \frac{R_{CC}}{R_{EE}} \qquad \text{(E 7.31)}$$

GM THE COMMON BASE AMPLIFIER

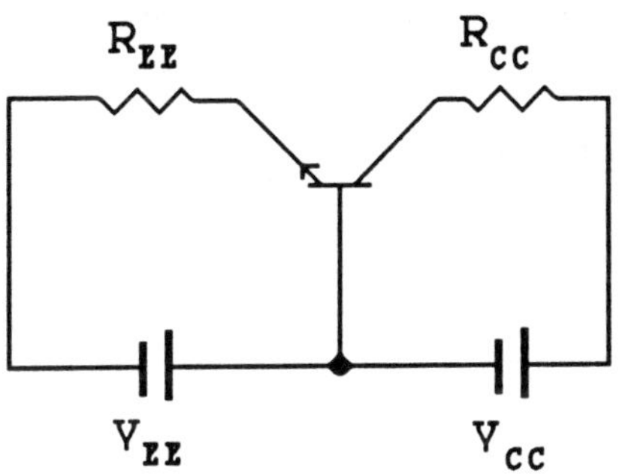

FIGURE 7.21 The simple dc common emitter circuit.

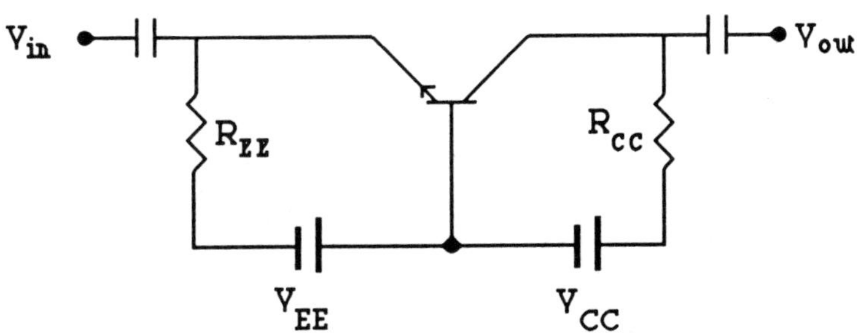

FIGURE 7.22 A dual power supply CB ac amplifier.

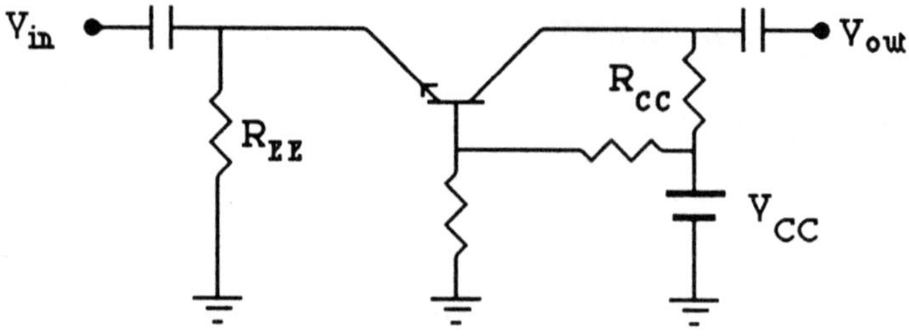

FIGURE 7.23 A single power supply ac CB amplifier.

7.10 THE COMMON COLLECTOR AMPLIFIER

TM THE COMMON COLLECTOR AMPLIFIER

The common collector amplifier is usually called the emitter follower amplifier. This transistor circuit makes the ideal isolation amplifier. The input resistance can be made quite high, which means minimal circuit loading and the output resistance is usually quite low, which means it approaches the ideal constant voltage source.

MM THE COMMON COLLECTOR AMPLIFIER

$$I_E R_E + V_{CE} = V_{CC} \quad \text{(E 7.32)}$$

$$I_C \approx \frac{V_{CC} - V_{CE}}{R_R} \text{ if it is assumed that } I_C \approx I_E \quad \text{(E 7.33)}$$

$$R_{in\ ac} = \beta\,(r'_e + R_E) \quad \text{(E 7.34)}$$

$$R_{out\ ac} = r'_e + \frac{R \parallel R2}{\beta} \text{ looking back into the emitter} \quad \text{(E 7.35)}$$

$$V_{in\ ac} = i_e\,(r'_e + R_E) \quad \text{(E 7.36)}$$

$$V_{out} = i_e R_E \quad \text{(E 7.37)}$$

$$A = \frac{R_E}{r'_e + R_E} \approx \frac{R_E}{R_E} \approx 1 \quad \text{(E 7.38)}$$

GM **THE COMMON COLLECTOR AMPLIFIER**

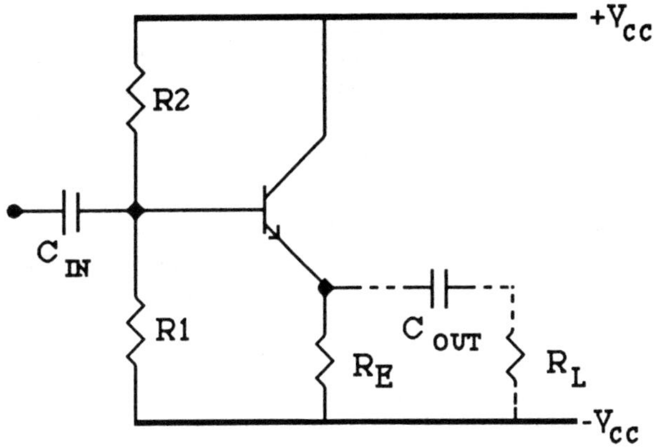

FIGURE 7.24 The classic emitter follower. Note that the resistors R1 and R2 may determine the major part of the input resistance (R_{in} = R1 || R2 || β (r'_e + R_E)).

7.11 JUNCTION TRANSISTOR EQUIVALENT CIRCUITS

The development of the junction transistor, starting in the 1950s, included the development of many equivalent circuits—some lasting and some very forgettable. Attempts were made to modify network, filter, and transmission-line models to explain and analyze transistors. Some of the equivalent circuits were primarily mathematical, while others concentrated on the physical actions inside the transistor. At first, both approaches were compromises, and neither gave an accurate mathematical model or an accurate physical model. Often a graphic transistor model might give some insight into transistor action but the parameters used either had no direct relationship to measurable quantities or the measurements were so esoteric that they had no value for normal design and analysis. The proliferation of computers made it possible to develop more accurate models containing many more parameters. Likewise, it became practical to store customized models of individual transistors rather than to use a few "universal" models for low-frequency, high-frequency, and power transistors. The models that follow include some historical configurations, including one that is still used, plus a sampling of computer transistor models that can be used for circuit design and analysis.

TM JUNCTION TRANSISTOR MODELS

Early transistor models included z, y, and s networks (see Chapter 1). These models take the "black box" approach to transistor analysis. S parameters are still used for some high-frequency modeling. However, the following graphics include models that primarily reflect the physical operation of the transistor. The T models were simplistic, but were easy to comprehend and included parameters that could be measured. The Ebers-Moll model presented a significant advancement in the understanding of the operation of junction transistors. The hybrid or h parameter models, while somewhat difficult to grasp, provided a way to include the wide variety of measurements from low-frequency, high-frequency, and power transistors in a coherent but expandable transistor model. It should be noted that much of the early literature on transistor models was inconsistent on terminology and assumptions. This made it difficult to follow then, and also now presents the problem of being historically inaccurate. The h parameters have proved to be the most valued parameters. They are still used to describe transistor performance and are included in most transistor data sheets. Likewise, most of the computer models either use h parameters or use measurable quantities that are similar or can be derived from them.

GM JUNCTION TRANSISTOR MODELS

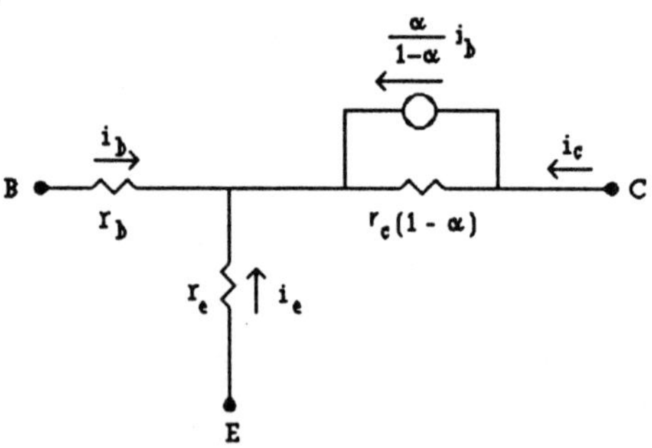

FIGURE 7.25 An early CE T-equivalent circuit.

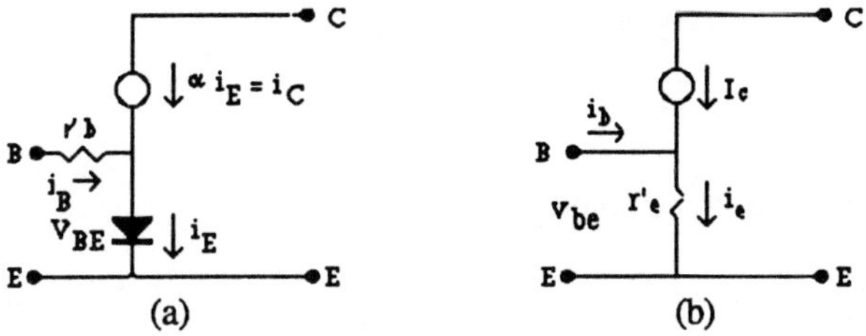

FIGURE 7.26 (a) The Ebers-Moll dc model; (b) The Ebers-Moll ac model.

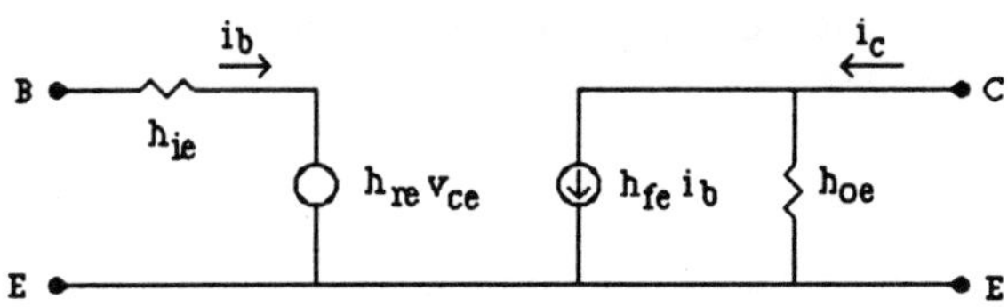

FIGURE 7.27 A simple hybrid-Pi equivalent circuit.

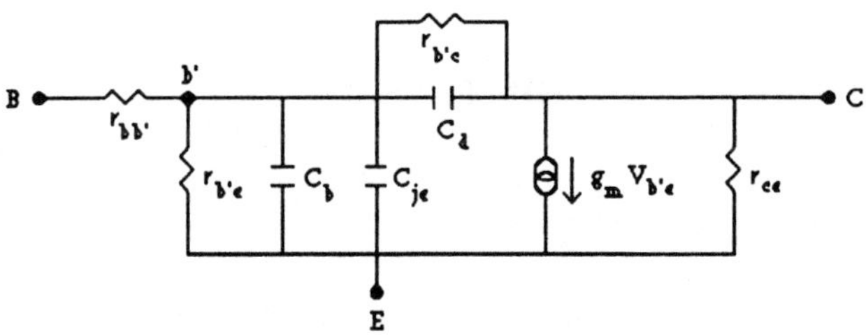

FIGURE 7.28 The normal h-parameter model for a CE transistor.

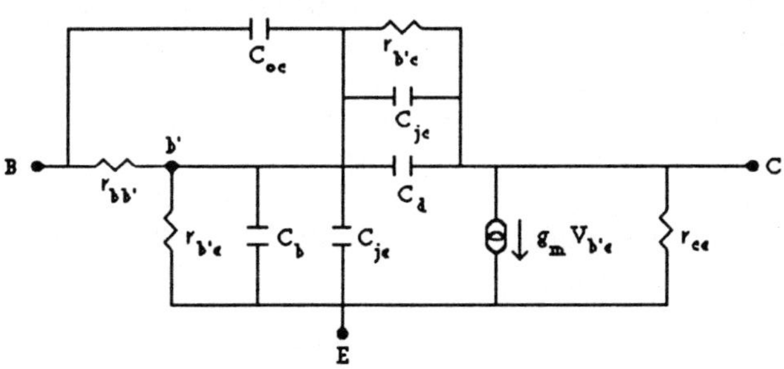

FIGURE 7.29 A hybrid-Pi model with high-frequency parameters included.

THE HYBRID-PI PARAMETERS SHOWN IN FIGURE 7.29

$r_{b'b}$ the base spreading resistance

C_b $= C_{ib}$ the CE input capacitance modeled with C_{oc}

$r_{b'e}$ the equivalent base input resistance

$r_{b'c}$ a collector-base resistance dependent

C_d a collector-base capacitance dependent upon the collector ac voltage upon the collector ac voltage

C_{je} the base-emitter junction space-charge capacitance

C_{jc} the collector-base junction space-charge capacitance

C_{oc} the overlap capacitor of the collector-base junction—used at high frequencies

COMPUTER PROGRAM JUNCTION TRANSISTOR EQUIVALENT CIRCUITS

TM COMPUTER PROGRAM MODELS

Engineers who design and analyze transistor circuits have a choice of simple or sophisticated computer transistor models to work with. Most programs give the user several choices. They usually give both dc and ac models. Likewise, there are generic models, where the parameters can be tailored to the transistor in use, or a number of specific transistors have been modeled and are available in library form. It should be remembered that each element in the model must be able to be represented as a computer number. For instance, the dc Ebers-Moll model, shown in Figure 7.26, could not be used unless an additional submodel was added for the diode. Usually, the schematics of the transistors given in the program instruction books are "examples" and are not necessarily the exact configuration the computer would use.

GM COMPUTER PROGRAM MODELS

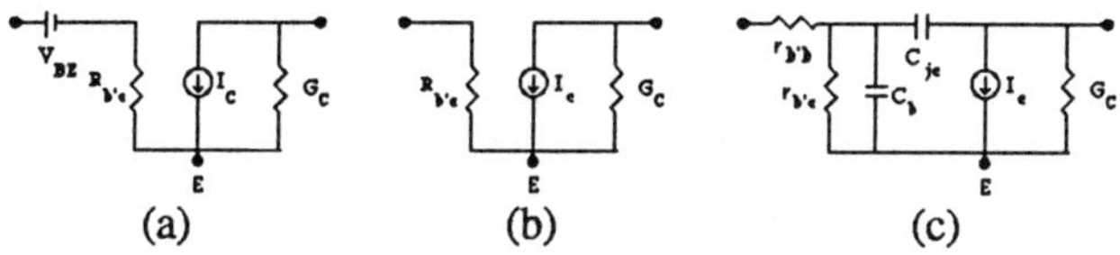

FIGURE 7.30 Prototype transistor models used in BV Engineering programs: (a) A dc model; (b) A low-frequency ac model; (c) A high-frequency ac model.

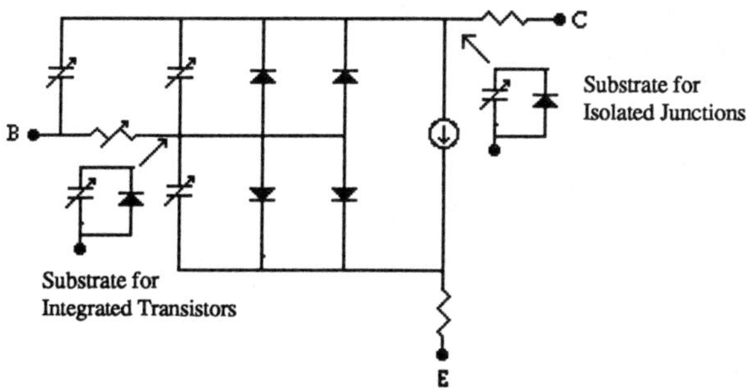

FIGURE 7.31 The transistor model described in PSpice. The program gives the user 50 parameters that can be modified for a specific transistor or a specific use.

7.12 JUNCTION FIELD-EFFECT TRANSISTORS, OR JFETs

AN INTRODUCTION

The junction transistors previously discussed operate by the flow of both electrons and holes and are termed *bipolar*. JFETs, because of their unique configuration, operate using only electrons or holes (n-channel or p-channel) and are termed *unipolar*. Likewise, the controlling element, the gate, which can be compared with the base, uses voltage rather than current. Also, JFETs can be operated in two different ways: as a controlled resistor or as a controlled amplifier. Nevertheless, even with these distinct differences, several of the basic modeling parameters that were used for bipolar transistors can be used as a foundation for building unipolar JFET models.

JFETs are three-terminal networks with their source, gate and drain roughly analogous to the emitter, base, and collector in bipolar transistor. The source-gate diode (roughly equivalent to the emitter-base diode) is reverse biased as is the drain-gate diode. In the JFET, the source and drain current is the same and is controlled or gated by a depletion field produced in the current path (channel) from the source to the drain.

The amplifier operation represented by the drain characteristic curves is similar to the operation represented by collector curves, with one major exception, the "ohmic" or saturation region (see Figure 7.34). For a bipolar transistor, the collector saturation voltage is usually only a few tenths of a volt. For a JFET, the analogous saturation region will show curves operating with a voltage across the FET of up to two volts. This operating area, where the characteristic curves are about vertical, represents a linear, low-resistance region that can be reliably used in switching and control circuits.

TM JUNCTION FIELD-EFFECT TRANSISTORS—CHARACTERISTICS

JFET semiconductor amplifiers are characterized by a (1) high input resistance, (2) a somewhat lower voltage gain than bipolar transistors (circuits) and (3) two output operating regions. One mode of operation can produce a very low resistance output. A second mode of operation gives a high-resistance, almost constant current drain output, similar to a bipolar collector. Like bipolar transistors, many JFET characteristics vary with temperature and from unit to unit.

V_{GS} = the dc (reverse bias) voltage from the gate to the source.

$V_{GS(cutoff)}$ = the dc (reverse bias) bias voltage that will cut off or reduce the drain current to zero.

$V_p = V_{pinch\text{-}off}$ = the gate source bias voltage that indicates the start of the drain constant-current operating region and the end of the voltage-controlled resistive (near-vertical) region.

$V_{GS(sat)}$ = a special definition for the particular knee point on each drain curve where it goes into the constant-current or pinch-off region.

$V_{DS(max)}$ = the maximum allowable source to gate voltage—the breakdown voltage.

I_{DSS} = the drain current when the gate is shorted to the source. This is usually the upper limit for the drain current in a JFET.

R_{DS} = the resistance of the JFET in the ohmic or vertical region on the characteristic curves.

R_{IN} = very high, the resistance of a low leakage current, reverse-biased diode.

R_{OUT} = similar to the collector output of a bipolar transistor.

MM JUNCTION FIELD-EFFECT TRANSISTORS—CHARACTERISTICS

$$V_{GS(off)} = -V_P \qquad \text{(E 7.39)}$$

$$R_{DS} = \frac{V_P}{I_{DSS}} \text{ in the resistive region} \qquad \text{(E 7.40)}$$

$$I_D = I_{DSS}\ (1 - \frac{V_{GS}}{GD_{(off)}})^2 \text{ in the active region} \qquad \text{(E 7.41)}$$

GM JUNCTION FIELD-EFFECT TRANSISTORS—CHARACTERISTICS

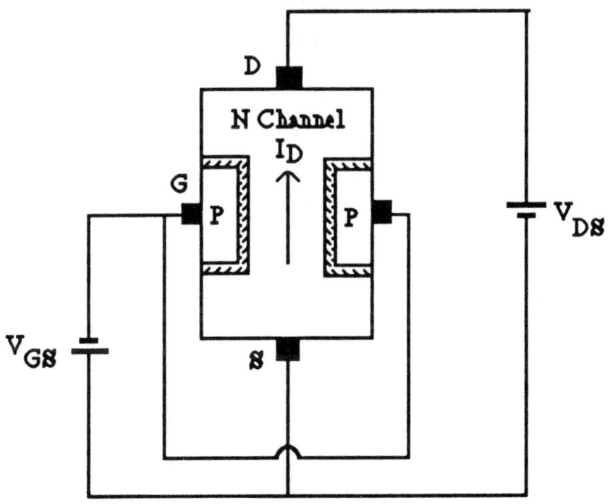

FIGURE 7.32 The JFET showing the depletion around the P regions. The gate-source diode is reverse biased and V_{DS} is very small.

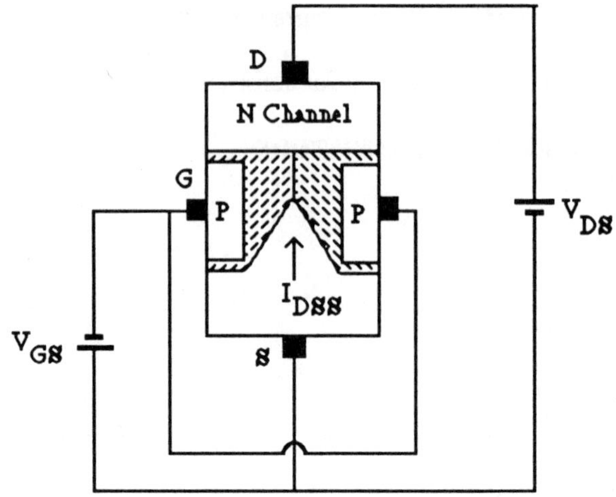

FIGURE 7.33 **By increasing V_{DS}, the depletion region is expanded in the heavily doped N region. When V_{DS} equals the pinch-off voltage V_p, the depletion regions meet and change the mode of operation from ohmic to active.**

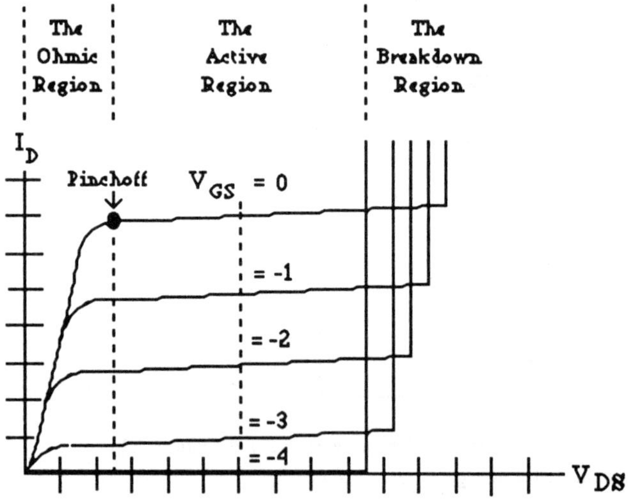

FIGURE 7.34 **An example of drain characteristic curves showing the ohmic, active, and breakdown areas as a function of source-to-drain voltage.**

TM JUNCTION FIELD-EFFECT TRANSISTOR BIASING

Common source JFET amplifiers are usually self-biased. Although voltage-divider bias or two-supply bias can be used successfully, the fact that the gate is a voltage controller, rather than a current controller, usually makes self-bias the most attractive method. Self-bias is accomplished by inserting a bias resistor R_S between the source and common (ground) node. This will make the source slightly positive in an N-channel JFET. If the gate is now connected to the common through a resistor R_G, the resulting gate voltage will be negative with respect to the source. Since the gate draws no current, there is no voltage drop in R_G.

MM JUNCTION FIELD-EFFECT TRANSISTOR BIASING

$V_S = +I_S\,R_S$ which is the dcvoltage drop across the source resistor.

(E 7.42)

$V_{GS} = -\,(I_S\,R_S)$ which is a NEGATIVE source-to-gate bias voltage.

(E 7.43)

GM JUNCTION FIELD-EFFECT TRANSISTOR BIASING

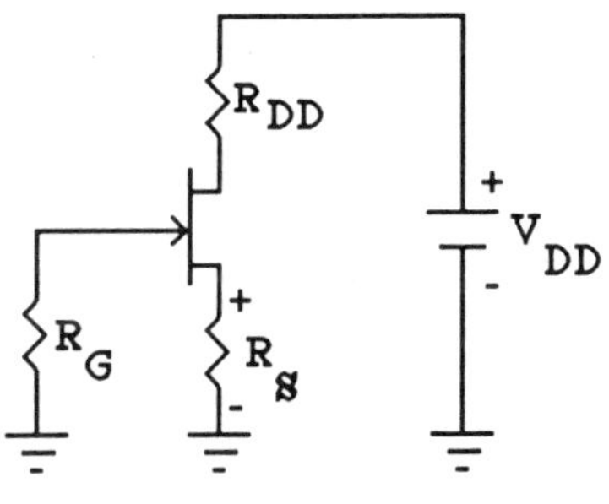

FIGURE 7.35 dc self-bias for an N-channel JFET.

TM JUNCTION FIELD-EFFECT TRANSISTOR TRANSCONDUCTANCE

The gain of a field-effect transistor is analyzed and computed differently than a bipolar transistor. The bipolar is essentially a "current-in–current-out" device, which gives a simple out/in gain ratio. If the voltage gain of a bipolar transistor is desired, it can be found using the fundamental current ratio plus the circuit resistors (and reactors) involved. An FET is essentially a "voltage-in–current-out" amplifier. It is easiest to use the transconductance—the change of output (drain) current divided by the change of input (gate) voltage—to calculate the stage gain.

MM JUNCTION FIELD-EFFECT TRANSISTOR TRANSCONDUCTANCE

$$g_m = \frac{\Delta I_D}{\Delta V_{GS}} \text{ or, for ac, } = \frac{i_d}{v_{gs}} \text{ (some data sheets use } g_{fs} \text{ for } g_m\text{)} \quad \text{(E 7.44)}$$

$$V_{GS\ off} = \frac{-2I_{DSS}}{g_{m\ max}} \text{ (some data sheets use } g_{m0} \text{ for } g_{m\ max}\text{)} \quad \text{(E 7.45)}$$

$$g_m = g_{m\ max}\left(1 - \frac{V_{GS}}{V_{GS\ off}}\right) \text{ note the parabolic form} \quad \text{(E 7.46)}$$

GM JUNCTION FIELD-EFFECT TRANSISTOR TRANSCONDUCTANCE

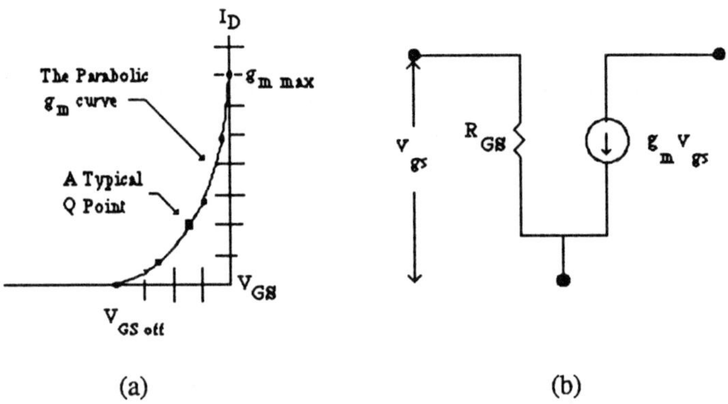

FIGURE 7.36 JFET transconductance: (a) The transconductance curve for $\Delta I_D/\Delta V_{GS}$; (b) A simple transconductance circuit model.

TM JUNCTION FIELD-EFFECT TRANSISTOR AMPLIFIERS

JFET circuits fall into two general classes: signal amplifiers and analog switches. The most common amplifier uses a common source operating in the active region, which is analogous to the bipolar common emitter amplifier. Second, a JFET can be used in the ohmic or control region where it appears as a voltage-controlled resistor.

MM THE JFET COMMON SOURCE AMPLIFIER

$$r_{in} \approx R_G \quad \text{(E 7.47)}$$

$$r_{out} = R_{DD} \parallel R_L \quad \text{(E 7.48)}$$

$$v_{out} = v_{in}\, g_m\, r_d \quad \text{(E 7.49)}$$

$$\frac{v_{out}}{v_{in}} = \text{voltage gain} = g_m\, r_d \quad \text{(E 7.50)}$$

GM THE JFET COMMON SOURCE AMPLIFIER

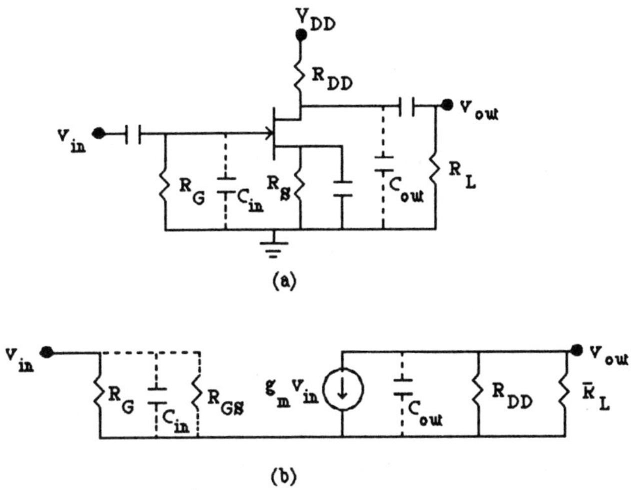

FIGURE 7.37 The JFET signal amplifier: (a) The circuit diagram of a CS JFET amplifier. The components R_{GS}, C_{in} and C_{out} are shown dashed because not all data sheets list them in this exact form; (b) Using superposition, the ac equivalent circuit can be simplified. The coupling and bypass capacitors are assumed as ac short circuits.

TM THE JFET AS ANALOG SWITCH

The JFET can be used as a series or shunt element with a low ON resistance and a high OFF resistance.

MM THE JFET AS ANALOG SWITCH

$$R_{DS} = \frac{V_{GSoff}}{I_{DSS}} \text{ remembering that } V_{GSoff} = -V_P \qquad \text{(E 7.51)}$$

GM THE JFET AS ANALOG SWITCH

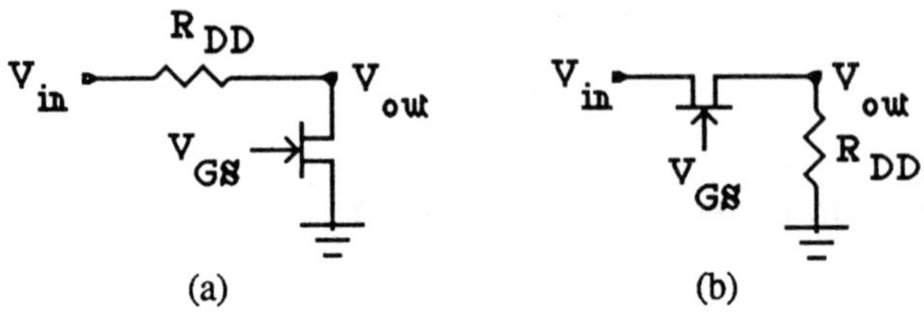

FIGURE 7.38 (a) The JFET as a shunt switch; (b) The JFET as a series switch.

7.13 METAL-OXIDE SEMICONDUCTOR FIELD EFFECT TRANSISTORS (MOSFETs)

The two-diode structure that was used as the foundation of the bipolar and the unipolar JFET transistors must be modified for the MOSFET. Although the MOSFET contains both N and P regions, the control element is not a semiconductor but a conductive plate. It is insulated from the rest of the transistor by a glass-type insulator of silicon dioxide (SiO_2). Also, there is another region, similar to the gate in a JFET, which is termed the *substrate*. MOSFETs, or, as they are sometimes called, insulated gate transistors, are divided into two classes: the depletion mode and the enhancement mode. Both depletion mode and enhancement mode MOSFETs are produced as N-channel and P-channel.

Special Note: It should be pointed out that, whether it be a bipolar, a JFET or a MOSFET, the graphical block models are used ONLY to convey an understanding of how the transistors operate in circuits and are NOT intended to infer any information about their exact physical layout or construction. The construction and integration varies greatly from type to type and manufacturer to manufacturer and is not treated in this work.

TM THE DEPLETION-MODE MOSFET

The depletion-mode MOSFET has a narrow channel of N or P semiconductor. In N-channel units, the channel is fabricated to a P substrate and the insulated gate. In most cases, the substrate is internally connected to the source. As in a JFET, the gate-to-source voltage controls the width of the channel and thus the drain current. However, for the depletion-mode MOSFET, a POSITIVE gate voltage will affect the channel width as well as a negative voltage. These transistors, like JFETs, can operate in an ohmic and an active region.

GM THE DEPLETION-MODE MOSFET

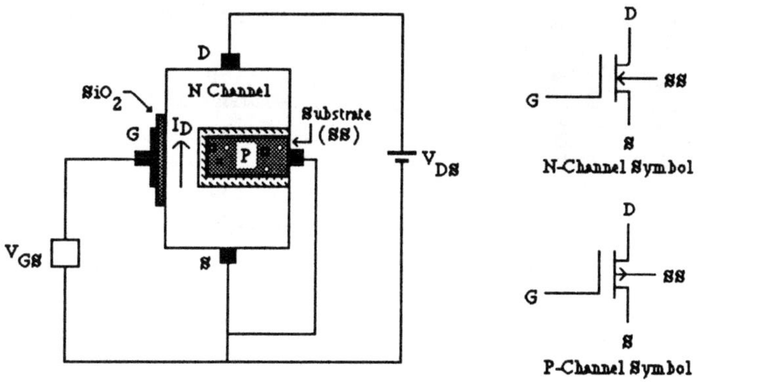

FIGURE 7.39 The depletion mode MOSFET. Note that V_{GS} can be + or –.

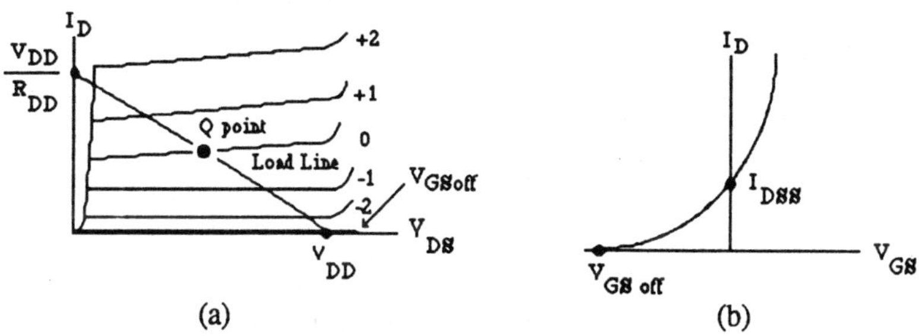

FIGURE 7.40 (a) Depletion mode MOSFET drain curves; (b) Depletion mode transconductance curve.

TM THE ENHANCEMENT-MODE MOSFET

The enhancement-mode MOSFET is much more common than the depletion-mode MOSFET—it is the building block for most VLSI integrated circuits. The model of the semiconductor structure resembles a bipolar transistor with the addition of an insulated gate. The drain characteristic curves somewhat resemble the collector curves of a bipolar transistor when gate voltage is substituted for base current. Drain current is created (enhanced) when a positive gate-source voltage exceeds the threshold voltage $V_{GD\ th}$ and creates an N-type inversion layer.

GM THE ENHANCEMENT-MODE MOSFET

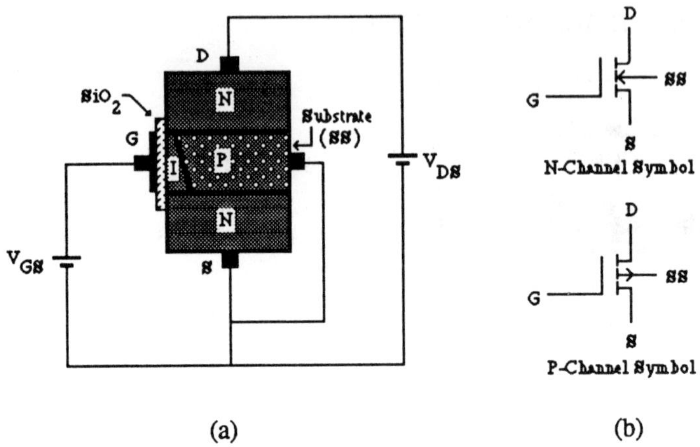

FIGURE 7.41 (a) The enhancement-mode MOSFET (note the N-type inversion layer created next to the insulated gate); (b) Enhancement-mode MOSFET symbols.

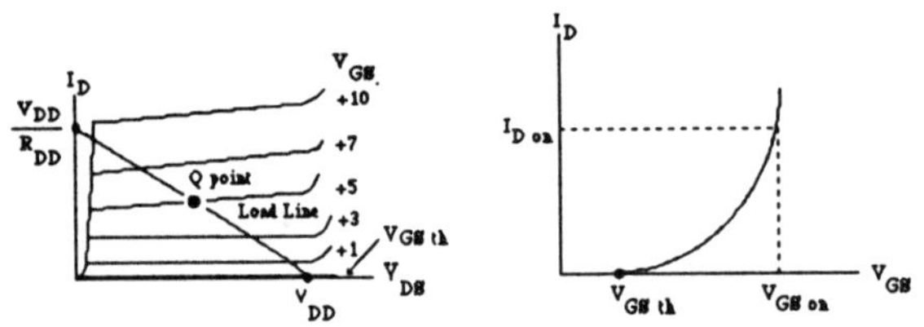

FIGURE 7.42 The gate and transconductance curves as a function of V_{GS}.

MOSFET ACTIVE LOAD SWITCHES

Enhancement-mode MOSFETs can be used as signal amplifiers, followers, etc., as can JFETs and depletion-mode MOSFETs. The circuit configurations and the design and analysis equations are similar. However, one of the most used and most interesting circuit configurations includes two MOSFETs working together as a digital switch. In this circuit pair, one of the transistors operates as a common source (inverting) amplifier and the other transistor operates as a two-terminal resistance load for the amplifier.

TM MOSFET ACTIVE LOAD SWITCHES

The prototype enhancement-mode active switch uses two MOSFETs, one acting as a common-source amplifier and the other as an active drain load resistor. The active load transistor does *not* operate in the normal ohmic region, but becomes a higher (load) resistance because its gate is connected to its drain.

MM MOSFET ACTIVE LOAD SWITCHES

$$V_{GS} = V_{DS} \text{ for the load MOSFET} \tag{E 7.52}$$

$$R_{DS} = \frac{\Delta V_{DS}}{\Delta I_D} \text{ for the load MOSFET} \tag{E 7.53}$$

GM MOSFET ACTIVE LOAD SWITCHES

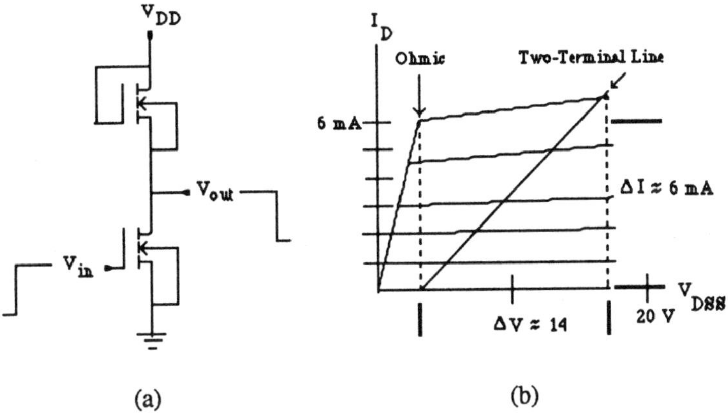

FIGURE 7.43 (a) A MOSFET switch with an active load; (b) The two-terminal active-load curve. As an example, ΔV/ΔI ≈ 14V/6mA ≈ 2.33 k ohms for the active resistance.

7.14 JFET AND MOSFET EQUIVALENT CIRCUITS

JFETs are widely used for analog circuits and, although their parameters vary widely from unit to unit, most textbooks and programs present equivalent circuits. MOSFET parameters also vary widely from unit to unit. For analog circuits, it is necessary to use feedback compensation with source resistors, etc., to partially compensate for the variations. Even so, digital circuits are really their forte. Most textbooks and computer programs present a universal ac "FET" equivalent circuit rather than a specific MOSFET equivalent.

GM A COMMON UNIVERSAL "FET" EQUIVALENT CIRCUIT

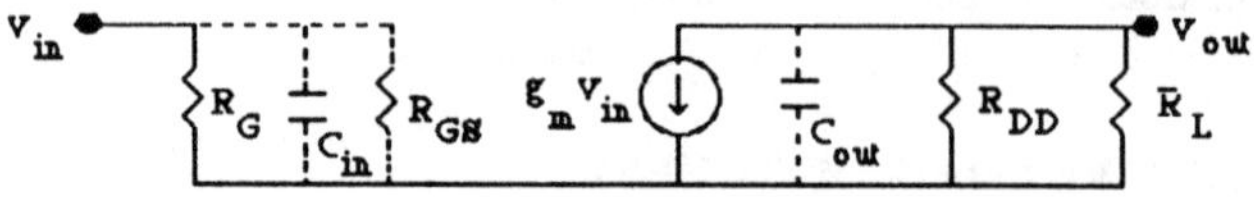

FIGURE 7.44 An ac equivalent circuit that can be used for JFET and MOSFET transistors.

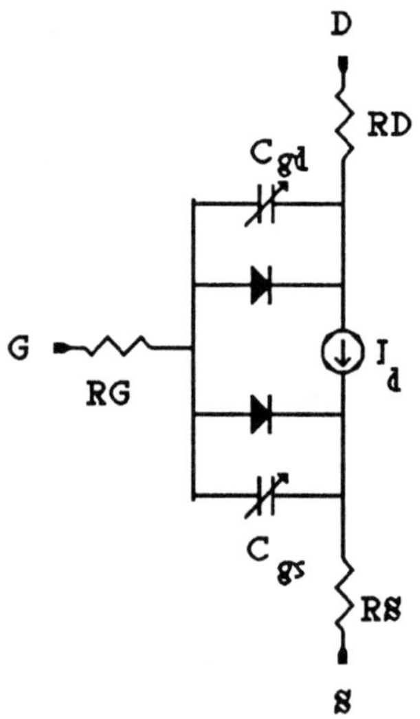

FIGURE 7.45 The JFET model given by PSpice. (The program gives the user 21 model parameters.)

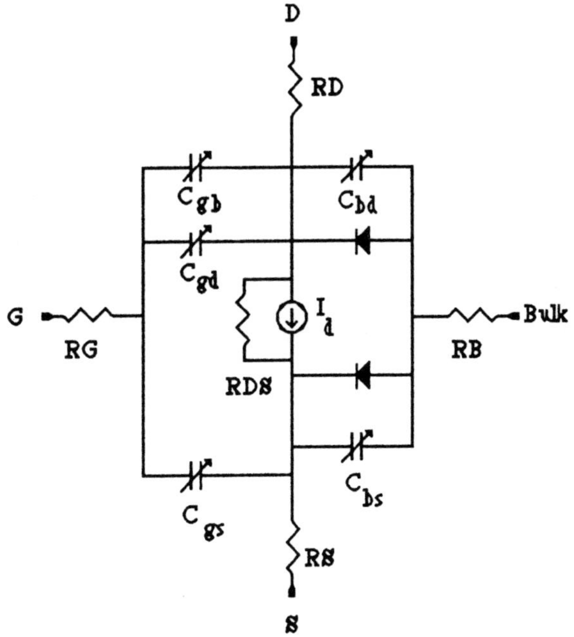

FIGURE 7.46 The PSpice MOSFET model. (This model is deceptively simple. PSpice presents four different variations on this model with up to 24 parameters in each variation.)

8

Transformer and Coil Equivalent Circuits

8.1 THE THEORETICALLY IDEAL VERSUS THE PRACTICAL, REALIZABLE TRANSFORMER

If an ideal transformer, with a turns ratio of one-to-one, is inserted in a simple ac series circuit, it will not change the values of the input or output voltage or current. For example, assume an ideal constant voltage generator, with an ac output of ten volts, feeding an ideal load of ten ohms. The generator and load voltage is, of course, ten volts and the series current is one ampere. If an ideal transformer, with a turns ratio of exactly one-to-one, is inserted between the generator and the load, the generator and load voltage and current will remain the same.

If a real-world transformer is inserted in the aforementioned circuit, things would change. The magnitude of the output voltage and current would be less. They may be only a fraction of one percent less or they might be 50 percent less, but always somewhat less. The object of this chapter is to list the reasons why transformers are less than ideal, and to derive the numbers and equivalent circuits that define and illustrate the operation of several types of transformers.

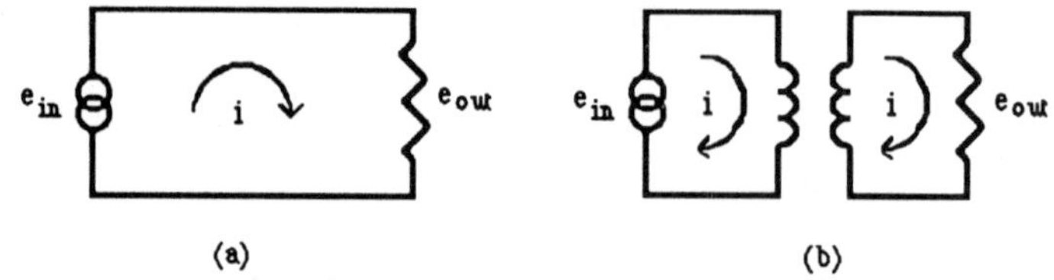

FIGURE 8.1 (a) A simple ac series circuit with a voltage generator and load; (b) The same circuit with an ideal transformer added.

8.2 THE GENERAL THREE–CIRCUIT MODEL OF AN ELECTRICAL TRANSFORMER

TM THE THREE–CIRCUIT TRANSFORMER MODEL

Any realizable transformer, with its losses, can be divided into three general circuits:

1. *The electric circuit of the coils.* The parameters include the resistance of the winding, the inductance of the coil, and any stray capacity or inductance. The degree of involvement of the stray parameters will usually be a function of the frequency range of the transformer. There is also a secondary resistance effect termed *skin effect* that makes the effective resistance of the wire increase as a function of frequency.
2. *The magnetic circuits of the coils.* When an electric current flows in a conductor, a magnetic field is produced. The magnitude and the location of this field is a function of the current, the geometry of the conductor and the losses in the circuit.
3. *The magnetic circuits of the core.* An additional medium, other than air or a non-magnetic support material, may be used inside or around the coil to concentrate the magnetic field flux. This will produce a more efficient transformer action. However, there are a number of physical characteristics that can produce losses in this magnetic core.

The two dominant core losses are termed *hysteresis* and *eddy currents.* Hysteresis losses result from the peculiar, very non-linear way that iron-derived magnetic materials react to being magnetized. The theories state that the composition of the material changes with different levels of magnetization. As a result, the output magnetic field is not a linear function of the input (amp-turns) magnetizing force. Likewise, a graph of the input-output relationship will indicate a loop, not a curve, because the path that the magnetization takes as the force is increasing is not the same as it is when the force is reversed and decreasing. Calculating the resulting hysteresis loss is very difficult.

Eddy-current losses result from I^2R losses in the core material. The flux generated in the primary will produce a current or currents in the core material. The core can be considered as one or more conductors just as the secondary. The current thus generated in the core will flow in the iron and produce I^2R_{iron} power losses. The classic way to reduce the eddy-current losses is to use a core of insulated iron layers (laminations) rather than a solid chunk of material.

A third core loss is termed *leakage inductance.* The portion of the magnetic field lines of force that is generated in the primary and does NOT link or pass through the secondary winding is called *leakage flux*, and an equivalent winding that created this lost or leakage flux can be postulated and called the leakage inductance. The "loss" from leakage inductance is not a power loss, as are the hysteresis and eddy-current losses, since the reactive current in the equivalent leakage inductance is not in phase with the winding currents.

A fourth core loss results from the unloaded *magnetizing current.* When the secondary is open-circuited, there may still be some primary current. In most equivalent circuits, a shunt inductance is created that requires this magnetizing current. The loss from the magnetizing current is also a reactive loss since this current is also not in phase with the normal winding currents.

The *permeability* of the core material is the very non-linear ratio of the output flux density **B** (flux output) to the input amp-turns **H** (forced input). This non-linear ratio changes with the magnitude and direction of the input **H**. The permeability (μ) of a material is defined as rate of change of the flux density with respect to the magnetizing force. The *relative permeability* of air may be assumed to be unity, whereas the relative permeability of iron and steel cores may range from a few hundred to well over 100,000.

MM THE THREE-CIRCUIT TRANSFORMER MODEL

1. The electric circuit of the coils—R_{dc}, X_L, and skin effect.

For a single coil:

$$Z_{in} = R_{coil\text{-}dc} + j\omega L_{coil} \qquad \text{(E 8.1)}$$

$$R_{skin} \approx \frac{261 * 10^{-9}}{P}\sqrt{f} \qquad \text{(E 8.2)}$$

where: R_{skin} is the effective resistance of a copper wire in ohms per cm
P is the perimeter (circumference) of the copper wire
f is the frequency in hertz

Note: The problem of skin effect can be materially reduced by the use of multistrand (litz) wire (see the S. Butterworth reference).

$$L_{coil} = \mu * N^2 * \frac{A}{l} \quad \text{(the } L_{coil} \text{ is self-inductance in henrys)} \qquad \text{(E 8.3)}$$

where: μ is the the permeability of the core medium.
N = the turns in a coil. Note the inductance is proportional to the turns squared!
A = the area of the core material (other than air)
l = the length of the core material

For two coils (L_{pri} and L_{sec}):

$$L_{pri} = \frac{M^2}{k^2 L_{sec}} \text{ or } L_{sec} = \frac{M^2}{k^2 L_{pri}} \text{ and, therefore } M = k\sqrt{L_{pri} L_{sec}} \qquad \text{(E 8.4)}$$

where: M is the *mutual inductance* produced by the mutual flux that links the "other" winding when L_{pri} and L_{sec} are in close proximity.
k is the *coefficient of coupling*. k is a function of the proximity of L_{pri} and L_{sec} and of the μ of the medeum. It states the amount (usually the per cent) of the total flux produced by a given winding L_1 that couples or "cuts" an adjacent winding L_2. With an iron core, up to 0.999% of the flux produced can link the "other" winding.

$$L_{Tot} = L_{pri} + L_{sec} + 2M \text{ or } L_{pri} + L_{sec} - 2M \qquad \text{(E 8.5)}$$

where: the L_{Tot} is measured by both connecting L_{pri} and L_{sec} in series flux adding and in series flux opposing (reversing the leads on one coil). Note that the inductance of either L_{pri} or L_{sec} can be measured separately if the other one is open circuited.

2. The magnetic circuit of the multiple coils—the magnetizing force.

$$\mathbf{F} = NI \qquad \text{(E 8.6)}$$

where: the force **F** to produce magnetic flux is proportional to N turns and I amperes. Note that the magnetizing force is completely independent of the core material!

$$\Phi = \frac{\mathbf{F}}{\Re} \qquad \text{(E 8.7)}$$

where: the *flux* Φ that is produced is proportional to the force **F** and inversely proportional to the core material *reluctance* $\Re$.

The reluctance $\Re$ is the magnetic "resistance" to the creation of flux by the force **F** just as the electrical resistance r opposes the flow of current i produced by the electromotive force e (Ohm's law).

$$H = \frac{\mathbf{F}}{l} = \frac{NI}{l} \qquad \text{(E 8.8)}$$

where: H is the *magnetomotive force* per unit length of the core medium, l is the total or mean (average) length of the magnetic medium.

3. The resulting magnetic flux:

$$B = \mu H \text{ for an air core} \qquad \text{(E 8.9)}$$

where: B is the *flux density* per unit area—usually in webers per square meter.

4. The magnetic circuits with a non-linear iron core.

$$B = m^{\kappa} H \qquad \text{(E 8.10)}$$

where: the "κ" might be some ethereal value that is very, very difficult to define since it is a function of the material and both the magnitude and the polarity of the input force (current). The linear equation E 8.9 usually does NOT describe the true "input force-output field" relationship for most iron-based materials.

There are two problems:

1. The B-H in-out relationship "droops" or saturates as the input current and thus the NI force increases. This non-linear behavior is somewhat analogous to the saturation characteristics in semiconductors—see Chapter 7.
2. When the sinusoidal input NI force peaks and then decreases, the decreasing output (flux) does not reverse and decrease along the same path it traveled when it was increasing. The paths form a loop rather than

a curve. This (hysteresis) loop represents power lost. The lost power is proportional to the (integrated) area of B-H loop.

$P_h \approx K_h f B^n$ is the Steinmetz approximation for transformer hysteresis power loss (E 8.11)

where: K_h and n are empirically determined material coefficients
$n \approx 1.6$
f is the frequency in hertz

$P_e \approx K_e f^2 B^2$ is the approximate transformer eddy-current power loss (E 8.12)

where: K_e is a material coefficient.

GM THE THREE-CIRCUIT TRANSFORMER MODEL

1. The electric circuit of the coils.

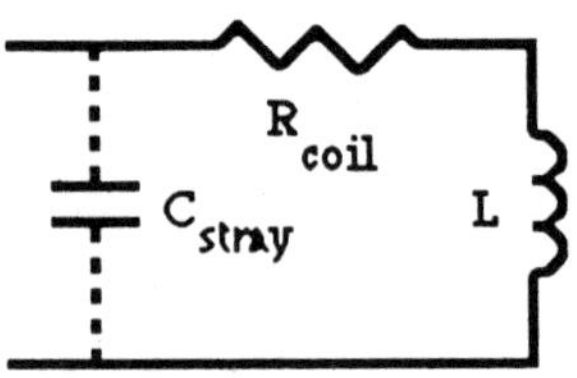

FIGURE 8.2 The low-frequency equivalent circuit of a transformer winding.

2. A realistic transformer equivalent circuit including an ideal turns ratio of 1:N.

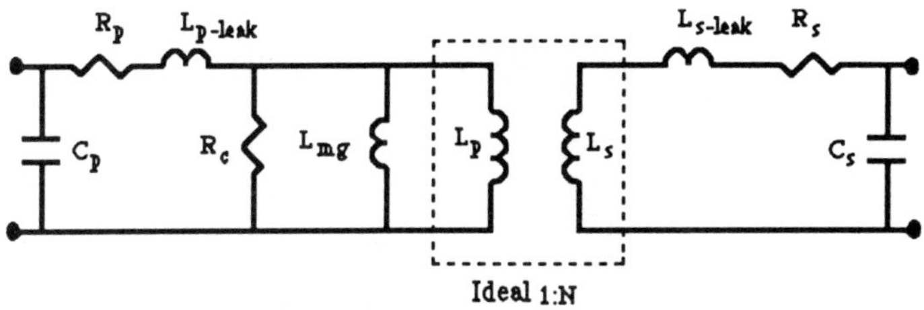

FIGURE 8.3 One transformer equivalent circuit: C_p = the stray capacitance of the primary, R_p = the series resistance of the primary (this may include skin effect), $L_{p\text{-leak}}$ = the primary leakage inductance, R_c = an equivalent resistor used to simulate the total core losses, L_{mg} = an equivalent inductor for magnetizing loss, L_p = the effective (non-loss) inductance of the primary, $L_{s\text{-leak}}$ = the effective (non-loss) inductance of the secondary, R_s = the series resistance of the secondary (this may include skin effect), C_s = the stray capacitance of the secondary.

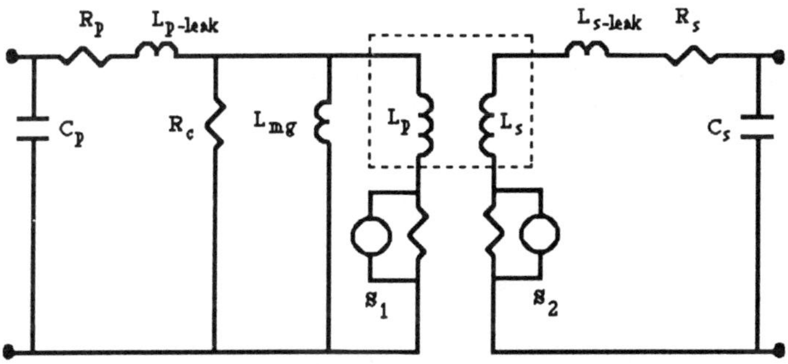

FIGURE 8.4 A transformer equivalent circuit that includes computer-generated constant-voltage or constant-current sources (S_1 and S_2) to accurately simulate the turns ratio.

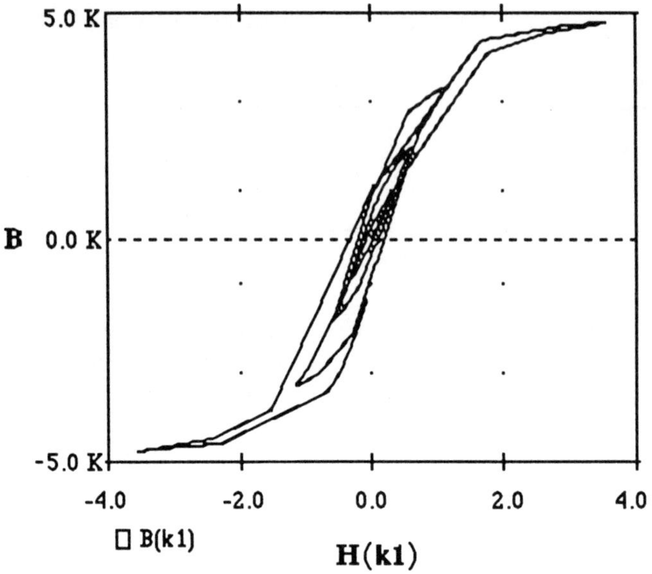

FIGURE 8.5 A reduced graph of a hysteresis as presented in PSpice. k1 is the computer program node (see the section on computer transformer analysis).

8.3 IRON-CORE TRANSFORMERS

TM IRON-CORE TRANSFORMERS

Power and audio transformers typically use iron-derived core materials. *Iron, steel,* and sometimes *ferrites,* provide a very efficient way to transfer the primary flux to the secondary. Usually, over 99 percent of the flux is utilized as an "ideal" transformer. The disadvantages of ferromagnetic cores are the core losses and the non-linear, in-out response with changing signal amplitude and frequency. For instance, audio transformers should, of course, be as frequency independent as possible. As with other components in electronics, the most useful equivalent circuit depends upon the general application. As an example, the leakage reactance and stray capacitance can often be ignored in the equivalent circuit of a power transformer. Sometimes the results of the "best" equivalent are not used. There are times when the user is willing to sacrifice some efficiency to get a particular shape or weight. In power transformers, the heating caused by the power losses must always be considered. If a given design will not properly radiate the heat caused by the power losses, the transformer must be made larger—with more surface area.

The input and output impedance is often particularly important in iron-core transformers. Obviously, it is undesirable for a power transformer to unnecessarily "load" the line. Likewise, audio transformers are very often specifically designed to match a particular input or output impedance. For either one, it is important that the equivalent inductive reactance be as high as possible compared to the winding resistance.

MM IRON-CORE TRANSFORMERS

$$Z_{pri} = R_p + j\omega L_p \text{ and } Z_{sec} = R_s + j\omega L_s \qquad \text{(E 8.13)}$$

$$Z_{in} = Z_p N^2 Z_{load} \qquad \text{(E 8.14)}$$

If Z_{load} is resistive, Z_{in} will be resistive. If Z_{load} contains both a resistance and a reactance, often a *conjugate match* is desired. A conjugate match is obtained when a series reactance is inserted in series with the input generator. The reactance of this series reactance should be equal to the reactance of the load but of opposite polarity. (Make the input a capacitive reactance if the load contains an inductive reactance, etc.)

8.4 AIR-CORE TRANSFORMERS

TM AIR-CORE TUNED TRANSFORMERS

One of the most interesting and challenging problems is the analysis and design of the much-used tuned radio frequency (RF) and intermediate frequency (IF) transformers. These are especially interesting for two general reasons. First, the voltage gain of this type of transformer has almost NOTHING to do with the turns ratio. Second, the calculation of the voltage gain, or loss, is dependent on several factors, including the "Q" of the primary and secondary coils, the tuned frequency of the primary and secondary resonant circuits and the coefficient of coupling (mutual inductance).

The very early designs of these RF-type transformers relied heavily upon graphs of data that were derived from both calculations and experiment. Contemporary computer-aided design not only accelerates the design and analysis, but also gives the investigator the ability to understand alternative approaches much better. Several equivalent network approaches are listed in the references. The mathematical models give two of these. The first, previously unpublished, equation is quite complete and shows the detail that can be used for extreme accuracy and understanding. The second equation is more simplified, but still illustrates the principles involved.

MM AIR-CORE TRANSFORMERS

The Murakami Equations (see reference):

Special Note: The following three equations, plus the equivalent circuit, represent a rather detailed approach to RF tuned transformers. These equations are presented to show some of the details that can be considered in the design of RF and IF transformer-coupled circuits. Used by kind permission from T. Murakami (see also the Terman *Handbook* Reference).

$$\frac{e_{out}}{i_{in}} = Z_{Trans} = \frac{k\sqrt{L_1L_2}\,s}{\left[\frac{s^2}{\omega_0^2} + \frac{s}{Q_1\omega_0} + 1\right]\left[\frac{s^2}{\delta^2\omega_0^2} + \frac{s}{\delta^2 Q_2\omega_0} + 1\right] - \frac{s^4k^2}{\omega_0^4\delta^2}}$$

(E 8.15)

where: $s = j\omega$

$\omega_0^2 = \frac{1}{L_1C_1}$ $\quad Q_2 = \frac{\omega_0 L_2}{R_2}$

$\delta^2\omega_0^2 = \frac{1}{L_1L_2}$ $\quad k^2 = \frac{M^2}{L_1L_2}$

$Q_1 = \frac{\omega_0 L_1}{R_1}$ $\quad Q_n = \sqrt{Q_1Q_2}$

$$\text{Amp-norm} = \left|\frac{Z_T}{Z}\right| = \frac{a_1\omega_n}{\sqrt{(b_4\omega_n^4 - b_2\omega_n^2 + b_0)^2 + (b_1\omega_n - b_3\omega_n^3)^2}}$$

(E 8.16)

$$\phi(\omega_n)_{phase} = \frac{\pi}{2} - \tan^{-1}\frac{b_1\omega_n - b_3\omega_n^3}{b_4\omega_n^4 - b_2\omega_n^2 + b_0} \qquad \text{(E 8.17)}$$

where: $\omega_n = \frac{\omega}{\omega_0}$

$a_1 = \frac{k\delta^2}{Q_n}$ $\quad b_2 = 1 + \delta^2 + \frac{1}{Q_n^2}$

$$b_0 = \delta^2 \qquad b_3 = \frac{1}{Q_1} + \frac{1}{Q_2}$$

$$b_1 = \frac{\delta^2}{Q_1} + \frac{1}{Q_2} \qquad b_4 = 1 - k^2$$

A tuned transformer including the mutual inductance:

$$Z_{in} = Z_{pri} + \frac{(\omega M)^2}{Z_{sec} + Z_{load}} \tag{E 8.19}$$

where: $m = k\sqrt{L_{pri} L_{sec}}$ and k is the coefficient of coupling.

GM AIR-CORE TRANSFORMERS

The Murakami Equivalent Model:

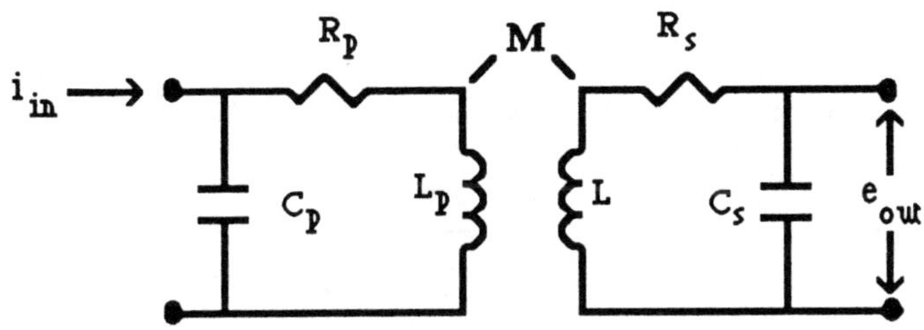

FIGURE 8.6 The circuit model for E 8.15.

A Tuned Transformer Including the Mutual Inductance:

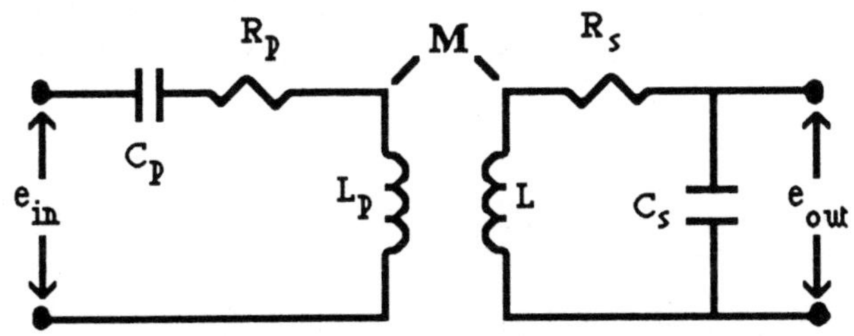

FIGURE 8.7 A circuit model for E 8.19.

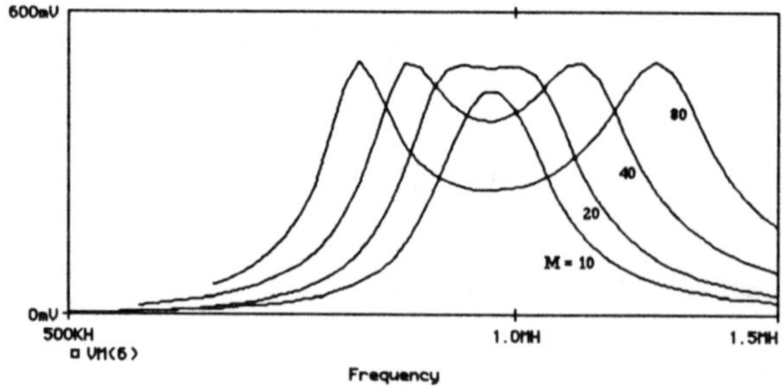

FIGURE 8.8 The normalized response of the circuit in Figure 8.7 using E 8.19. C_p = 250 pF, R_p = 100 ohms, L_1 = 0.1 mH, C_s = 250 pF, R_s = 100 ohms, L_s = 0.1 mH, M = $k\sqrt{L_{pri} L_{sec}}$ where the M shown in the figure is in µH.

8.5 COMPUTER-AIDED TRANSFORMER ANALYSIS AND DESIGN

The sophisticated circuit analysis program and the spreadsheet have both greatly reduced the time required for transformer design and improved the precision of the end product. The transformer can either be considered by itself or as a component that is part of a more complicated system.

Iron-core transformer analysis and design program alternatives now include non-linear models of many commonly available iron-derived and ferrite materials. As an example, the models in PSpice® not only include the B–H characteristics of the material, but give the user the ability to select a core area, a core path length, a stacking factor, etc. Some manufacturers use the results of their information and experience in the form of a scientific spreadsheet that includes some or all of the above factors plus information on core weight and weight vs. standard core or price, etc. For air-core transformer design, it is now quite easy to match such things as air gap and voltage breakdown to coefficient of coupling.

9

The Operational Amplifier and an Introduction to Operational Mathematics

9.1 ABOUT THIS CHAPTER—THE IDEAL OPERATIONAL AMPLIFIER

An operational amplifier is much more than just a handy circuit device that will give the user a predetermined fixed-gain amplifier with a minimum of design effort. A (near) ideal operational amplifier will not only provide this predetermined gain, but (1) will provide the gain over a wide range of life, load, temperature, etc., (2) will not oscillate or radically change its frequency or transient response with load, temperature, life, etc., and (3) will provide (almost) zero dc and ac output for zero dc and ac input.

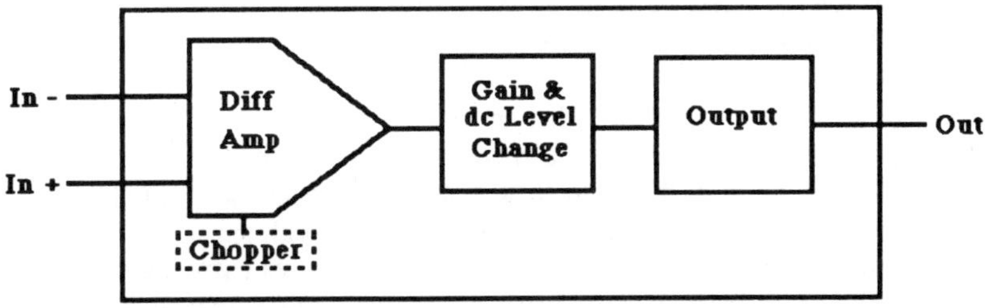

FIGURE 9.1 The elements of a prototype operational amplifier.

The operational amplifier is the major theoretical building block for electrical analog computers and analog control systems such as servo mechanisms and phase locked loops. Operational amplifier circuits that can add and subtract, multiply and divide, integrate and differentiate, etc., can be combined to solve complex differential equations or they can be combined to simulate the operation of complex natural physical systems. The study and use of analog simulating and computing systems can give the student and practicing engineer an understanding of the "real world" that is difficult to obtain by any other method.

There are limitations, however, especially in accuracy, to the use of operational analog circuits. This chapter will catalog the development of operational amplifiers, the analysis of their circuits and limitations, and outline the analysis of operational configurations that can be used for calculation and system simulation.

9.2 THE DIFFERENTIAL AMPLIFIER—THE INPUT STAGE OF THE OPERATIONAL AMPLIFIER

The ideal operational amplifier, with its input differential amplifier, would possess an infinite input impedance, an infinite bandwidth, starting at dc, and circuits that would have no dc drift with temperature or age. It especially would accept either a non-inverting (+) signal or an inverting (–) signal relative to its output. While no input differential amplifier will meet all of these (and more) requirements, it is important to understand the analysis of the circuits so that judgments can be made concerning what errors are potentially detrimental and what errors may not be significant. A review of Chapter 7 will provide a background review of transistor parameters that can change with temperature, β, etc.

TM A PROTOTYPE DIFFERENTIAL AMPLIFIER USING TWO DIFFERENTIAL TRANSISTORS AND TWO COLLECTOR RESISTORS

The generic differential amplifier (diff amp) contains two transistors and three resistors. It has both balanced inputs and outputs. Its gain is primarily dependent upon transistor β, r'_e and the collector resistors. The use of the two opposing base-emitter diodes cancels most of the circuit dc drift with temperature (see E 7.1). The emitter currents are supplied by a (nearly) constant current source R_S. When R_S is large, it does not significantly affect the signal gain since the input signal path is into one transistor base, down its emitter, and then up the emitter of the other transistor and out the other base.

GM A PROTOTYPE DIFFERENTIAL AMPLIFIER USING TWO DIFFERENTIAL TRANSISTORS AND TWO COLLECTOR RESISTORS

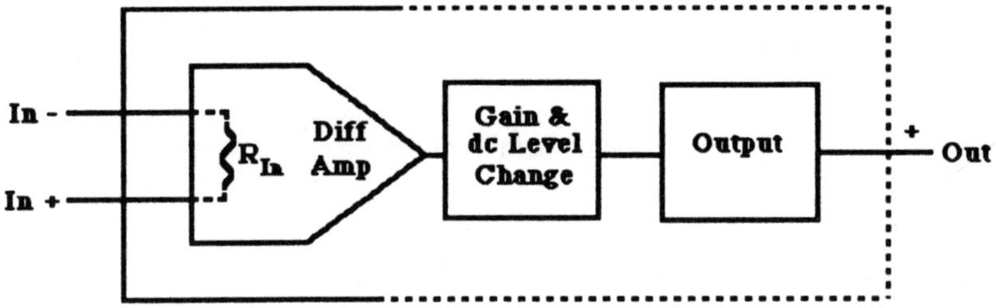

FIGURE 9.2 The differential amplifier as part of a prototype operational amplifier.

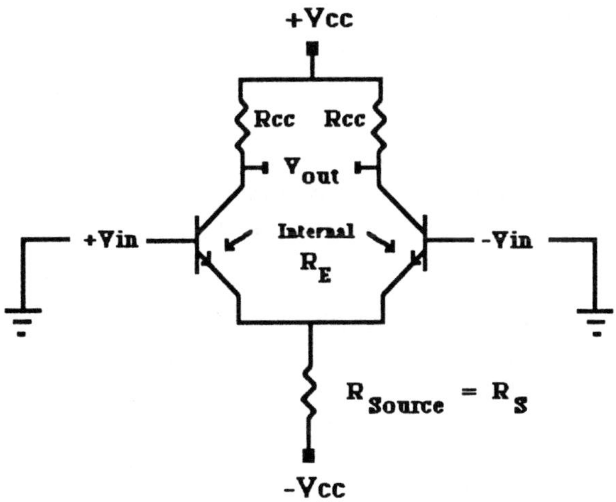

FIGURE 9.3 The prototype differential amplifier.

MM A PROTOTYPE DIFFERENTIAL AMPLIFIER USING TWO DIFFERENTIAL TRANSISTORS AND TWO COLLECTOR RESISTORS

$$I_{R_S}\,dc = \frac{V_{CC} - V_{BE}}{R_S}$$ where V_{CC} is the magnitude of $-V_{CC}$ and

V_{BE} is typically 0.7 volts (E 9.1)

$I_{R_{cc}}\,dc \approx \frac{1}{2} I_{R_S}\,dc$ assuming both transistors have the same β_{dc} and V_{BE} (E 9.2)

$$I_{Base}\,dc \approx \frac{1}{2} I_{R_S}\,dc \,/\, \beta_{dc} \quad \text{(E 9.3)}$$

$E_{out}\,dc \approx \frac{1}{2} I_{R_S}\,dc\, R_{CC}$ assuming both transistors have the same β_{dc} and V_{BE} (E 9.4)

$$R_{In} \approx 2\, R_E * \beta \quad \text{(E 9.5)}$$

$$A_V \approx V_{out}/V_{in} \approx (R_{CC}/R_E) \quad \text{(E 9.6)}$$

TM A PROTOTYPE DIFFERENTIAL AMPLIFIER USING TWO DIFFERENTIAL TRANSISTORS AND ONE COLLECTOR RESISTOR

Differential amplifiers that are used in op amps rarely, if ever, use the circuit in Figure 9.3 because the balanced outputs are not convenient for interfacing to the remainder of the circuit. A more practical design uses just one collector resistor although this reduces the gain by a factor of two.

GM A PROTOTYPE DIFFERENTIAL AMPLIFIER USING TWO DIFFERENTIAL TRANSISTORS AND ONE COLLECTOR RESISTOR

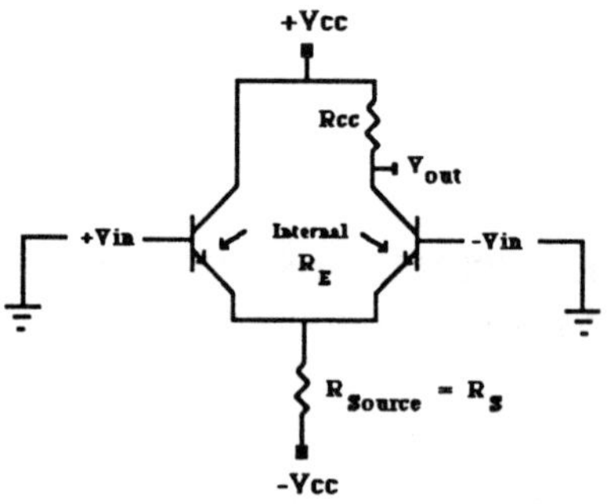

FIGURE 9.4

MM A PROTOTYPE DIFFERENTIAL AMPLIFIER USING TWO DIFFERENTIAL TRANSISTORS AND ONE COLLECTOR RESISTOR

The dc equations are the same as for the two-collector resistor model shown in Figure 9.1.

TM A PROTOTYPE DIFFERENTIAL AMPLIFIER USING TWO DIFFERENTIAL TRANSISTORS, ONE COLLECTOR RESISTOR AND A SEPARATE TRANSISTOR CONSTANT-CURRENT GENERATOR

An improvement in performance can be obtained if the emitter current resistor R_S is replaced with a transistor constant-current generator. The operation is improved for three reasons: (1) In the above explanation, it was assumed that R_S was "large." In practical discrete circuits, the value of R_S is often limited by $-V_{CC}$; (2) The current in R_S is a direct function of $-V_{EE}$, which is undesirable; (3) It will be shown that the resistor R_S limits the ac *common-mode rejection.* If R_S is replaced by the collector of a constant-current transistor, the effective resistance is usually increased many times—see Chapter 7.

GM A PROTOTYPE DIFFERENTIAL AMPLIFIER USING TWO DIFFERENTIAL TRANSISTORS, ONE COLLECTOR RESISTOR AND A SEPARATE TRANSISTOR CONSTANT-CURRENT GENERATOR

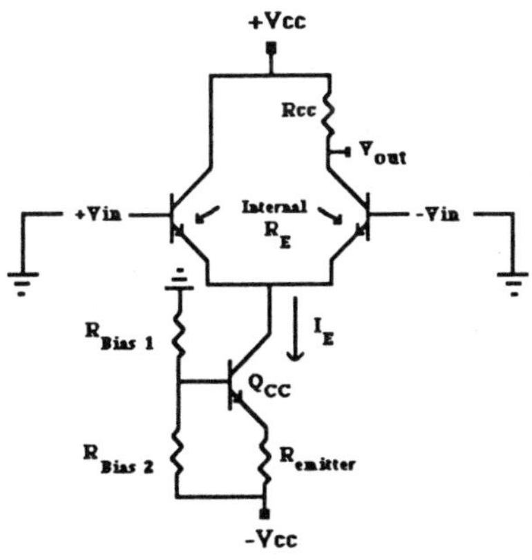

FIGURE 9.5 A common diff amp circuit when discrete components are used.

MM A PROTOTYPE DIFFERENTIAL AMPLIFIER USING TWO DIFFERENTIAL TRANSISTORS, ONE COLLECTOR RESISTOR AND A SEPARATE TRANSISTOR CONSTANT-CURRENT GENERATOR

$$V_{base\ Qcc} \approx -V_{CC}\frac{R_2}{R_1 + R_2} = \text{the voltage from the } Q_{CC} \text{ base to } -V_{CC} \tag{E 9.7}$$

$$I_{C\ Qcc} \approx I_{E\ Qcc} \approx \frac{V_{base\ Qcc} - 0.7}{R_{emitter}} \tag{E 9.8}$$

The differential currents are as before.

9.3 ERRORS IN THE NON-IDEAL DIFFERENTIAL AMPLIFIER

TM COMMON MODE REJECTION RATIO

The use of a physical source resistor usually produces a problem in the ac operation of the differential amplifier. While it is assumed that the inputs to the diff amp are balanced and out of phase, this is not always true. Unwanted signals such as ac hum and noise will often be picked up by the diff amp inputs. Since these unwanted signals usually act equally and in phase at both the + and – inputs, they are termed common or "common mode" signals. In theory, an ideal balanced amplifier would completely cancel these in-phase signals and they would not produce a resulting output. However, a differential amplifier with a low source resistance will let some of the common signal get through. To have good common mode performance, the equivalent source resistance must be a high value. The ratio of the normal differential gain to the common mode gain is termed the common mode ratio and is usually expressed in dB.

MM COMMON MODE REJECTION RATIO

$$A_{v\ diff} = V_{out}/V_{in} \approx R_{CC}/R_E \tag{E 9.9}$$

$$A_{com\ mod} \approx R_{CC}/R_S \tag{E 9.10}$$

$$CMRR = \frac{|A_v|}{|A_{com\ mod}|} = 20\log\frac{|A_v|}{|A_{com\ mod}|} \text{ in dB} \tag{E 9.11}$$

TM DIFFERENTIAL INPUT CURRENT AND VOLTAGE OFFSET

For the ideal operational or differential amplifier, it is assumed that with no dc or ac signal input, there will be no dc or ac signal output. Even in a typical single-output differential amplifier (see Figure 9.5), where the output may be above ground, it is assumed that this is a pre-designed offset (a Q-point) and, with no input signals, this specific dc Q-point will not change with time and temperature. Unfortunately, mismatch in differential transistors can cause quite serious offset errors. For instance, if the mismatch in base currents is only 1 μA, a differential amplifier current gain of 100 (a typical value) would produce an unwanted voltage offset of 0.68 volts across an R_{CC} of 6.8k ohms. Likewise, for a voltage gain of 100, a mismatch in V_{BE} of 10 μV would produce an offset of 1 mV. If the voltage gain were 10^6, which is typical for instrumentation amplifiers, the resulting offset might be more than the normal signal output or, more commonly, this would drive the operational amplifier out of its working range. The specifications of commercial operational amplifiers list the offset specifications, and the prudent investigator should know the possible limitations they impose on the unit's operation or on custom designs.

MM DIFFERENTIAL INPUT CURRENT AND VOLTAGE OFFSET

$I_{in\ bias} = \frac{I_{B1} + I_{B2}}{2}$ where the *input bias current* is the average of the two, usually at 25°C (E 9.12)

$I_{in\ off} = I_{B1} - I_{B2}$ where the *input offset current* is the difference in the base currents (E 9.13)

$V_{in\ off} = V_{BE1} - V_{BE2}$ where the *input voltage offset* is the differential V_{BE} unbalance (E 9.14)

TM THE DC LIMITATIONS OF DISCRETE-COMPONENT OPERATION AMPLIFIERS

The diff amp shown in Figure 9.5 does not operate as an ideal amplifier when it is constructed with real-world, physical parts. The assumption that the two differential transistors possess identical characteristics is NEVER true! This means that the gain (β_{dc}), the input current (I_B), the emitter-base voltage (V_{BE}), and the emitter-base diode dc drift is not the same in both transistors. Thus, (1) the input characteristics of the two transistors are not the same, (2) the gains are not the same, (3) the dc drifts are not the same, and (4) because these items are not identical, the V_{OUT} will not be the value that was calculated assuming symmetrical transistors. Likewise, Q_{CC} will usually produce a non-ideal constant collector current.

These problems are defined as input offset voltage and offset current and, in some manufacturers' specifications, the average of the base currents is given as input bias current.

MM THE DC LIMITATIONS OF DISCRETE-COMPONENT OPERATION AMPLIFIERS

$V_{in\ drift}$ = a manufacturer's specification, usually in µv/°C, of approximate voltage drift.

$V_{in\ off}$ = the average, approximate residual offset voltage when both inputs are grounded.

R_{in} = the average, approximate resistance as seen from Input 1 to Input 2.

$I_{in\ off} = I_{B1} - I_{B2}$ = how closely are the base currents matched. (E 9.15)

$I_{in\ bias} = \frac{I_{B1} - I_{B2}}{2}$ = the average of the two input bias currents. (E 9.16)

$E_{out\ dc} \approx \frac{1}{2} I_{R_E\ dc} R_{CC}$ assuming both transistors have the same β_{dc} and V_{BE}. (E 9.17)

TM VOLTAGE OFFSET AND DRIFT—CHOPPER STABILIZATION

In many operational amplifier applications, especially in instrumentation and analog computers, even an input voltage offset or drift of a few millivolts is not acceptable. By using a special modulated amplifier in series with the differential amplifier, the equivalent input offset and drift can be reduced to about a microvolt. The chopper circuits originally used a mechanical reed switch, called a vibrator or chopper, to first modulate the incoming signal and, after amplifying it, to demodulate it. The amplified signal is really an ac signal. The vibrating chopper switch regularly shorts the input and the output to ground and thus maintains their average potential near ground. Mechanical choppers usually operate at 60 or 400 Hz. More modern FET choppers may operate at higher frequencies. The chopper amplifier is a "modulate-demodulate" communication system and thus requires input and output low-pass filters. Its bandwidth, following the Nyquist criterion, must be less than one-half of the chopper frequency.

Chopper stabilized operational amplifiers have exceptional offset and drift performance, but usually have a very limited bandwidth. They are also limited to one (inverting) input. The improved performance of some modern integrated op amps makes them competitive for most, but definitely not all, applications.

GM CHOPPER STABILIZED OPERATIONAL AMPLIFIERS

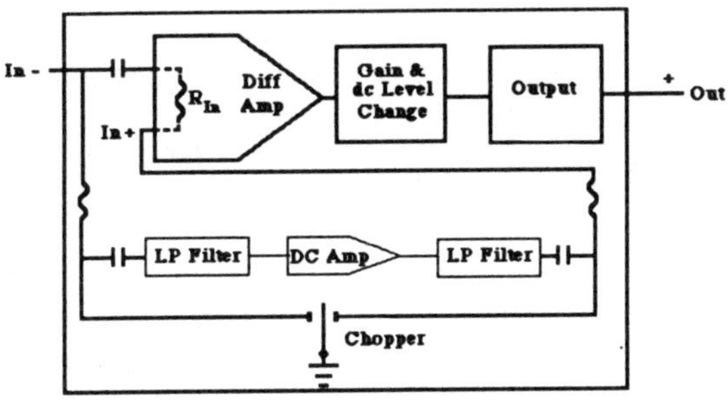

FIGURE 9.6 A chopper stabilized dc operational amplifier.

The stabilized signal going into the + input has been amplified enough to override any drift in the diff amp. The dc and very low frequency gain of the complete operational amplifier is the gain of the chopper amplifier, which is usually in the order of 1000, multiplied by the gain of the rest of the system. The voltage gain of complete chopper stabilized operational amplifiers is usually 10^6 to 10^7.

9.4 INTEGRATED (IC) DIFFERENTIAL AMPLIFIERS—THE CURRENT MIRROR AND THE ACTIVE LOAD

The diff amp shown in Figure 9.5 uses the constant current transistor Q_{CC} to simulate the physical source resistor R_S in Figure 9.4. This improvement in I_E constancy is still limited because the base voltage, and thus the emitter voltage and current, are a function of $-V_{CC}$ and the Q_{CC} emitter-base voltage is still very temperature dependent. Likewise, in discrete circuits such as Figures 9.4 or 9.5, simple circuit analysis usually assumes the ideal. The circuit analysis usually assumes that the β, V_{BE}, r'e, etc., of both transistors are identical—THIS IS NEVER TRUE! However, if these circuits are produced in integrated form, their operation is much more ideal. Integrated circuits provide many circuit advantages besides reduced circuit size. Not the least of these is the improvement in transistor parameter consistency.

Creative designers found that the radical improvement in IC component reproducibility provided many new tools for improving circuit performance. Two examples of these new circuit techniques are the current mirror constant-current generator and the active load.

TM THE CURRENT MIRROR CONSTANT-CURRENT GENERATOR

Using the much improved integrated component consistency, a constant current generator composed of two transistors and one resistor can be fabricated in an integrated circuit. In the following Figure 9.7, it is assumed that the dc emitter-base characteristics of Q_1 and Q_2 are identical—not a bad assumption for integrated circuits. The current through the resistor R is divided between the transistor Q_1, connected as a diode, and the base of Q_2. If it is assumed that the base current of Q_2 is very low ($\beta_{dc} > 100$), then the current in the diode Q_1 is very close to the emitter and collector current in Q_2. Thus, the collector current in Q_2 is very close (within one percent) to the current in the resistor R. So, the Q_2 collector current is said to be mirrored in the resistor-R current.

GM THE CURRENT MIRROR CONSTANT-CURRENT GENERATOR

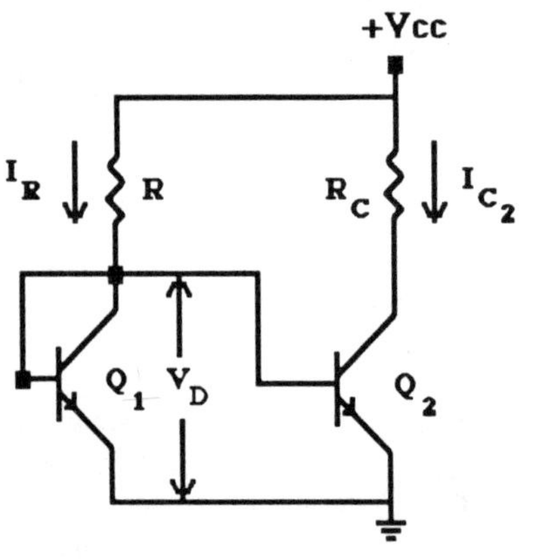

FIGURE 9.7 The IC current mirror.

MM THE CURRENT MIRROR CONSTANT-CURRENT GENERATOR

$$I_R = \frac{V_{CC} - V_{BE}}{R} = \frac{V_{CC} - V_D}{R}$$ where the left transistor is

connected as a diode (E 9.18)

$I_{Q1} = I_R - I_{B\ Q2}$ where $I_{B\ Q2}$ is a very small part of I_R (< 1 %) (E 9.19)

$I_{C2} \approx I_{E2} \approx I_{Q1}$ because of high transistor β and the emitter-base diodes are alike (E 9.20)

$I_{C2} \approx I_R$ within about 1%—thus I_{C2} is mirrored by I_R (E 9.21)

TM THE CURRENT MIRROR ACTIVE LOAD

In the model differential amplifiers shown so far, the dc (and ac) gain is a function of the collector resistor R_C. However, the upper value of R_C is limited by the value of $+V_{CC}$. By using the current mirror circuit in a slightly different form, a synthesized "collector resistor" can be produced by the collector of a transistor, which makes the effective value of this resistor very high (usually > one megohm).

GM A PROTOTYPE INTEGRATED DIFFERENTIAL AMPLIFIER USING TWO DIFFERENTIAL TRANSISTORS, A CURRENT-MIRROR, CONSTANT-CURRENT GENERATOR, AND A CURRENT-MIRROR ACTIVE LOAD

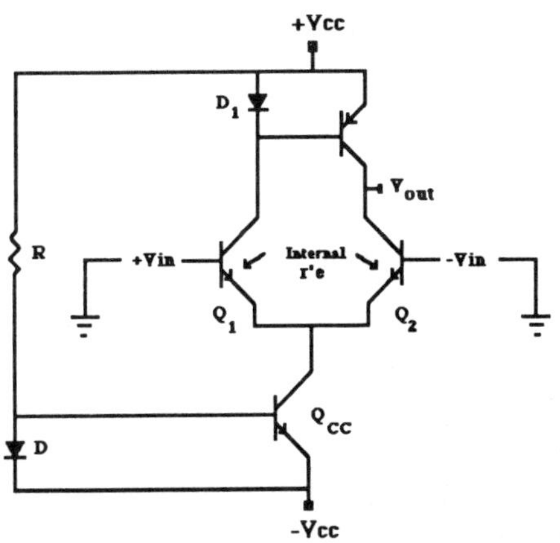

FIGURE 9.8 An IC differential amplifier using current mirrors for the constant-current generator and the transistor load. (Note that the active load transistor is a PNP and the diodes D and D_1 are synthesized from transistors.) The current in D_1 is established by Q_1 and, because of the current mirror, the current in the active load PNP transistor will be the same as in Q_2.

9.5 VOLTAGE LEVEL SHIFTERS AND THE OUTPUT STAGE—THE COMPLETE OP AMP

TM THE COMPLETE OP AMP

Most definitions of operational amplifiers include the requirement that, with no dc or ac input there will be no dc or ac output—zero out for zero in. Usually, there must be some additional circuitry after the differential amplifier to add more gain and to shift the dc level back to zero (ground) potential. Discrete circuits use a combination of transistors, resistors, and sometimes zener diodes to add desired gain and shift the output level. ICs use several similar techniques plus current mirrors and active loads.

GM THE COMPLETE OP AMP

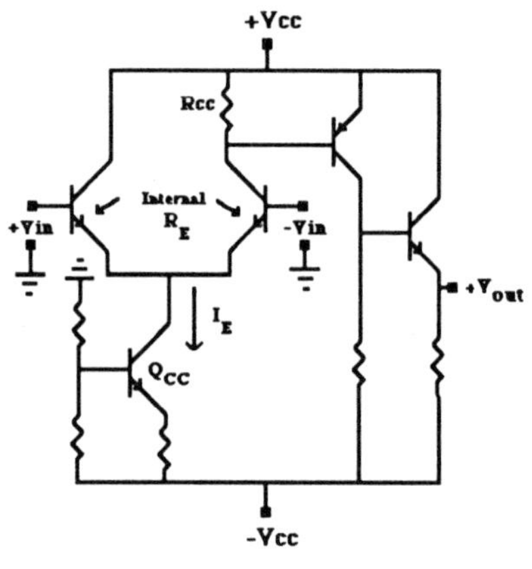

FIGURE 9.9 An elementary discrete operational amplifier.

9.6 THE OPEN LOOP CHARACTERISTICS OF OPERATIONAL AMPLIFIERS

TM THE AC SMALL-SIGNAL RESPONSE OF OPERATIONAL AMPLIFIERS

For *small signal* operation, the ideal operational amplifier will amplify very small signals with a constant gain from dc to a single high-frequency cutoff pole. Beyond this cutoff pole, the gain will gracefully roll off at a near constant slope of about 6 dB per octave (20 dB per decade) until the response (V_{out}/V_{in}) reaches a value of one—the unity-gain frequency. For most commercial op amps, the dc and midband gain is in the order of 10^5 to 10^7. The phase response (phase shift) is zero at dc, starts to change and go negative in the midband (see one pole low-pass filters in Chapter 5), is –45° at the pole frequency, and approaches –90° at the high-frequency unity gain.

Real-world op amps deviate from this ideal definition in several ways. Chapter 6 showed that there are input and output capacitors, including the equivalent Miller capacitors, for all transistors. When many transistors are cascaded in an amplifier, be it discrete or integrated, the high-frequency region can contain a number of cutoff poles. Many times a manufacturer will include one special capacitor that will roll off the amplifier at a new, slightly lower frequency. Since this is designed to override (swamp out) the effects of all the transistor high-frequency capacitors, the entire operational amplifier operates as a single pole, wide band, low-pass filter. If the amplifier includes a chopper stage, at least one pole is added at very low frequencies. The addition of a chopper stage also complicates the phase and transient response.

GM THE AC SMALL-SIGNAL RESPONSE OF OPERATIONAL AMPLIFIERS

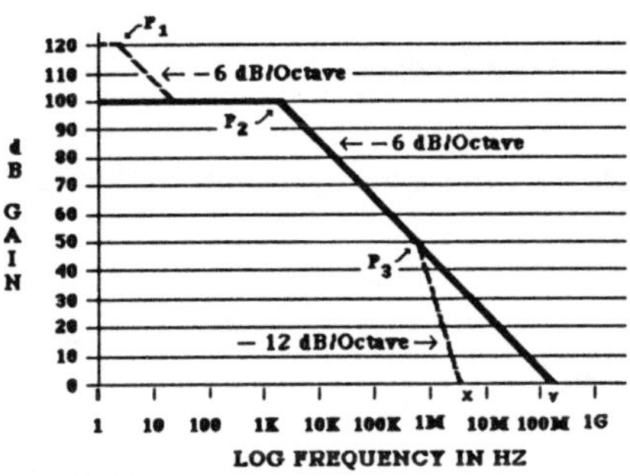

FIGURE 9.10 A composite graphic showing the effects of several circuit configurations on open-loop operational amplifier frequency response: (a) The bold line from 100 dB, breaking at P_2 and sloping to the unity-gain point v, is a typical open-loop response for a high-gain operational amplifier; (b) If a chopper stage is included, the very low frequency gain is increased, as illustrated, to 120 dB and a pole is added at P_1; (c) If an amplifier does not swamp out all of the high-frequency capacitive effects, the amplifier may roll off at a steeper rate as shown with the -12 dB line from P_3 to the unity-gain point x. (The log frequency is not necessarily to scale.)

MM THE SMALL-SIGNAL SINE-WAVE RESPONSE OF OPERATIONAL AMPLIFIERS

$A = V_{out}/V_{in} = K\,H(s)$ where K is the dc gain and H(s) is the complex transfer function. (E 9.22)

$$A(s) = V_{out}(s)/V_{in}(s) = K\frac{1}{1+RC(s)} = K\frac{1}{1+j\omega RC}$$

$$= K\frac{1}{\sqrt{1+\omega^2R^2C^2}}e^{j\omega t}$$ for a single pole where $s = j\omega$. Also, (E 9.23)

$$A(s) = K\frac{1}{1+j\omega/\omega_c} \text{ or } K\frac{1}{1+jf/f_c}$$ where ω or f is normalized to the pole frequency (E 9.24)

$$K\,H(j\omega) = K\frac{X_c}{\sqrt{R^2+X_c{}^2}}$$ compare with an RC filter in Chapter 5 (E 5.6)

$$\phi = -\tan^{-1}\frac{X_c}{R}$$ where φ is the phase shift—see Chapter 5 (E 5.6)

If the op amp has more than one pole, the sine wave gain and phase are calculated:

$$A_{Total} = K_1\frac{1}{1+j\omega R_1C_1} * K_2\frac{1}{1+j\omega R_2C_2} * \cdots \quad \text{(E 9.25)}$$

$$\phi_{Total} = -\tan^{-1}\frac{X_{c1}}{R_1} + -\tan^{-1}\frac{X_{c2}}{R_2} + \cdots \quad \text{(E 9.26)}$$

The single-pole phase shift starts at zero, goes to –45° at the pole frequency and approaches –90° at about five times the pole frequency. For two separated poles, the phase shift will approach –180° above the frequency of the second pole and, for three or more poles, the phase shift approaches –180° quite rapidly.

9.7 THE CLOSED LOOP (FEEDBACK) CHARACTERISTICS OF OPERATIONAL AMPLIFIERS

TM FEEDBACK GAIN

The most common feedback connection uses two resistors configured for a fixed gain. However, even with this simple circuit, there are some potential errors that must be considered for "real-life" operation. (1) The tolerance of the resistors, or any feedback components, must be considered if precise gain, such as in analog computers, is desired. (2) The exact gain of the system also involves the open-loop gain of the amplifier—is the open-loop gain large enough? (3) The system stability must be considered. If a particular feedback configuration forces the amplifier near the limits of stable operation, a slight change in temperature, power-supply voltage, etc., can produce intermittent or steady oscillations. (4) The output capability must be considered. Both the power output and the amplifier slew rate must be reviewed to be sure that the input wave form is reproduced correctly at the gain predicted by the feedback ratio.

MM SIMPLE FEEDBACK AMPLIFIER GAIN

$$A_{fb} = \frac{A}{1+AR_1/R_2} = \frac{A}{1+A\beta} \text{ where } \beta = R_1/R_2 \qquad \text{(E 9.27)}$$

An example: Assume A = 10000, 1000, 100, 20 and $\beta = 0.1$

A_{fb} will = 9.9999+, 9.9990, 9.990, and 9.950

GM SIMPLE FEEDBACK AMPLIFER GAIN

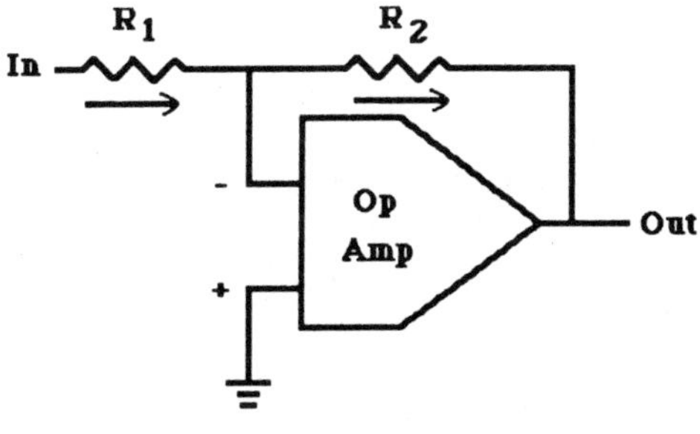

FIGURE 9.11 An ideal operational amplifier with ideal resistor feedback.

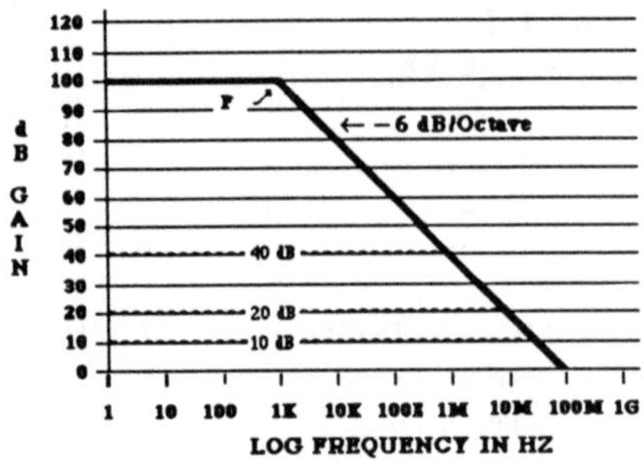

FIGURE 9.12 The frequency response of an ideal operational amplifier with and without feedback. The open-loop gain is shown as 100 dB and the closed-loop gains are 10, 20, and 40 dB. Note that the amplifier bandwidth increases as the closed-loop gain decreases.

TM GAIN AND PHASE MARGINS

All practical operational amplifiers contain one or more capacitors that cause the frequency response to be reduced at some higher frequencies. The capacitors will also cause a phase shift from input to output. If this phase shift equals or exceeds 180°, the complete circuit becomes an oscillator rather than an amplifier. When feedback is used around any amplifier, the designer must be certain that not only will the amplifier not oscillate under certain "normal" or defined circumstances, but that changes in temperature, power-supply voltages, age, etc., will not produce oscillations. Any feedback amplifier should have built-in margins of safety for changes of gain and phase.

GAIN MARGIN is the gain change, usually in dB, that can be allowed before the amplifier becomes unstable. On a Bode plot, the *gain margin* is the distance, in dB, from the horizontal unity-gain line to a point on the gain plot corresponding to the frequency where the phase plot is −180°.

PHASE MARGIN is the allowable phase-shift change, usually in degrees, that can be allowed before the amplifier becomes unstable. On a Bode plot, the *phase margin* is the distance, in degrees, from the horizontal line where the phase shift is −180° up to the point on the phase plot corresponding to the unity-gain frequency on the gain plot.

MM GAIN AND PHASE MARGINS

The gain and phase can be calculated with equations E 9.25 and E 9.26. Since graphical techniques actually "show" the relationship between gain and phase and their margins, they are an excellent way to analyze a feedback-amplifier system.

GM GAIN AND PHASE MARGINS

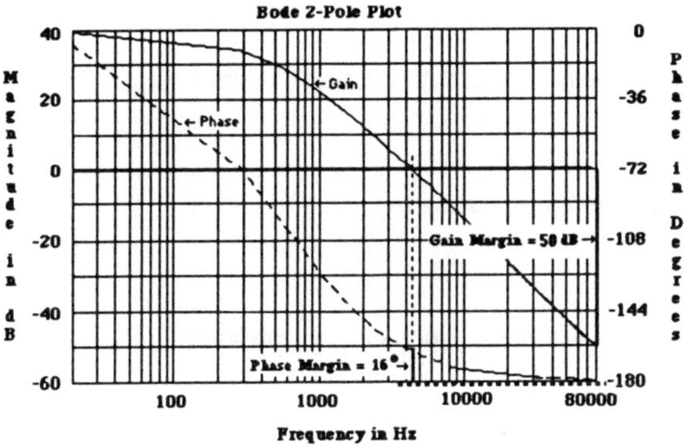

FIGURE 9.13 A 2-Pole, 40-dB amplifier gain and phase plot showing the gain and phase margins. The form of the transfer function would be

$$H(j\omega) = \frac{K}{(1+j\frac{\omega}{\omega_0})\,(1+j\frac{\omega}{5\omega_0})}$$

where ω_0 is the first pole at about 1 kHz and $5\omega_0$ is the second pole at about 5 kHz and K is a gain of 100 or 40 dB. The 0 dB line and the gain and phase margins are shown in bold for clarity. Note that this amplifier will NEVER be unstable because the phase shift approaches, but never reaches or exceeds, -180°.

TM R_{IN}, R_{OUT} AND DISTORTION WITH FEEDBACK

The proper use of feedback can improve and stabilize several amplifier parameters, including the input and output impedances. The ideal op amp would have an infinite input impedance (it would not load the previous circuit at any frequency) and a zero output impedance (it would be a constant voltage generator at all frequencies). Stable feedback can raise the input impedance and lower the output impedance.

Amplifier distortion is also reduced with feedback. However, it should be remembered that only the AMPLIFIER distortion is reduced—feedback will NOT decrease the distortion of the incoming signal!

MM Z_{IN}, Z_{OUT}, AND DISTORTION WITH FEEDBACK—USING THE INVERTING AND NON-INVERTING INPUTS

$R_{in\ inv} = \frac{R_2}{1+A\beta}$ = the input resistance at the inverting (–) input—not including R_1 (E 9.28)

$R_{in\ non} = R_{in}\ (1+A\beta)$ for the non-inverting (+) input and R_{in} is the manufacturers listing (E 9.29)

$R_{out\ inv} = \frac{R_{out}}{1+A\beta}$ where R_{out} is the open-loop output resistance. A is the open loop gain. (E 9.30)

$R_{out\ non} = \frac{R_{out}}{1+A\beta}$ where β is the feedback ratio (E 9.31)

$\%Dist_{inv} = \frac{\%Dist}{1+A\beta}$ where amplifier distortion is usually given in percent (E 9.32)

$\%Dist_{non} = \frac{\%Dist}{1+A\beta}$ where amplifier distortion is usually given in percent (E 9.33)

GM Z_{IN}, Z_{OUT} AND DISTORTION WITH FEEDBACK

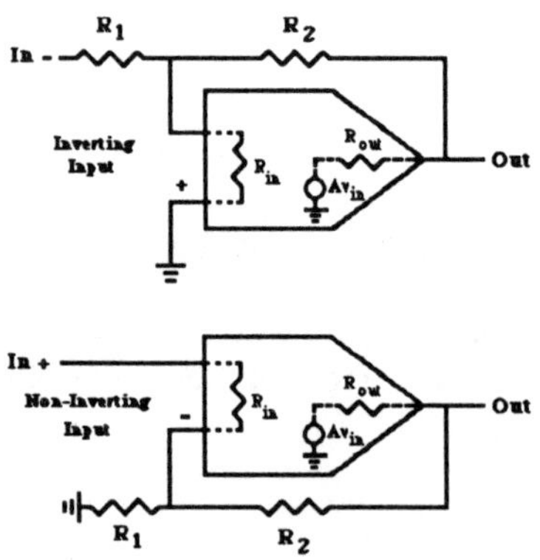

FIGURE 9.14 The impedance equivalent circuits of operational amplifiers.

TM GAIN BANDWIDTH PRODUCT

For a stable feedback system, as the gain is decreased, the bandwidth is increased. For a simple, one-pole, feedback amplifier, the gain times the bandwidth is a constant—the *gain bandwidth product.*

MM GAIN BANDWIDTH PRODUCT

$$f_{PCL} = (1+A\beta)\, f_{POL} \quad \text{(E 9.34)}$$

where $f_{P\,OL}$ is the open-loop position (frequency) of the pole and $f_{P\,CL}$ is the closed loop position. Also,

$$A_{CL} = \frac{A}{1+A\beta} \text{ and} \quad \text{(E 9.27)}$$

$$1+A\beta = \frac{A}{A_{CL}} \quad \text{(E 9.35)}$$

combining E 9.34 and E 9.35,

$$f_{P\,CL} = \frac{A}{A_{CL}} f_{POL} \text{ or}$$

$$A_{CL}\, f_{PCL} = A\, f_{POL}$$

where the left side of the equation is the closed-loop, gain bandwidth product and the right-hand side of the equation is the open-loop, gain bandwidth product. (E 9.36)

GM GAIN BANDWIDTH PRODUCT

See Figure 9.12.

An example: In Figure 9.12, the open-loop pole is at 1 KHz (10^3 Hz) and the gain is 10^5 (100 dB). Thus, the open-loop gain bandwidth is 10^8. At the 20 dB dashed line, the gain is 10 and the pole is at 10 MHz (10^7), so the gain bandwidth is 10 times 10 million or, again, 10^8.

TM POWER BANDWIDTH—SLEW-RATE DISTORTION

Using calculus (see Chapter 14), it can be shown that the maximum rate of change for a sine wave is at the middle point(s), where it crosses the horizontal axis. To avoid signal distortion, the rise time of the amplifier must be equal to or greater than the maximum rate of change of the sine wave. By using some mathematical manipulation, this maximum possible rate of change can also be equated to the *power bandwidth* of the amplifier.

MM POWER BANDWIDTH—SLEW-RATE DISTORTION

$S_{S\text{-}Max} = S_R = 2\pi f V_{pk}$ where $S_{S\text{-}Max}$ is the maximum slope of a sine wave and V_{pk} is the sine wave peak voltage. (E 9.37)

Thus, E 9.37 can be rewritten,

$$f_{max} = \frac{S_R}{2\pi V_{pk}}$$ where S_R is the *slew rate* and f_{max} is the *power bandwith.* (E 9.38)

GM POWER BANDWIDTH—SLEW-RATE DISTORTION

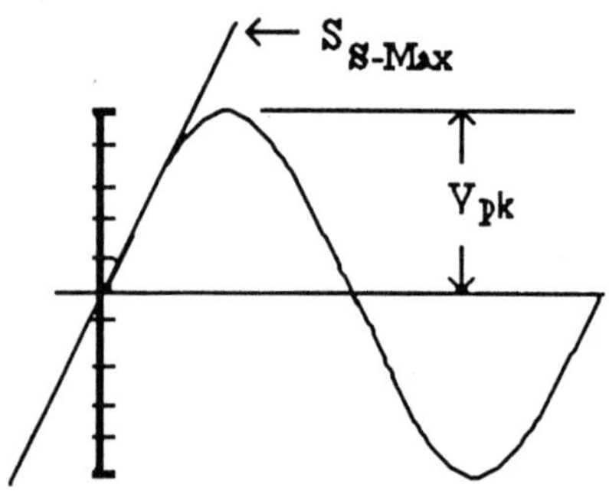

FIGURE 9.15 The maximum slope of a sine wave. The slope $S_{S\text{-}Max}$ must be equal to or less than the amplifier slew rate to prevent slew-rate distortion.

TM RISE TIME AND BANDWIDTH

For a one-pole amplifier, the *rise time* of a step wave can be specified using an approach simular to the analysis of an RC low-pass filter. For an amplifier, the rise time is DEFINED as the elapsed time from the point where the amplifier output has risen to 10% of its final output to the point where the output has risen to 90% of its final output.

Since both the rise time and the frequency response are determined by the RC time constant, it is also possible to develop a close relationship between the TRANSIENT rise time and the STEADY STATE sine wave response. This is a fundamental relationship and is the starting point for developing more advanced relationships between the analog steady state and transient response and the responses of digital signal-processing circuits.

MM RISE TIME AND BANDWIDTH FOR A ONE-POLE CIRCUIT

$T_R = 2.2RC$ where T_R is the rise time and RC is the pole time constant—both in seconds. (E 9.39)

$f_P = \frac{1}{2\pi RC}$ where f_P is the steady-state pole frequency—see Figure 9.10. (E 9.40)

$f_P = \frac{0.35}{T_R}$ where the same f_P is also defined in terms of the transient rise time. (E 9.41)

GM RISE TIME AND BANDWIDTH

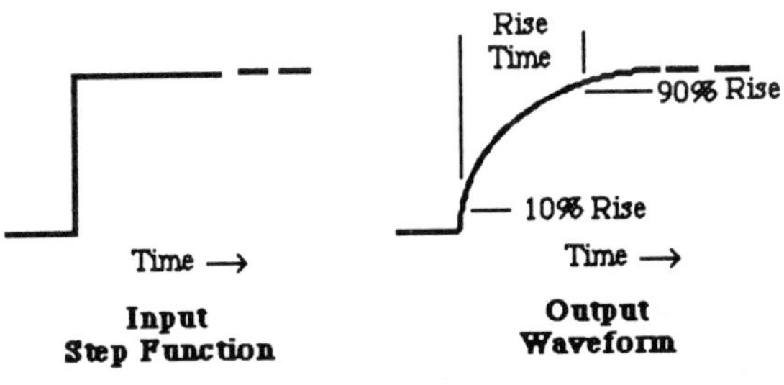

FIGURE 9.16 The relative input-output rise-time wave forms for an ideal amplifier driven by a step function.

9.8 AN INTRODUCTION TO ANALOG COMPUTING—ANALOG COMPUTING CIRCUITS

TM ADDERS AND SUBTRACTORS

Operational amplifier circuits can be configured to perform many mathematical operations including addition and subtraction. Their accuracy will be limited by the accuracy of the external component networks as were the input and output impedances, offsets, drifts, etc., of the op amps. For example, 0.1% resistors, which are typically used, cause approximately the same errors as one would get in a 10-bit digital circuit. However, well-designed operational circuits can very closely simulate the operation of mechanisms and processes and also can be used to solve complex functions, such as the differential equations describing electric circuits and servo mechanisms.

The fundamental adder circuit is a variation on the elementary feedback-amplifier circuit shown in Figure 9.11. The most common adder is connected to the inverting input, but it is also possible to configure non-inverting input. If the op amp is chopper stabilized, the inverting input must be used. A simple subtractor circuit can be configured using the non-inverting input for the (+) minuend and the inverting (–) input for the subtrahend.

MM ADDERS AND SUBTRACTORS

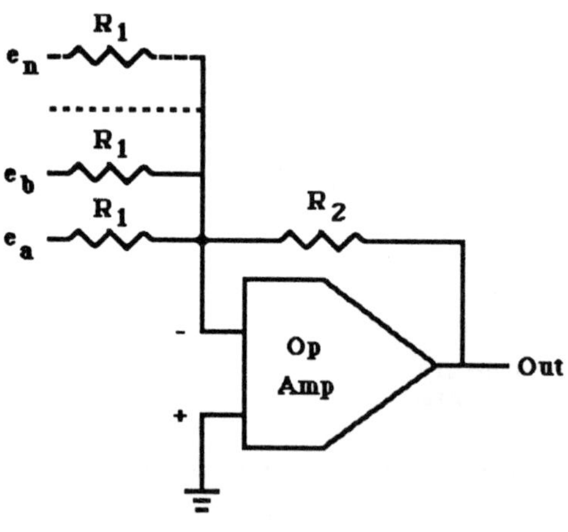

FIGURE 9.17 An adder using the inverting input.

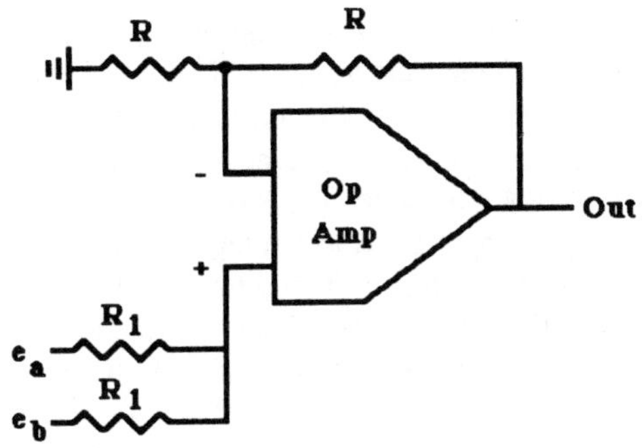

FIGURE 9.18 An adder using the non-inverting input.

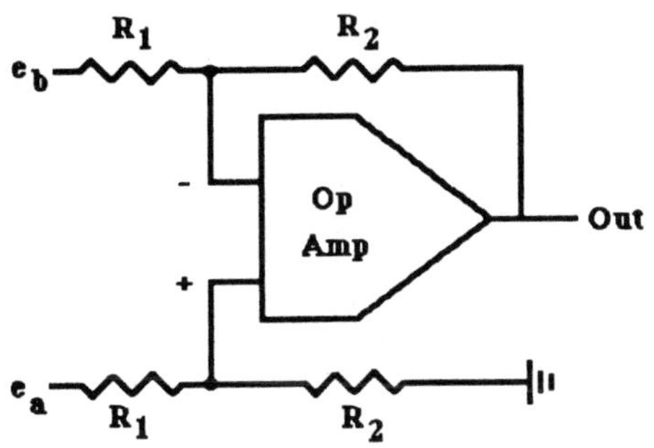

FIGURE 9.19 A subtractor.

MM ADDER (SUMMERS) AND SUBTRACTORS

$$E_{out} = -\frac{R_2}{R_1}(e_a + e_b + \cdots e_n) \text{ for the adder in Figure 9.17} \quad \text{(E 9.42)}$$

$E_{out} = e_a + e_b$ for the adder in Figure 9.18 if all resistors are equal (for unity gain). (Gain can be obtained by increasing the value of the feedback resistor.) (E 9.43)

$E_{out} = e_a - e_b$ for the subtractor in Figure 9.19 where $R_1 = R_2$. (Increase the R_2 resistors for a gain other than unity.) (E 9.44)

TM MULTIPLIERS AND DIVIDERS

Building an accurate analog multiplying or dividing circuit is a difficult task. An amplifier is really a multiplier, but it multiplies by a constant. To obtain a usable multiply or divide operation between two variables, it is necessary to have a controllable gain. Referring to Figure 9.11, a variable-gain amplifier (multiplier) could be obtained if it were possible to change the value of R_2 in a linear, convenient manner. The many attempts to provide a linear, controllable resistive element, before the introduction of FET transistors and integrated circuits, were less then spectacular. Many times in analog computers, R_2 was a rheostat that was set to the appropriate value for each problem. Using integrated circuits, several quite linear multipliers have been developed, including the Motorola MC1494 and MC1495 four-quadrant multipliers. These chips can be used as electronic gain controls as well as being adaptable for the functions of multiply, divide, square and square root.

MM MULTIPLIERS AND DIVIDERS

a ∗ b = c where a is considered the input and b is the multiplying factor

(E 9.45)

GM MULTIPLIERS AND DIVIDERS

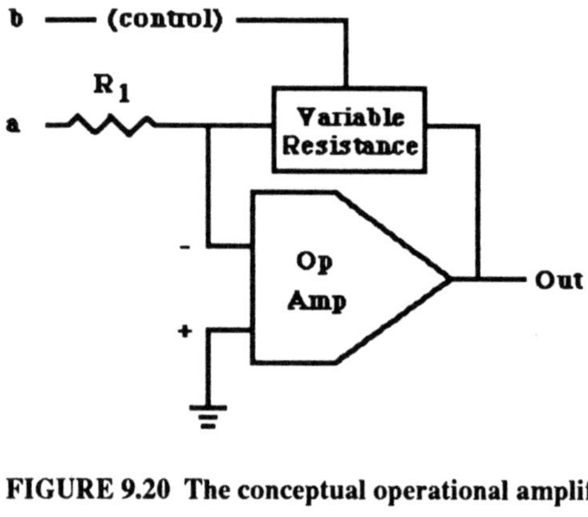

FIGURE 9.20 The conceptual operational amplifier multiplier or divider.

TM OPERATIONAL SQUARE AND SQUARE ROOT FUNCTIONS

The square function is relatively easy if a suitable multiplier can be obtained. If a quantity is multiplied by itself, its square is obtained. Producing the square root is more difficult. In the past, non-linear (usually diode) circuits were developed that would simulate in-out relationships corresponding to squares, square roots, logs, etc. These were satisfactory, but had some of the same problems that will be seen in integrators and differentiators. The advent of the four-quadrant multiplier has made the root, square and log function somewhat easier.

GM SQUARE ROOT FUNCTION

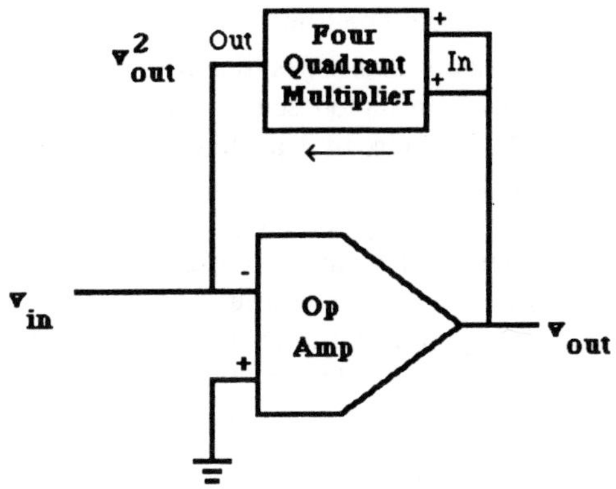

FIGURE 9.21 A basic square root circuit.

MM SQUARE ROOT FUNCTION

$v^2_{out} - v_{in} = 0$ Kirchhoff's Voltage at the (–) input, or (E 9.46)

$v^2_{out} = v_{in}$ by rearranging, again at the (–) input, therefore, (E 9.47)

$v_{out} = \sqrt{e_{in}}$ (E 9.48)

(See "Specifications and Applications Information for the MC1495L/MC1595L" Motorola Semiconductors, Phoenix, Arizona.)

TM INTEGRATORS

Analog computers, which are famous for solving differential equations, of course must have the best possible integrators and differentiators. It is also interesting to compare these circuits to low-pass and high-pass filters.

A true integrator circuit will produce the integral of an input signal. As an example, the integral of a step function is a ramp and the integral of a sine is a –cosine. For sine waves, the output (the integral) is INVERSELY proportional to the frequency of the input. If an RC low-pass filter is driven by a constant-voltage step function, the output starts at zero volts and rises, for a short time, in a straight, diagonal line. However, it soon deviates from this linear rise and follows the familiar exponential rise. For the first moments, the linear output voltage is proportional to current. In fact, if the RC were driven by a CURRENT step, the (output) voltage across the capacitor would be a truly linear ramp. If a capacitor replaces R_2 in the prototype feedback circuit in Figure 9.11, the op amp will drive a current through the capacitor and the output will be a true integration—a voltage ramp. However, even this supposedly ideal integrator may have three practical problems: (1) The circuit may have inadequate amplifier gain at very low frequencies; (2) It may lack sufficient loop gain at high frequencies; and (3) Once the capacitor is charged for a given operation, it may keep that charge and produce an (offset) error for the next operation.

These problems have several solutions. If a given amplifier has inadequate very low frequency gain, change to a high-gain, chopper-stabilized op amp. If the high-frequency loop gain is too low, change the feedback factor β or likewise increase the amplifier gain—again usually with a chopper-stabilized amplifier. If the capacitor charge is a problem, add a large-value parallel "leak" resistor or use an FET to discharge the capacitor between operations.

MM INTEGRATORS

$e_o/e_{in} = -Z_2/Z_1$ is the fundamental transfer function for a feedback (operational) amplifier. (E 9.49)

$e_o/e_{in}(s) = -1/RC(s)$ where $R = Z_1$, $1/C(s) = Z_2$ and $s = j\omega$ (E 9.50)

$$v_o = \frac{-1}{R_1 C} \int_{t=0}^{t} v_{in}\, dt \quad \text{(E 9.51)}$$

An example: Using an operational (amplifier) integrator with a sine wave input:

$i_{in} + i_c = 0$ states Kirchhoff's law for an operational integrator. This assumes a virtual ground at the (-) amplifier input and *no* amplifier input or offset current. The i_{in} goes in R_1 and i_c is fed back through the capacitor.

$i_c = C\dfrac{dv_o}{dt}$ is the current in the capacitor as a function of the changing output voltage.

$\dfrac{v_{in}}{R_1} + C\dfrac{dv_o}{dt} = 0$ or $\dfrac{dv_o}{dt} = \dfrac{-1}{R_1 C} v_{in}$, and now integrating,

$v_0 = \frac{-1}{R_1C} \int_{t=0}^{t} v_{in}\, dt$ which is the model op amp integrator equation! Now, for a sine wave,

$v_0 = \frac{-1}{R_1C} \int_{t=0}^{t} (A \sin \omega t\,)\, dt = \frac{-A}{\omega R_1C}(-\cos \omega t)$ (note that the output is inversely proportional to ω)

$$= \frac{A}{\omega R_1C} \cos \omega t \qquad \text{(E 9.52)}$$

GM INTEGRATORS

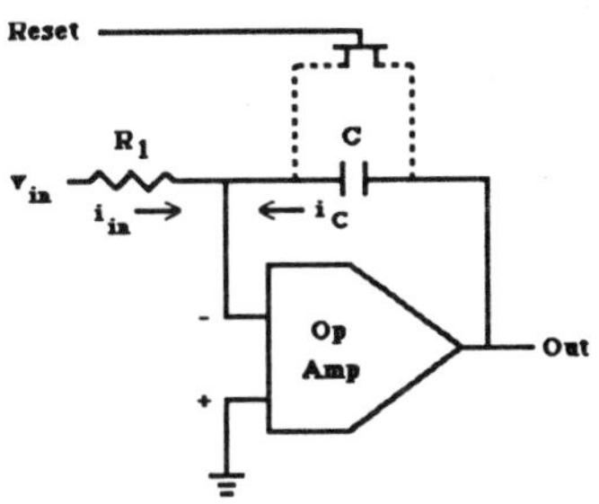

FIGURE 9.22 The fundamental operational integrator circuit. The FET can be used to dischange, and thus reset, C for the next operation.

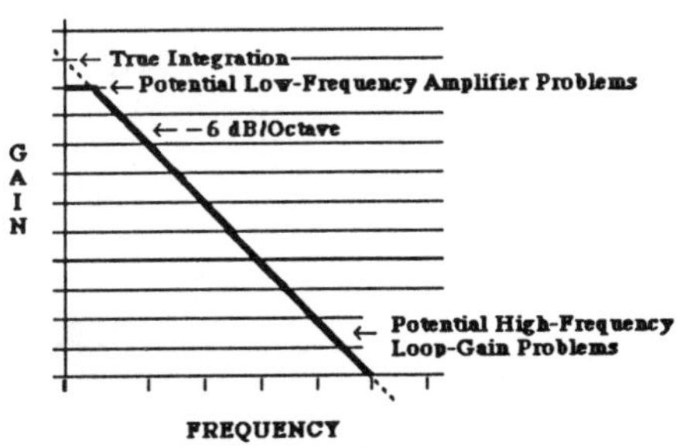

FIGURE 9.23 A general Bode plot of an operational integrator showing potential amplifier gain and loop-gain problems. A true integrator would have no pole crossover points.

TM DIFFERENTIATORS

The model operational differentiator circuit can be formulated by exchanging the position of the resistor and the capacitor in Figure 9.22. However, the accuracy and usefulness of electronic differentiators is always limited by high-frequency noise. Just as a true integrator resembles a low-pass filter, a true differentiator resembles a high-pass filter. High-pass filters always have noise problems. As the response, and thus the output, keeps going up in frequency, the signal energy usually decreases, leaving the residual noise. Likewise, ALL high-pass filters, including operational differentiators, eventually start to roll off (eventually there is a high-frequency pole) and the differentiator becomes an integrator! Practical operational control and computer circuits solve these problems in the following ways. First, it is sometimes possible to limit the frequency range of operation and scale (amplifying) the input so the output will be above the noise. Second, very exotic differentiators, which are far more intricate than the operational prototype design, have helped. Third, many times the mathematical problem can be restated so that integration operations can be substituted for the differentiations. Notwithstanding these problems, the differentiator is an interesting and fundamental circuit.

MM DIFFERENTIATORS

$$v_o = -R_2 C \frac{dv_{in}}{dt} \qquad \text{(E 9.53)}$$

An example: Using an operational (amplifier) differentiator with a sine wave input.

$$i_C + i_{R2} = 0 \text{ Kirchhoff's current law} \qquad \text{(E 9.54)}$$

$$i_C = C\frac{dv_{in}}{dt} \text{ and } i_{R2} = \frac{v_o}{R_2} \qquad \text{(E 9.55)}$$

$$C\frac{dv_{in}}{dt} + \frac{v_o}{R_2} = 0 \text{ or rearranging,} \qquad \text{(E 9.56)}$$

$$v_o = -R_2C\frac{dv_{in}}{dt} \text{ which is the model op amp differentiator}$$

equation! Now, for a sine input, (E 9.57)

$v_O = -R_2C\frac{d(A\sin\omega t)}{dt} = -A\omega R_2C\cos\omega t$ where the $-A$ is from the inverting input, or (E 9.58)

$v_O = A\omega R_2C\cos\omega t$ which is in the cos form that is found in integral tables and where the differentiated output is proportional to ω (frequency) (E 9.59)

GM DIFFERENTIATORS

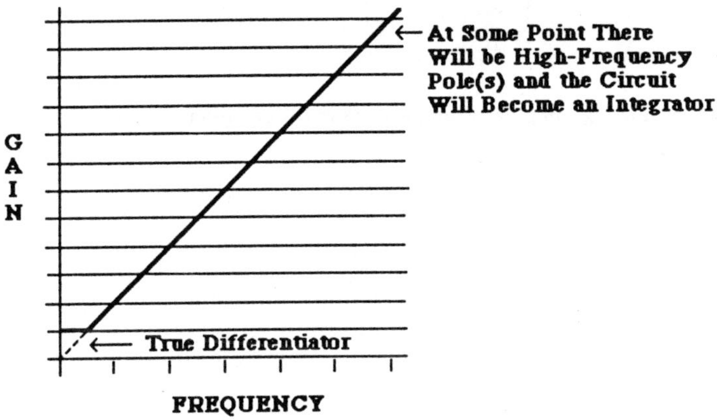

FIGURE 9.24 A general Bode plot of an operational differentiator. The true differentiator is similar to a high-pass filter, but it does not have its characteristic low-frequency crossover.

AN EXAMPLE OF PROBLEM SOLVING USING AN ANALOG COMPUTER MODEL:

A number of second order differential equation problems found in nature, including RLC resonance problems, can be reduced to this general form:

$$x = \frac{-b \pm \sqrt{b^2 - 4ac}}{2a}$$

How can one solve for x using an analog computer? The following graphic illustrates one of many methods.

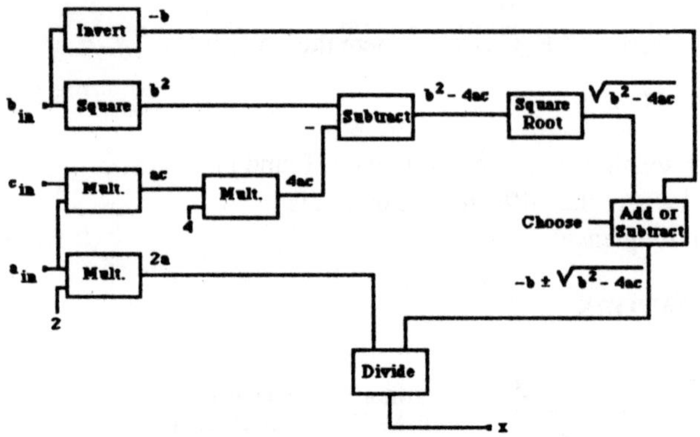

FIGURE 9.25 An analog computer example to solve the equation

$$x = \frac{-b \pm \sqrt{b^2 - 4ac}}{2a}$$

10

Fourier Series and the Algebra of Growth

10.1 ABOUT THIS CHAPTER—ANALYSIS, SYNTHESIS, COMPOSITION, AND DECOMPOSITION

Circuit analysis is based upon signal analysis. It is assumed that investigators understand the characteristic outputs of a battery or a sine-wave generator before they are used as signals to analyze a circuit. If the signal is more complex, as a narrow pulse, the complete operation of the circuit can only be understood if, first, the time and frequency composition of the pulse are known and understood.

Most signals can be broken into a series of elemental component parts. For example, a square wave usually can be reduced to a dc term, plus a series of sine or cosine waves. Knowing the frequency-domain composition of this time-domain signal can be of much help in understanding any problems caused by its transmission through a filter circuit. Conversely, knowing the time information about a frequency-domain signal can aid in determining any transient problems. Signals can also be created if the proper component parts are known. Music synthesizers often rely upon the ability to combine wave forms that are simple, easy to define and produce, to create complex wave shapes.

Mathematicians have invented several number series that conveniently “fit” most electrical signals. In fact, these series play an important part in the analysis of most inputs and processes in science and technology, social science, business, or finance. These series help us understand the growth of tree rings, of population, and the interest on money, as well as the charge on a capacitor. The following is a general review of some of the series that are used in electronics and circuit analysis. They can be used for background on the use of the Fourier series.

10.2 INFINITE SERIES

TM SIMPLE INFINITE SERIES

The number series 1 + 2 + 3 + 4 + ··· continually adds to a bigger and bigger number—the more terms, the larger the final sum. However, the number series 1 + .1 + .01 + .001 + ··· gets closer and closer to the number 10/9. As the number of terms increases to infinity, the first series is said to diverge (keep increasing) and the second series is said to converge (finally come to a single value). Series can contain several types of numbers such as the simple ones above, fractions, exponents (called powers), complex numbers, trigonometric functions, etc.

A group of numbers that are related to each other by some function such as an addition, a ratio, etc., is called a progression or a series. If the numbers continue on and on ad infinitum, the progression is called an *infinite series*.

ARITHMETIC SERIES

A group of numbers that are characterized by the continual addition of some number related to the term in the series is called an *arithmetic series*. It should be remembered that there are many arithmetic series since there are many numbers that can be used in the progression of a series. The general form of the series is a, a + k, a + 2k, a + 3k, ···. The calculation of simple, yearly interest on money produces an arithmetic series. For one example, what is the yearly value of one dollar invested at (r) rate of simple interest?

End of year:	0	1	2	3	4
Value at end of year:	1	1+k	1+2k	1+3k	1+4k
If r = 0.1 = 10%:	\$1	\$1.1	\$1.2	\$1.3	\$1.4

GEOMETRIC SERIES

A group of numbers where each term bears a constant ratio to the preceding term is called a *geometric series*. The ratio can be a whole number or a fraction. As with other series, there are many number sequences that form geometric series. The general form, with r standing for ratio, is a, ar, ar^2, ar^3 ···.

For a geometric series example, find the interest on one dollar invested if the interest is paid (compounded) yearly.

End of year:	0	1	2	3	4
Value at end of year:	1	1+r	$(1+r)^2$	$(1+r)^3$	$(1+r)^4$
If r = 0.1 = 10%:	\$1	\$1.1	\$1.21	\$1.33	\$1.46

HARMONIC PROGRESSION

A group of numbers that form a sequence that is the reciprocal of the numbers in an arithmetic progression is called a *harmonic progression or series.* Related numbers having to do with wavelengths in science and music often form a harmonic series. The ratio of the length of pipes for a pipe organ or the ratio of the lengths of the strings in a piano forms a reciprocal arithmetic or harmonic series. Likewise, the curve in the cabinet of a grand piano follows the ratios of a harmonic series. The signal output of non-linear amplifiers (often considered distortion) usually follows a mathematical harmonic progression.

LOGARITHMS AND LOGARITHMIC PROGRESSION

Logarithms and their antilogs can be written as arithmetic and geometric series.

Log (arithmetic)	1	2	3	4	5	6
Base number	2	2	2	2	2	2
Antilog (geometric)	2	4	8	16	32	64

To use the logarithms, here with a base of 2, select any two antilog numbers to multiply but, instead, add the corresponding logs and then find the corresponding answer in its antilog. As an example, to multiply the numbers 4 times 8, add the corresponding logs 2 and 3 for a total of 5. Now, take the antilog of 5 which is 32.

Log (arithmetic)	1	2	3	4	5	6
Base number	10	10	10	10	10	10
Antilog (geometric)	10	100	1000	10k	100k	1M

For a second example, to multiply 100 times 1000, add the corresponding logs 2 and 3 and see the antilog answer 100k under the log value of 5. If the intervening decimal numbers are used, log and antilog lookup tables can be constructed.

The logarithm of a number is the power to which the base must be raised to equal the number!

EXPONENTIAL SERIES

$$e = 1 + \frac{1}{1!} + \frac{1}{2!} + \frac{1}{3!} + \cdots. \quad \text{(E 10.1)}$$

$$e^x = 1 + x + \frac{1}{2!} + \frac{1}{3!} + \cdots \quad \text{(E 10.2)}$$

$$a^x = 1 + x \log a + \frac{(x \log a)^2}{2!} + \frac{(x \log a)^3}{3!} + \cdots \quad \text{(E 10.3)}$$

TRIGONOMETRIC SERIES

$$\sin x = x - \frac{x^3}{3!} + \frac{x^5}{5!} - \frac{x^7}{7!} + \cdots \text{ (where x is in radians)} \quad \text{(E 10.4)}$$

$$\cos x = 1 - \frac{x^2}{2!} + \frac{x^4}{4!} - \frac{x^6}{6!} + \cdots \quad \text{(E 10.5)}$$

$$\cos x + j \sin x = 1 + j\theta - \frac{\theta^2}{2!} - j\frac{\theta^3}{3!} + \frac{\theta^4}{4!} + j\frac{\theta^5}{5!} - \cdots = e^{j\theta}$$ the complex Euler identity (E 10.6)

BINOMIAL (TWO NUMBER) SERIES

$$(1+x)^k = 1 + kx + \frac{k(k-1)}{2!}x^2 + \frac{k(k-1)(k-2)}{3!}x^3 + \cdots \quad \text{(E 10.7)}$$

TAYLOR AND MACLAURIN—GENERAL "FRAMEWORK" SERIES

Many of the above series as well as others not covered here fit under the umbrella of either a Taylor or Maclaurin series. The Maclaurin is a special case of the Taylor series and requires that x be defined at zero. As an example, the cotangent of x is not defined at x = 0 and, therefore, it can only be expanded with the more universal Taylor series. Unfortunately, these series are quite overwhelming in their general form, but they are included here for reference.

TAYLOR SERIES

$$f(x) = f(a) + f'(a) * \frac{(x-a)}{1!} + f''(a) * \frac{(x-a)^2}{2!} + f'''(a) * \frac{(x-a)^3}{3!} + \cdots \tag{E 10.8}$$

where f ' is the first derivative, f " is the second derivative, etc.

MACLAURIN SERIES

$$f(x) = f(0) + f'(0)\frac{x}{1!} + f''(0)\frac{x^2}{2!} + f'''(0)\frac{x^3}{3!} + \cdots \tag{E 10.9}$$

10.3 AN INTRODUCTION TO FOURIER SERIES

The fundamental principle of a Fourier series is that (almost) any real, physical signal can be represented as a combination of sine and cosine wave forms. In general, the only conditions limiting the use of a Fourier series for either analysis or synthesis are that the signal must be periodic and physically realizable (a musical sound, an electrical signal, a mechanical vibration, etc.) and the components must be harmonically related (ω, 2ω, 3ω, etc.). Thus, it is possible to assemble a series of sinusoidal *frequency-domain* wave shapes into a composite wave that can be, and usually is, represented in the *time domain.* Likewise, a time-domain signal can be reduced (decomposed) to a series of sinusoids in the frequency domain.

The frequency domain can be defined in several ways. The equations sin ω or cos ω or the graph of these equations bring to mind a clear picture of a "frequency" signal. However, in addition, a frequency-domain signal (or many signals) can be represented in a frequency *spectrum.* A spectrum is a series of frequencies usually, but not necessarily, related. The notes on a piano, as represented by the keys, compose a spectrum from about 33 Hz to 8372 Hz. This is a *discrete spectrum* since the piano cannot produce all of the frequencies in this range. An FM receiver, with its range of 88 Mhz to 108 Mhz, represents a *continuous spectrum* since, at least in theory, it could receive an FM signal transmitting on any frequency in this spectrum range.

The two major parameters of each term in a Fourier series are the *spectral components*, i.e., the sines or cosines of ω for the harmonics involved and the *amplitude coefficients* for each of these spectral components. It should be understood that not all wave forms contain both sine and cosine components. Likewise, sometimes a component will add (be +) and sometimes it will subtract (be –) from the other sine or cosine terms. The following models will show the classical method of obtaining some of the components. After this method is shown, the values of the components and their coefficients for more wave forms will be listed, without derivation. Finally, the techniques of finding the components and their coefficients with computer programs will be illustrated.

TM FOURIER SERIES

A Fourier Series is a trigonometric series representing the components, including their frequencies and amplitudes, of a periodic wave shape as a series of sine and cosine functions.

TM THE CLASSICAL FOURIER SERIES

For a function f (t), with a period $T = \frac{1}{f}$ and $\omega_0 = \frac{1}{T}$ and a_0 = the average dc or $a_{average}$,

$$f(t) = \frac{1}{2} a_0 + \sum_{n=1}^{\infty} a_n \cos n\omega_0 t + \sum_{n=1}^{\infty} b_n \sin n\omega_0 t \qquad \text{(E 10.10)}$$

$$f(t) = \frac{1}{2} a_0 + a_1 \cos \omega_0 t + a_2 \cos 2\omega_0 t + \cdots$$
$$+ b_1 \sin \omega_0 t + b_2 \sin 2\omega_0 t + \cdots$$
$$= \frac{1}{2} a_0 + \sum_{n=1}^{\infty} (a_n \cos n\omega_0 t + b_n \sin n\omega_0 t)$$

an infinite number of terms is assumed (E 10.11)

$$= \frac{1}{2} a_0 + \sum_{n=1}^{n} (a_k \cos n\omega_0 t + b_k \sin n\omega_0 t) \text{ only terms from 1 to n}$$

are assumed (E 10.12)

$$a_0 = \frac{1}{T} \int_0^{T=2\pi} f(t)\, dt = 1/2 \text{ the average value of the wave form} \qquad \text{(E 10.13)}$$

$$a_n = \frac{2}{T} \int_0^{T=2\pi} f(t) \cos n\omega_0 t\, dt \quad n = 0, 1, 2, 3, \cdots \qquad \text{(E 10.14)}$$

$$b_n = \frac{2}{T}\int_0^{T=2\pi} f(t)\sin n\omega_0 t\,dt \quad n = 1, 2, 3, \cdots \qquad \text{(E 10.15)}$$

Note: If the wave form is symmetrical about the y axis, the integration is from $-\pi$ to $+\pi$.

TM WAVE SYMMETRY AND THE FOURIER SERIES

A Fourier series can contain only sine terms, only cosine terms, or a combination of the two. Likewise, the series may or may not include a dc term. If the wave shape is symmetrical about the vertical (y) axis, it is termed *even* and contains only cosine terms. If it starts at the y axis, it is termed *odd* and it contains only sine terms. If the wave form is unsymmetrical about the y axis but starts at a place other than the y axis, it will contain both sine (b_n) and cosine (a_n) terms. If the area is not balanced about the horizontal (x) axis, it contains a dc (a_0) term.

Special Note: In the analysis, synthesis, and operation of wave shapes and their circuits, it is usually extremely important to know the precise position of a wave shape relative to some time reference. The precise position of the start of a signal, such as a pulse, relative to a time reference, may determine whether or not the circuit is operating properly.

MM WAVE SYMMETRY AND THE FOURIER SERIES

If f(t) is even, i.e., if $f(-t) = f(t)$, then $b_n = 0$ and the series contains only cosine terms.

If f(t) is odd, i.e., if $f(-t) \neq f(t)$, then $a_n = 0$ and the series contains only sine terms.

GM ODD AND EVEN FUNCTIONS

(See Figure 10.1, top of page 230.)

EXAMPLE 10.1: Using the Classical Fourier Series, E 10.10 and E 10.11, derive the coefficients for the square wave in Figure 10.1 (a).

By inspection, this wave form is NOT symmetrical about the x axis and thus there is a dc offset A_0. A_0 is the average or dc value of the wave form from 0 to 2π. Since the wave is in two discrete parts, to find A_0 requires two integrals,

a_0 = the average value from 0 to 1/2T + the average value from 1/2T to T.

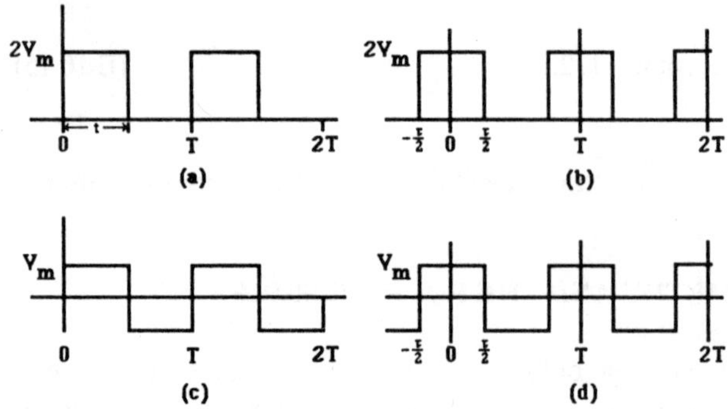

FIGURE 10.1 Even and odd Fourier wave forms:

(a) $f(t) = V_m + \frac{4}{\pi} V_m (\sin \omega t + \frac{1}{3} \sin 3\omega t + \frac{1}{5} \sin 3\omega t + \cdots\cdots\cdots)$

an odd function **(E 10.16)**

where the V_m (a_0) term represents the +dc offset—the average value of the wave $\neq 0$

(b) $f(t) = V_m + \frac{4}{\pi} V_m (\cos \omega t - \frac{1}{3} \cos 3\omega t + \frac{1}{5} \cos 3\omega t - \cdots\cdots\cdots)$

an even function **(E 10.17)**

where the V_m (a_0) term represents the +dc offset—the average value of the wave $\neq 0$

(c) $f(t) = \frac{4}{\pi} V_m (\sin \omega t + \frac{1}{3} \sin 3\omega t + \frac{1}{5} \text{sine } 3\omega t + \cdots\cdots\cdots)$

odd, no dc offset **(E 10.18)**

(d) $f(t) = \frac{4}{\pi} V_m (\cos \omega t - \frac{1}{3} \cos 3\omega t + \frac{1}{5} \cos 3\omega t - \cdots\cdots\cdots)$

even, no dc offset **(E 10.19)**

$$= \frac{1}{T} \int_0^{\frac{1}{2}T} 2V_m dt + \frac{1}{T} \int_{\frac{1}{2}T}^{T} 0 dt = \frac{V_m}{T} [\, 2 * \frac{1}{2} T - 0 \,]_0^{1/2T} + \frac{V_m}{T} [\, 0 - 0 \,]_{1/2T}^{T} =$$

$$V_m \frac{1T}{1T} = 1V_m$$

$a_n = 0$ because it is an odd function—there are no cosine functions

$$b_n = \frac{2}{T} \int_0^T f(t) \sin n\, \omega\, t\, dt \text{ and, when } \omega = 2\pi f = 2\frac{\pi}{T}$$

$$= \frac{2}{T} \int_0^T f(t) \sin n\, 2 \frac{\pi}{T}\, dt$$

$$= \frac{2}{T} \int_0^T f(t) \sin 2\pi n \frac{t}{T}\, dt$$

$$= \frac{2}{T} \int_0^{T/2} f(t) \sin 2\pi n \frac{t}{T}\, dt - \frac{2}{T} \int_{T/2}^{T} f(t) \sin 2\pi n \frac{t}{T}\, dt$$

$$= \frac{2}{T} \frac{T}{2\pi n} \left[-\cos 2\pi n \frac{t}{T} \right]_0^{1/2T} - \frac{2}{T} \frac{T}{2\pi n} \left[\cos 2\pi n \frac{t}{T} \right]_{1/2T}^{T}$$

$$= \frac{1}{\pi n} [-\cos n\pi + 1 + \cos 2n\pi - \cos n\pi]$$

$= \frac{1}{\pi n} [2 - 2 \cos n\pi]$ and, for n odd, $\cos n\pi = -1$, so

$b_n = \frac{1}{\pi n} [2 - (-2)] = \frac{4}{n\pi}$ and, finely, for the indicated square wave,

$$f(t) = V_m + \frac{4}{\pi} V_m \left[\sin 2\pi \frac{t}{T} + \frac{1}{3} \sin 6\pi \frac{t}{T} + \frac{1}{5} \sin 10\pi \frac{t}{T} + \cdots \right]$$

If the area of the wave form is balanced about the x axis—as in Figure 10.1 (c) — there is no dc offset and the a_0 term is zero.

EXAMPLE 10.2: Using the following graphic, find the Fourier coefficients and write the Fourier series equation.

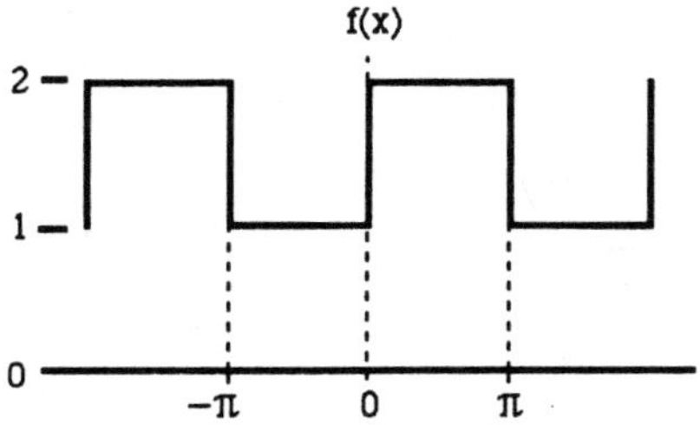

FIGURE 10.2 The graphic for EXAMPLE 10.2.

$$f(t) = \frac{1}{2} a_0 + \sum_{n=1}^{\infty} a_n \cos n\omega_0 t + \sin n\omega_0 t$$

a copy of (E 10.10)

$$a_0 = \frac{1}{2\pi} \int_{-\pi}^{0} dx + \frac{1}{2\pi} \int_{0}^{\pi} dx$$

$$= \frac{1}{2\pi} [x]_{-\pi}^{0} + \frac{1}{2\pi} [x]_{0}^{\pi} = \frac{1}{2\pi} ([\, 0 - (-\pi)\,] + [\, 2\pi - 0\,])$$

$$= \frac{1}{2\pi} [\, \pi + 2\pi\,] = \frac{3}{2} \qquad \text{(E 10.20)}$$

a_n = an odd function = 0

$$b_n = \frac{1}{\pi} \int_{-\pi}^{0} \sin nx \, dx + \frac{2}{\pi} \int_{0}^{\pi} \sin nx \, dx$$

$$= -\frac{1}{n\pi} [\, \cos nx\,]_{-\pi}^{0} - \frac{2}{n\pi} [\, \cos nx\,]_{0}^{\pi}$$

$$= -\frac{1}{n\pi} [\, 1 - (-\pi)^n\,] - \frac{2}{n\pi} [\, (-1)^n - 1\,] = \frac{1 - (-1)^n}{n\pi} \qquad \text{(E 10.21)}$$

$$a_1 = \frac{2}{\pi}$$

$$a_2 = 0$$

$$a_3 = \frac{2}{3\pi} \cdots$$

$$f(t) = f(x) = \frac{3}{4} + \frac{2}{\pi} (\frac{\sin x}{1} + \frac{\sin 3x}{3} + \frac{\sin 5x}{5} + \cdots) \qquad \text{(E 10.22)}$$

10.4 GRAPHICAL FOURIER ANALYSIS

Often the Fourier coefficients of special wave shapes are desired. The following method illustrates one graphical technique for finding these coefficients. This and similar methods are based upon the principle of sampling, and the accuracy is a function of the number of samples. For instance, if 12 intervals are used, because of the *theory of sampling*, it is impossible to determine coefficients for more than the first six (one-half the number of samples) harmonics. It is usually inconvenient to graph, by hand, more than 12 to 24 samples. Usually, an oscilloscope picture or a hand-drawn graph is used and the sample amplitudes are determined by dividers. It is interesting to note that computer Fourier analysis uses much the same idea but, typically, selects 512 samples, which give a reasonably accuracy (not perfect) value for the first 256 coefficients. It should be kept in mind that any graphical technique produces some inaccuracies because of the number of samples, the method of sample measurement and rounding errors in the calculations.

TM GRAPHICAL FOURIER ANALYSIS

To determine the Fourier coefficients, any periodic wave shape can be divided into k samples per period—the example uses k = 12. For each sample, there are m intervals such as m1, m2, ··· up to mk. Likewise, n harmonics can be calculated. No harmonic above n = k/2 can be calculated. Once the wave period is divided into k intervals, the peak value at the end of the period is measured and tabulated as an amplitude a_m. There will be m1, m2, ··· intervals with a1, a2, ··· peak values. The rules of even or odd symmetry, described earlier, also apply to graphical Fourier analysis. Thus, depending on where the axes are chosen and the type of wave shape, some of the harmonics and some of the sine or cosine terms may be inherently zero and need not be calculated.

MM GRAPHICAL FOURIER ANALYSIS

If there is a dc offset value, i.e., if the wave shape is not balanced about the x axis, the value of a_0 must be calculated.

$$a_0 = \frac{1}{k}\sum_{m=1}^{k} a_m \qquad \text{(E 10.23)}$$

$$a_n = \frac{2}{k}\sum_{m=1}^{k} a_m \cos m\frac{n\,360}{k} = \frac{2}{k}\sum_{m=1}^{k} a_m \cos m\,30 \text{ using 12 (k) samples} \qquad \text{(E 10.24)}$$

(Note that 360°/12 = 30°)

$$b_n = \frac{2}{k} \sum_{m=1}^{k} a_m \sin m \frac{n\,360}{k} = \frac{2}{k} \sum_{m=1}^{k} a_m \sin m\, 30 \text{ using 12 samples}$$

(E 10.25)

GM GRAPHICAL FOURIER ANALYSIS

EXAMPLE 10.3: Although graphical analysis is designed for special wave forms, the following example will use a square wave so that the results can be compared with the formulas given earlier.

Assume the wave forms given in Figure 10.3.

(a) Find the Fourier coefficients for Figure 10.3 (a) where the ordinate (y) scale is –1 to +1.

(b) Find the coefficients for Figure 10.3 (b) where the ordinate scale is 0 to +3.

(c) In 10.3 (c) the wave is not quite symmetrical. Discuss how this will alter the method of finding the coefficients.

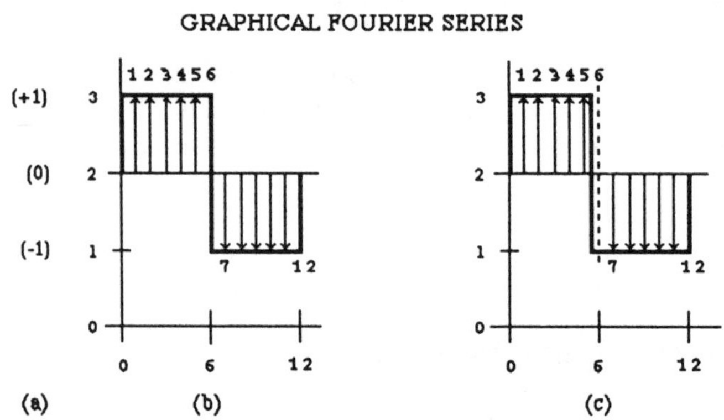

FIGURE 10.3

(a) Find the Fourier coefficients for Figure 10.3 (a) where the ordinate (y) scale is –1 to +1.

Using the (a) scale, the square wave is symmetrical about the 0 (x) axis and therefore there is no average dc offset, so $a_0 = 0$. Likewise, the wave is odd and there are no cosine (a_n) terms. Also, for a square wave, there are no even sine terms.

$a_0 = 0$

$b_n = 0$

To find the b_n coefficients of the fundamental:

$$b_1 = \frac{2}{12} \sum_{m=1}^{12} a_m \sin m\, n\, 30 \text{ where } n = 1$$

$$= \frac{2}{12} [1 \sin 1 * 30 + 1 \sin 2 * 30 + \cdots -1 \sin 7 * 30 + \cdots -1 \sin 12 * 30]$$

$$= \frac{2}{12} [\,.5 + .866 + \cdots .5 + \cdots 0\,] = 1.24402$$

To find the coefficient of the third harmonic:

$$b_3 = \frac{2}{12} \sum_{m=1}^{12} a_m \sin m\, n\, 30 \text{ where } n = 3$$

$$= \frac{2}{12} [\,1 \sin 1 * 3 * 30 + 1 \sin 2 * 3 * 30 + \cdots -1 \sin 7 * 3 * 30 + \cdots$$

$$-1 \sin 12 * 3 * 30\,]$$

$$= \frac{2}{12} [\,1 + 0 + -1 + \cdots 0\,] = 0.3333 \cdots$$

To find the coefficient of the fifth harmonic:

$$b_5 = \frac{2}{12} \sum_{m=1}^{12} a_m \sin m\, n\, 30 \text{ where } n = 5$$

$$= \frac{2}{12} [\,1 \sin 1 * 5 * 30 + 1 \sin 2 * 5 * 30 + \cdots -1 \sin 7 * 5 * 30 + \cdots$$

$$-1 \sin 12 * 5 * 30\,] = 0.0893$$

$$= \frac{2}{12} [\,.5 + -.866 + .5 + 0\,] = 0.0893$$

The numbers for the above calculations are tabulated in Table 10.1.

Some of the errors of the graphical method can be seen by referring back to E 10.15. That equation shows the coefficients as $b_1 = 1$, $b_3 = 0.3333$ and $b_5 = 0.2$ compared with the equivalent graphical values of 1.244, 0.3333, and 0.089.

HARMONIC CALCULATIONS
USING A SPREADSHEET

a = the amplitudes of the sampled wave
k = number of intervals = 12
m = this interval = 1 ····· 12
n = this harmonic = 1, 2, 3, ···

First Harmonic n = 1				
m	a values	m(n360/k)	SIN	a Sin (x)
1	1	30	0.5000	0.5000
2	1	60	0.8660	0.8660
3	1	90	1.0000	1.0000
4	1	120	0.8660	0.8660
5	1	150	0.5000	0.5000
6	1	180	0.0000	0.0000
7	-1	210	-0.5000	0.5000
8	-1	240	-0.8660	0.8660
9	-1	270	-1.0000	1.0000
10	-1	300	-0.8660	0.8660
11	-1	330	-0.5000	0.5000
12	-1	360	0.0000	0.0000
			SUM =	7.4641
			b1 = SUM/6 =	1.24402

Third Harmonic n = 3				
m	a values	m(n360/k)	SIN	a Sin (x)
1	1	90	1.0000	1.0000
2	1	180	0.0000	0.0000
3	1	270	-1.0000	-1.0000
4	1	360	0.0000	0.0000
5	1	450	1.0000	1.0000
6	1	540	0.0000	0.0000
7	-1	630	-1.0000	1.0000
8	-1	720	0.0000	0.0000
9	-1	810	1.0000	-1.0000
10	-1	900	0.0000	0.0000
11	-1	990	-1.0000	1.0000
12	-1	1080	0.0000	0.0000
			SUM =	2.0000
			b3 = SUM/6 =	0.3333

Fifth Harmonic n = 5				
m	a values	m(n360/k)	SIN	a Sin (x)
1	1	150	0.5000	0.5000
2	1	300	-0.8660	-0.8660
3	1	450	1.0000	1.0000
4	1	600	-0.8660	-0.8660
5	1	750	0.5000	0.5000
6	1	900	0.0000	0.0000
7	-1	1050	-0.5000	0.5000
8	-1	1200	0.8660	-0.8660
9	-1	1350	-1.0000	1.0000
10	-1	1500	0.8660	-0.8660
11	-1	1650	-0.5000	0.5000
12	-1	1800	0.0000	0.0000
			SUM	0.5359
			b5 = SUM/6 =	0.0893

TABLE 10.1 A tabulation of the calculations in the graphic Fourier series example.

Special Note: Programs such as *Excel* or *Lotus 1-2-3* are another example of the power that personal computers afford the engineer. Besides the vast array of financial functions, spreadsheets also offer trigonometic, exponential, statistical, and logic functions. The use of spreadsheets is only limited by the user's imagination. Operations that are the equivalent of "Fill Down" or "Fill Right" make calculations rapid and easy to duplicate.

(b) Find the coefficients for Figure 10.3 (b) where the ordinate scale is 0 to +3.

With the (b) ordinate scale of 0 to +3, the wave will have an average dc offset which says that $a_0 \neq 0$. The other a and b coefficients will be the same and need not be recalculated. Using Figure 10.2 (b) and E 10.19,

$$a_0 = \frac{1}{k} \sum_{m=1}^{k} a_m$$

$$a_0 = \frac{1}{12} [3 + 3 + 3 + 3 + 3 + 3 + 1 + 1 + 1 + 1 + 1 + 1] = \frac{24}{12} = 2$$

which is, by inspection, the mid-axis average dc value of the square wave in part (b).

(c) In 10.2 (c) the wave is not quite symmetrical. Discuss how this will alter the method of finding the coefficients.

The wave form in the (c) part is now not quite a true square wave; i.e., the duty cycle or on-off ratio is now not exactly 50 percent. This will change the amplitude of the sampled values and can affect all the coefficients. For instance, the sampled value of a_6 has changed from +1 to −1. The new value of a_0 would = (5*3 + 7*2)/12 or 1.833. Likewise, the calculations of a_n and b_n would also be changed. There are two significant points about a_n and b_n that should be remembered. If a "square wave" is not truly symmectrical—not really a true square wave—there may be cosine terms (a_n) as well as sine terms. In the (c) part of this example, the change is so small that accurate values for a_0, a_n and b_n would be impossible without using more samples.

10.5 THE FOURIER SERIES EQUATIONS FOR TYPICAL WAVE FORMS

Special Note: In the following graphs and equations, the time-domain dimension (T) and the angular or frequency-domain dimension (π) can be, and are, used interchangeably. Thus, $T = 2\pi$, $T/2 = \pi$, etc. In some cases, the values of a_n, b_n, and c_n are not listed because these are a rather complicated function of the length of the pulse relative to the length of the period.

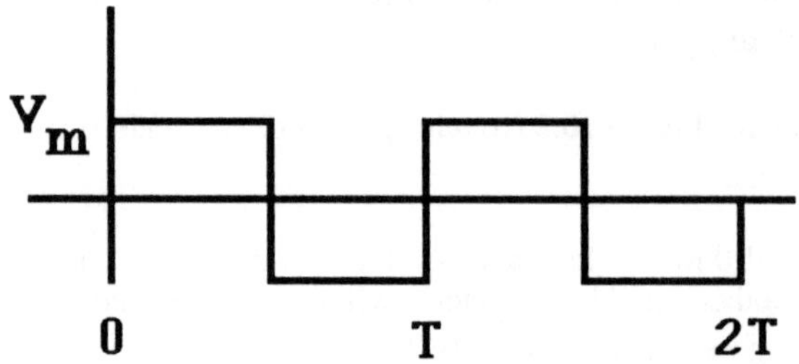

FIGURE 10.4 A true square wave odd with no dc offset:

$$f(t) = \frac{4}{\pi} V_m \left(\sin \omega t + \frac{1}{3} \sin 3\omega t + \frac{1}{5} \text{sine } 5\omega t + \cdots\cdots\right) \text{ derived from (E 10.16)}$$

$a_0 = 0$, $a_n = 0$, $b_n = 4/\pi n$ **where n is the order of the harmonic**

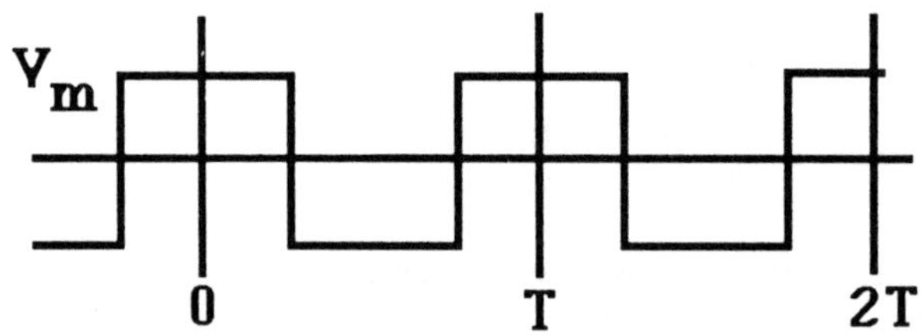

FIGURE 10.5 A true square wave even with no dc offset:

$$f(t) = \frac{4}{\pi} V_m \left(\cos \omega t - \frac{1}{3} \cos 3\omega t + \frac{1}{5} \cos 3\omega t - \cdots\cdots\right) \text{ derived from (E 10.17)}$$

$a_0 = 0$, $a_n = 4/\pi n$, $b_n = 0$

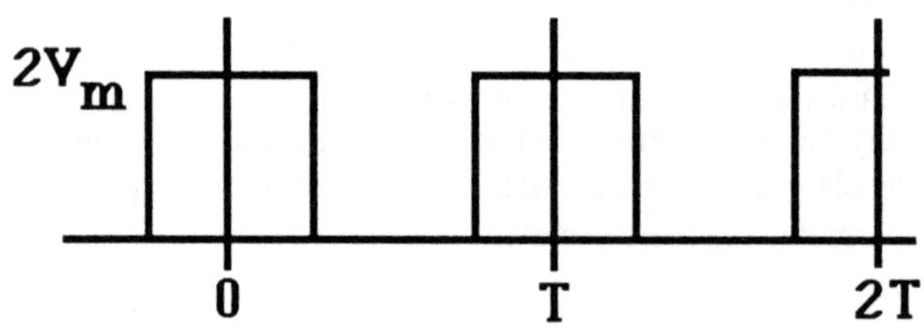

FIGURE 10.6 A true square wave even with a dc offset:

$$f(t) = V_m + \frac{4}{\pi} V_m \left(\cos \omega t - \frac{1}{3} \cos 3\omega t + \frac{1}{5} \cos 3\omega t - \cdots\cdots\right)$$

a repeat of (E 10.17)

$$a_0 = \frac{2v_m}{2},\ a_n = 2V_m \frac{0.5T}{T} \sin \frac{\sin \pi n}{n} \frac{0.5Tn}{T},\ b_n = 0$$

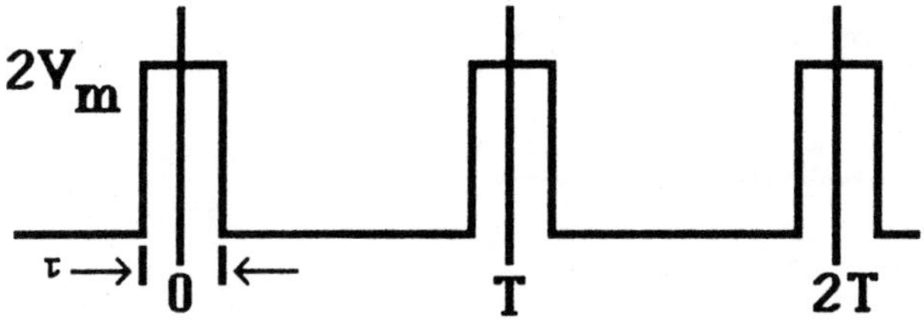

FIGURE 10.7 A rectangular pulse. Even with no dc offset:

$$f(t) = 2V_m \left[k + \frac{\pi}{2} \left(\sin k\pi \cos \omega_0 + \frac{1}{2} \sin 2k\pi \cos 2\omega_0 + \frac{1}{3} \sin 3k\pi \cos 3\omega_0 + \cdots \right) \right] \quad k = \frac{\tau}{T} \qquad \text{(E 10.26)}$$

Note: This formula will also work for a true square wave when $k = 0.5$. Also, the magnitude $2V_m$ was used for consistency. Some books or equations use V_m.

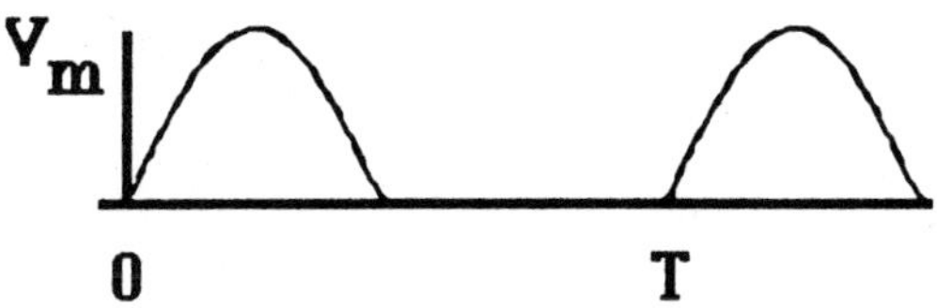

FIGURE 10.8 Half-wave rectifier output:

$$f(t) = \frac{2}{\pi} V_m \left[\frac{1}{2} + \frac{\pi}{4} \cos \omega t + \frac{1}{3} \cos 2\omega t - \frac{1}{15} \cos 4\omega t + \frac{1}{35} \cos 6\omega t \pm \cdots \right] \qquad \text{(E 10.27)}$$

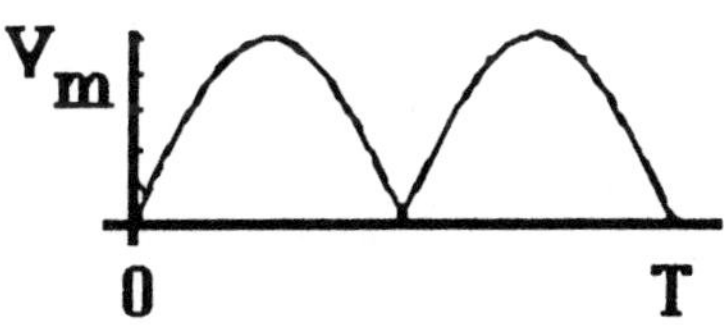

FIGURE 10.9 Full-wave rectifier output:

$$f(t) = \frac{4}{\pi} V_m \left[\frac{1}{2} + \frac{1}{3} \cos 2\omega t - \frac{1}{15} \cos 4\omega t + \frac{1}{35} \cos 6\omega t \pm \cdots \right] \qquad \text{(E 10.28)}$$

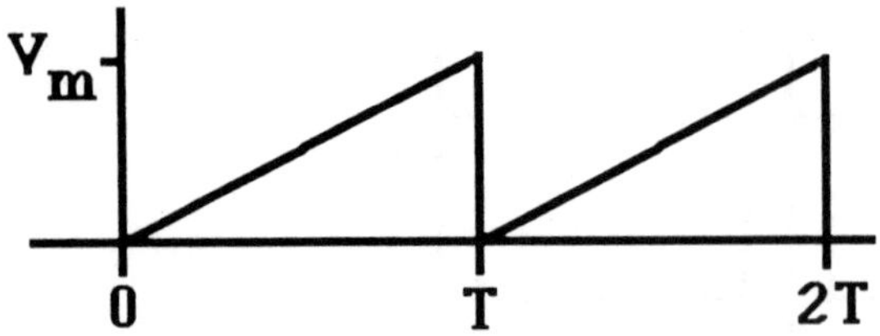

FIGURE 10.10 Sawtooth or ramp:

$$f(t) = \frac{A}{\pi}\left(\frac{\pi}{2} - \sin\omega t - \frac{1}{2}\sin 2\omega t - \frac{1}{3}\sin 3\omega - \cdots\right) \qquad \textbf{(E 10.29)}$$

10.6 THE COMPLEX FORM OF THE FOURIER SERIES

The FORM of the Fourier series can be presented in more than one way. Since the series involves sine and cosine terms, which means that wave shapes can be constructed (synthesized) from simple sine or cosine waves, it seems logical that it can be presented in terms of their equivalent exponentials. Chapter 4 introduced Euler's rule and showed that the series expansion of the sine and cosine function could be equated to exponential expansions. After some manipulation, it was shown that a rather complex combination of sine and cosine terms could be neatly reduced to one straightforward exponential expression. The same general technique can be used to simplify the form of the Fourier series. This somewhat abbreviated form finds special application in the development of the frequency (domain) spectrum to be discussed in a following section.

TM THE COMPLEX FORM OF THE FOURIER SERIES

An exponential form of the Fourier series uses Euler's rule to present the sine and cosine terms in a complex, exponential form. A new variable, c_n, is derived from the complex combination of the more familiar a_n and b_n. This new, abbreviated, complex variable gives the amplitude and phase of the harmonics that make up the wave form and is often used for signal spectral analysis. (In the exponential form, the integration of the terms is done from minus infinity to plus infinity because of the Euler expansion.)

MM THE COMPLEX FORM OF THE FOURIER SERIES

For review and comparison, the (original) trigonometric Fourier series:

$$f(t) = a_0 + \sum_{n=1}^{\infty} [\, a_n \cos k\omega_0 t + b_n \sin k\omega_0 t \,]$$

for reference see (E 10.10)

Now, the (new) exponential Fourier series:

$$f(t) = a_0 + \sum_{n=1}^{\infty} \frac{a_n}{2} [(e^{jn\omega_0 t} + e^{-jn\omega_0 t}) + \frac{b_n}{j2} (e^{jn\omega_0 t} - e^{-jn\omega_0 t})]$$

$$= a_0 + \sum_{n=1}^{\infty} (\frac{a_n - jb_n}{2}) e^{jn\omega_0 t} + (\frac{a_n + jb_n}{2}) e^{jn\omega_0 t}$$

$$= \sum_{-\infty}^{\infty} c_n e^{jn\omega_0 t} \qquad \text{(E 10.30)}$$

Define the new complex coefficient c_n:

$$c_n = \frac{1}{2} (a_n - jb_n) = \frac{A_n}{2} \angle\text{-}\theta_n \,,\; n = 1, 2, 3, \cdots$$

$$= \frac{1}{2} [\frac{2}{T} \int_{t_0}^{t_0+T} f(t) \cos n\omega_0 t \, dt \; - j \frac{2}{T} \int_{t_0}^{t_0+T} f(t) \sin n\omega_0 t \, dt \,]$$

$$= \frac{1}{T} \int_{t_0}^{t_0+T} f(t) (\cos n\omega_0 t \, dt - j \sin n\omega_0 t) \, dt$$

$$= \frac{1}{T} \int_{t_0}^{t_0+T} f(t) \, e^{-jn\omega_0 t} \, dt \qquad \text{(E 10.31)}$$

In E 10.31, if ω_0 is zero, then $f(t)\,e^{-jn\omega_0 t}$ goes to f(t), which produces the familiar offset or dc term, so

$$c_0 = \frac{1}{T}\int_{t_0}^{t_0+T} f(t)\,dt = a_0 \qquad \text{(E 10.32)}$$

GM THE COMPLEX FORM OF THE FOURIER SERIES

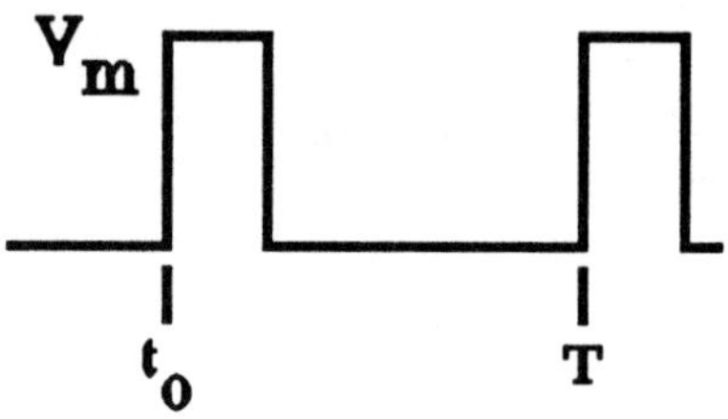

FIGURE 10.11 t_0 can be taken at the start, at the middle or at any place on the pulse.

EXAMPLE 10.4: Find the coefficients c_n for the wave form shown in Figure 10.12, starting at $-\tau/2$.

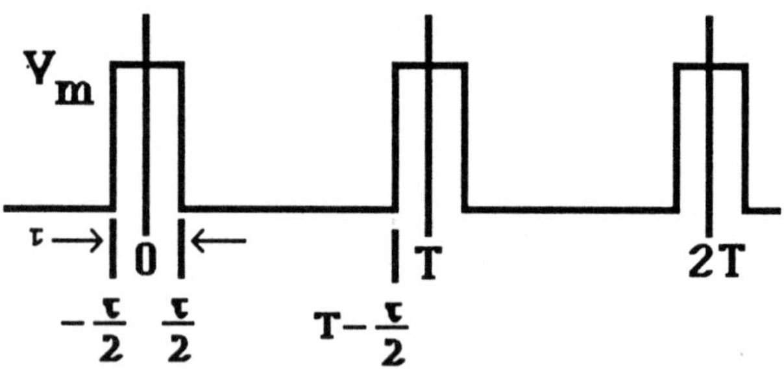

FIGURE 10.12 A pulse centered at T = 0, T = 1T, T = 2T, etc.

$$C_n = \frac{1}{T}\int_{t_0}^{t_0+T} f(t)\,e^{-jn\omega_0 t}\,dt$$

$$C_n = \frac{1}{T}\int_{-\tau/2}^{\tau/2} V_m\,e^{-jn\omega_0 t}\,dt + \frac{1}{T}\int_{\tau/2}^{T-\tau/2} V_m\,e^{-jn\omega_0 t}\,dt \text{ for Figure 10.12}$$

$$= \frac{1}{T} \int_{-\tau/2}^{\tau/2} V_m e^{-jn\omega 0t} dt + 0$$

$$= \frac{V_m}{T} \left[\frac{e^{-jn\omega 0t}}{-jn\omega_0 t} \right]_{-\tau/2}^{\tau/2}$$

$$= \frac{jV_m}{n\omega_0 T} \left[e^{-j\omega_0 \tau/2} - e^{j\omega_0 \tau/2} \right]$$

$$= \frac{V_m}{n\omega_0 T} \sin n\omega_0 \tau/2$$

$$= V_m \frac{\sin n\omega_0 \tau/2}{n\omega_0 \tau/2} = \text{the form of } 2V_m \frac{\sin nx}{nx} \text{ where n = the harmonic of } \omega_0$$

10.7 THE GENERALIZED SPECTRUM OF C_n USING THE COMPLEX FORM OF THE FOURIER SERIES

TM THE GENERALIZED SPECTRUM

The foregoing analysis of the TIME DOMAIN pulses was designed to obtain FREQUENCY DOMAIN information about their harmonic content. Several of the equations were manipulated to obtain the generalized form of sin x/x. This form, which is often called the *sinc* function, is often found in the analysis of both analog and digital wave forms. It is interesting to note that the amplitudes of the harmonics—the c_n values—always decrease as a sinc function.

A second observation is that the c_n amplitudes are inversely proportional to the width of the pulse. For instance, in a square wave which has only odd harmonics, the first harmonic has a value of 0.6366 V_m, the next or third harmonic is –0.2122 V_m, and the fifth is only 0.1273 V_m. If the pulse is narrowed to a (k =) T/t ratio of 10, the fifth harmonic is 0.6366, and if the T/t ratio is 100, the fifth harmonic has an amplitude of 0.9959.

MM THE GENERALIZED SPECTRUM OF C_n USING THE COMPLEX FORM OF THE FOURIER SERIES

$$c_n = V_m \frac{\sin n\omega_0 \tau/2}{n\omega_0 \tau/2} = \text{the form of } 2V_m \frac{\sin nx}{nx}$$

where n = the harmonic of ω_0 (E 10.33)

MM THE GENERALIZED SPECTRUM OF C_n USING THE COMPLEX FORM OF THE FOURIER SERIES

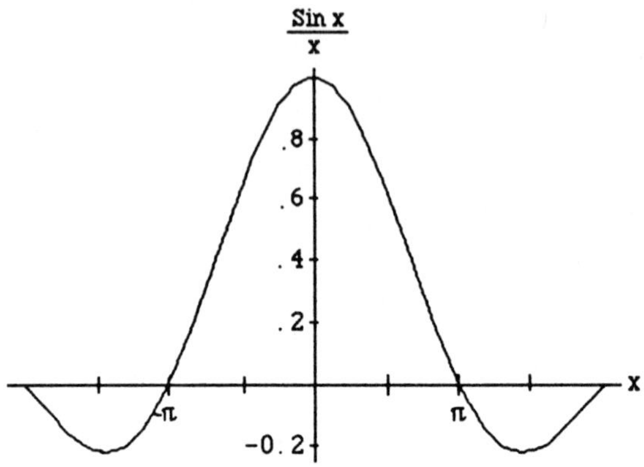

FIGURE 10.13 A plot of the harmonic amplitudes as a function of angular frequency. As the pulse time is made smaller by the ratio of T/t, the equation becomes $c_0 = V_m \frac{\sin n\pi/k}{n\pi/k}$.

THE AMPLITUDE OF HARMONICS AS A FUNCTION OF PULSE WIDTH

c = (Sin n π/k)/(n π/k)
n = the number of the harmonic = 1, 2, 3, ...
k = the ratio of the period T to the pulse time t.

	k = 2	k = 5	k = 10	k = 20	k = 100
n	c	c	c	c	c
1	0.6366	0.9355	0.9836	0.9959	0.9998
2	0.0000	0.7568	0.9355	0.9836	0.9993
3	-0.2122	0.5046	0.8584	0.9634	0.9985
4	0.0000	0.2339	0.7568	0.9355	0.9974
5	0.1273	0.0000	0.6366	0.9003	0.9959
6	0.0000	-0.1559	0.5046	0.8584	0.9941
7	-0.0909	-0.2162	0.3679	0.8103	0.9920
8	0.0000	-0.1892	0.2339	0.7568	0.9895
9	0.0707	-0.1039	0.1093	0.6986	0.9867
10	0.0000	0.0000	0.0000	0.6366	0.9836
11	-0.0579	0.0850	-0.0894	0.5716	0.9802
12	0.0000	0.1261	-0.1559	0.5046	0.9765
13	0.0490	0.1164	-0.1981	0.4363	0.9724
14	0.0000	0.0668	-0.2162	0.3679	0.9681
15	-0.0424	0.0000	-0.2122	0.3001	0.9634
16	0.0000	-0.0585	-0.1892	0.2339	0.9584
17	0.0374	-0.0890	-0.1515	0.1700	0.9531
18	0.0000	-0.0841	-0.1039	0.1093	0.9475
19	-0.0335	-0.0492	-0.0518	0.0524	0.9417
20	0.0000	0.0000	0.0000	0.0000	0.9355
21	0.0303	0.0445	0.0468	-0.0474	0.9290

TABLE 10.2 A tabulation of the harmonic amplitudes for pulses with T/t ratios of 2, 5, 10, 20, and 100.

11

Transient Analysis: The Laplace Transform

11.1 AN INTRODUCTION: THE METHODS OF CIRCUIT ANALYSIS AND THE LAPLACE TRANSFORM

This chapter is about creating and solving equations for (1) the inputs, (2) the transmission or intermediate processing functions, and (3) the outputs of electrical circuits and networks. It concentrates upon transient analysis and the solution of circuit equations with differential and integral terms using the Laplace Transform.

Circuit analysis can be accomplished by breadboarding, by classical mathematical analysis, or by using predesigned computer programs. These programs "look" at the network and produce nodal voltages or currents based upon the rules and calculations required by standard network theorems. Traditional breadboarding can sometimes give the investigator an intuitive feel for the operation of the circuit. However, it is quite limited when statistics are required such as those that come from component tolerance, circuit variations caused by temperature, power supply variations or delay changes. Likewise, it is often difficult to simulate all of the input functions that a circuit might encounter. Mathematical or automated computer analysis gives a much broader and more precise analysis of circuit behavior. However, it is the responsibility of the investigator to understand and always double check the mathematics or computer routines to be sure that the results truly are for the network that is intended and that the final results "make sense."

11.2 AN INTRODUCTION TO THE LAPLACE TRANSFORM

The Laplace Transform technique was first developed (Pierre Simon Marquis de Laplace, French mathematician and astronomer, 1749–1827) as a technique for solving complicated *differential equations*. Although the Laplace Transform is often taught simply as a method of solving electric circuit differential equations, its use and influence is much wider than that in the fields of electronics and communications. The use of the Laplace transform has produced a literature and a tradition that is the foundation of our transient analysis. For instance, the Fourier Transform and its Fast Fourier Transform (see Chapter 12) are a special case of the Laplace. Most of the fundamental ideas of poles and zeros come from Laplace analysis. Likewise, the general form of most transient analysis is also derived from the use of the Laplace Transform.

TM THE DEFINITIONS OF THE LAPLACE TRANSFORM

The *Laplace Transform* of a function f(t)—using the symbol $\mathcal{L}$—is the integral, from zero to plus infinity, of that function multiplied by "e" to the minus st, where s is the complex frequency, $\alpha + j\omega$, and t is time.

The Laplace Transform (LT) is a circuit-analysis technique that allows complex differential equations in TIME to be changed (sometimes called mapped) into algebraic equations in COMPLEX FREQUENCY. When these algebraic equations are solved for the parameters such as voltage, current or impedance, the resulting equations can be retransformed back to the time domain by the process called the Inverse Laplace Transform (ILT). Using the LT, differential equations, which represent complicated circuits driven by a variety of wave shapes, can be solved. Also, the nodal voltages and currents can be obtained as a function of time and, if need be, frequency.

Although the Laplace method is primarily used to obtain the transient response, the steady-state frequency response can also be obtained. There are transforms (and inverse transforms) for input forcing functions, circuit components such as resistors, capacitors, inductors and amplifier gain, as well as for the differential terms.

MM THE DEFINITIONS OF THE LAPLACE TRANSFORM

$$\mathcal{L}\{f(t)\} = \int_0^{\infty} f(t)\, e^{-st}\, dt \qquad \text{(E 11.1)}$$

$$\mathcal{L}\{f(t)\} = F(s) \qquad \text{(E 11.2)}$$

$\mathcal{L}^{-1}$ F(s) = f (t) is the Inverse Laplace Transform equation which indicates the process of going back to the time (t) domain from the complex frequency (s) domain. (E 11.3)

Note that in E 11.1, the integral includes f(t) times e^{-st}, where e^{-st} will make the transform in the FORM of a decreasing exponential. This indicates (and it is true) that the Laplace Transform of a function puts a function in this familiar decreasing exponential form.

11.3 THE PROCEDURE AND RULES OF LAPLACE TRANSFORM CIRCUIT ANALYSIS

The Laplace Transform and its associated Inverse Laplace Transform are quite straightforward if the following procedure and rules are followed. The historical analogy for the transform process has been the use and methods of logarithms and the logarithmic look-up process. This is unfortunate and confusing because, although there is some vague resemblance, the Laplace Transform requires much more adherence to form and rules than does the use of logarithms. To solve circuit differential equations using the LT:

LT RULE 1: The time-domain circuit equations should be written in a form following Kirchhoff's voltage or current laws (*KVL* or *KCL*). For clarity and convenience, the differentials are usually presented in reduced order as the equation proceeds from left to right.

As a convention, parameters in the time domain, such as v(t) or i(t), are written in lower case while parameters in the s-domain, such as V(s) or I(s), are written in upper case.

LT RULE 2: Each individual term of the differential equation, including the forcing function, must be transformed, one at a time, from the time (t) domain to the complex frequency (s) domain. Each one of these $\mathcal{L}$ transformations can be obtained by using E 11.1. However, there is a much easier way!

Most, if not all, of the terms of a circuit differential equation can be transformed to the s-domain using a look up table. Likewise, terms of equations in the s-domain—if they are in the proper form—can be re-transformed by using an inverse transform look up table. The transform of a function and its accompanying inverse transform are called *transform pairs*.

LT RULE 3: The resulting algebraic equation in the s-domain can now be solved or changed to obtain one or more of the following:

a. the desired response of E(s) or I(s)
b. the circuit transfer function
c. the natural response of the circuit
d. information for plotting a graph of the circuit equation's poles and zeros

LT RULE 4: To use the ILT look up tables to retransform equations back to the time domain:

a. the solved s-domain equations must be factored (the roots must be found)
b. in most cases, the factored equations must be partial-fraction expanded (This step is often overlooked or misunderstood in using the LT method!)

LT RULE 5: As a check of the validity and accuracy of the transform and algebraic calculations, in some cases, the Initial-Value Theorem (IVT) and/or the Final Value Theorem (FVT) can be used to check the response.

LT RULE 6: The s-domain equation can now be reverse transformed back to the time domain.

11.4 THE MODELS FOR THE LT RULES

MM FOR LT RULE 1

$v_{in} = v_L + v_C + v_R$ is an example of the Kirchhoff voltage FORM , and (E 11.4)

$$v_{in} = L\frac{di}{dt} + \frac{1}{C}\int_0^t i\,dt + Ri \text{ is an example of the differential form} \qquad \text{(E 11.5)}$$

MM FOR LT RULE 2—THE DERIVATION OF SOME COMMON LAPLACE TRANSFORMS

Laplace Transforms are usually found in tables that include both the forward, or standard, Laplace Transforms and the "going back," or inverse, Laplace Transform. Even so, it is advisable to understand how a few of the Transforms are obtained from the fundamental equation of the Laplace Transform.

The LT of the unit step function u(t):

$$\mathcal{L}[u(t)] = \int_0^\infty f(t)\, e^{-st}\, dt = \int_0^\infty 1\, e^{-st}\, dt = \frac{e^{-st}}{-s}\Big|_{0+}^{\infty} = \frac{1}{s}$$

Note: Since the LT of A, a constant, is A, the $\mathcal{L}[A\ u(t)]$ would be $A * \frac{1}{s} = \frac{A}{s}$ (E 11.6)

The LT of a unit impulse:

$$\mathcal{L}[\delta(t)] = \int_0^\infty \delta\, e^{-st}\, dt = 1\, e^{-s(0)} = 1 \qquad \text{(E 11.7)}$$

The LT of a sine:

$$\mathcal{L}[\sin(\omega t)] = \int_0^\infty \sin(\omega t)\, e^{-st}\, dt = \int_0^\infty \frac{e^{j\omega t} - e^{-j\omega t}}{2j}\, e^{-st}\, dt$$

$$= \int_0^\infty \frac{e^{-(s-j\omega)t} - e^{-(s-j\omega)t}}{2j}\, dt$$

$$= \frac{1}{2j}\left(\frac{1}{s-j\omega} - \frac{1}{s+j\omega}\right) = \frac{\omega}{s^2+\omega^2} \qquad \text{(E11.8)}$$

The LT of a REAL exponential:

$$\mathcal{L}[e^{-at}] = \int_0^\infty e^{-at}\, e^{-st}\, dt = \int_0^\infty e^{-(a+s)t}\, dt = \frac{e^{-(a-s)t}}{a-s}\Big|_{0+}^{\infty} = \frac{1}{s+a} \quad \text{or} \quad \frac{A}{s+a}$$

with a constant (E 11.9)

The LT of a COMPLEX exponent:

$$\mathcal{L}[e^{-(a+j\omega)t}] = \int_0^\infty e^{-(a+j\omega)t}\, e^{-st}\, dt = \frac{e^{-(a+j\omega-s)t}}{a-s}\Bigg|_{0+}^{\infty}$$

$$= \frac{1}{s+a+j\omega} \quad \text{or} \quad \frac{A}{s+a+j\omega} \qquad \text{(E 11.10)}$$

MM FOR LT RULE 2—A TABLE OF LAPALCE TRANSFORM PAIRS

Functional Transforms

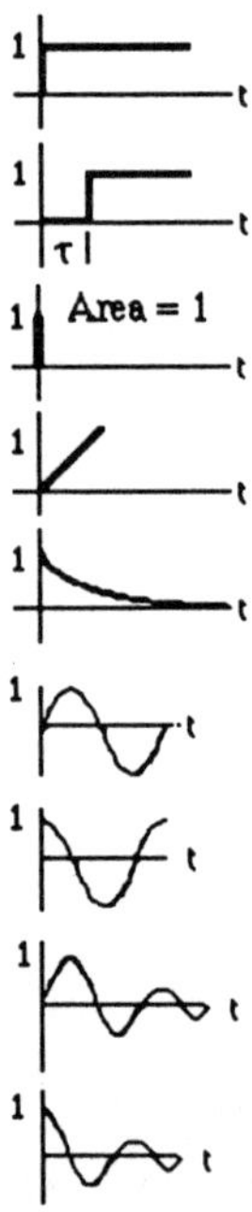

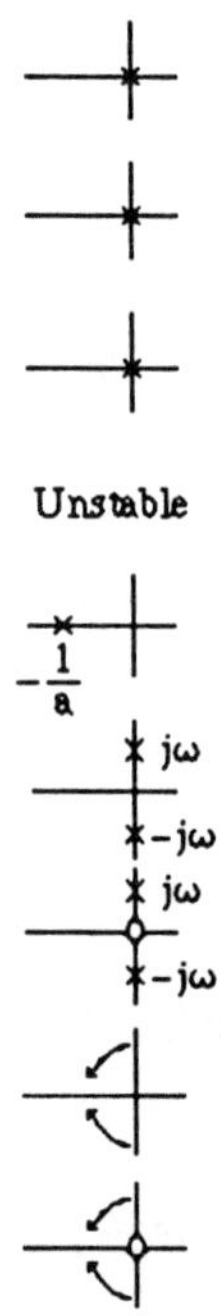

	f(t)		F(s)	s-Plane
1.	$u(t)$	Step	$\frac{1}{s}$	
2.	$u(t-\tau)$	Delayed Step	$\frac{1}{s}e^{-\tau s}$	
3.	$\delta(t)$	Impulse (Area = 1)	1	
4.	t	Ramp	$\frac{1}{s^2}$	Unstable
5.	e^{-at}	Exponential	$\frac{1}{s+a}$	
6.	$\sin \omega t$	Sine	$\frac{\omega}{s^2+\omega^2}$	
7.	$\cos \omega t$	Cosine	$\frac{s}{s^2+\omega^2}$	
8.	$e^{-at}\sin \omega t$	Damped Sine	$\frac{\omega}{(s+a)^2+\omega^2}$	
9.	$e^{-at}\cos \omega t$	Damped Cosine	$\frac{s+a}{(s+a)^2+\omega^2}$	

TABLE 11.1 Laplace transform pairs.

Operational Transforms

	$f(t)$	$F(s)$
1.	$Kf(t)$	$KF(s)$
2.	$f_1(t) + f_2(t) + f_3(t) \cdots$	$F_1(s) + F_2(s) + F_3(s) \cdots$
3.	$df/dt\,(t)$	$sF(s) - f(0^-)$
4.	$d^2f/dt^2\,(t)$	$s^2F(s) - sf(0^-) - df(0^-)/dt$
5.	$\int_0^t f(x)\,dx$	$F(s)/s$

TABLE 11.2 Operational transform pairs.

Partial Fraction Transforms

	$f(t)$	$F(s)$
1.	$K e^{-at}\,[u(t)]$	$\dfrac{K}{(s+a)}$
2.	$K t e^{-at}\,[u(t)]$	$\dfrac{K}{(s+a)^2}$
3.	$2\,\lvert K\rvert\, e^{-at} \cos(\beta t + \theta)\,[u(t)]$	$\dfrac{K}{(s+a-j\beta)} + \dfrac{K^*}{(s+a+j\beta)}$
4.	$2t\,\lvert K\rvert\, e^{-at} \cos(\beta t + \theta)\,[u(t)]$	$\dfrac{K}{(s+a-j\beta)^2} + \dfrac{K^*}{(s+a+j\beta)^2}$
	$\lvert K\rvert$ is the absolute value of K	* where K is a complex qunaty

TABLE 11.3 Partial-fraction pairs.

MM FOR LT RULE 3

See examples later in this chapter.

MM FOR LT RULE 4A—FACTORING THE S-DOMAIN EQUATION

Most often the solved s-domain equation will be in a form:

$$F(s) = \frac{\text{Numerator }(s)}{\text{Denominator }(s)} = \frac{a_n s_n + a_{n-1}s^{n-1} + a_{n-2}s^{n-2} + \ldots + a_0}{b_m s^m + b_{m-1}s^{m-1} + b_{m-2}s^{m-2} + \ldots + b_0} \quad \text{(E 11.11)}$$

Component Transforms

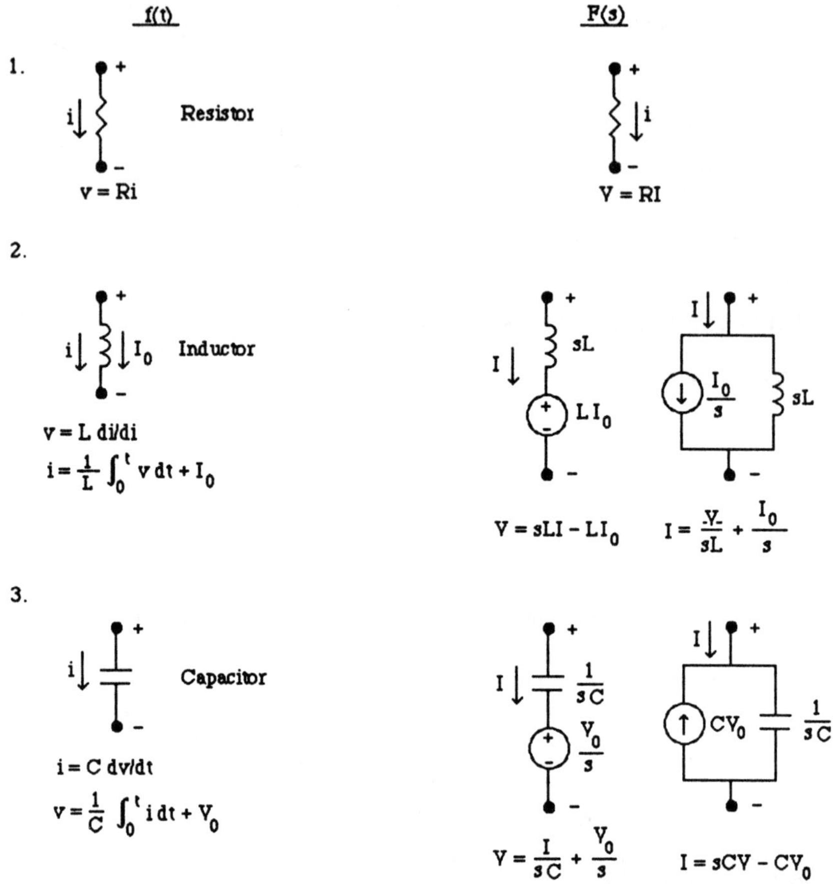

TABLE 11.4 Component transform pairs.

But, to use the Inverse Laplace Transforms, the equation must have factored terms as:

$$F(s) = \frac{K_0\,(NT_1)\,(NT_2)\,(NT_3)\,\cdots}{(DT_1)\,(DT_2)\,(DT_3)\,\cdots}$$ where K_0 is a constant, NT is a numerator term, etc. (E 11.12)

Very simple first and second degree equations can be factored by inspection with a little practice. A few models are given below:

$$x^2 + 2x + 1 = (x + 1)(x + 1) \qquad x^2 - 1 = (x + 1)(x - 1)$$

$$x^2 + 3x + 2 = (x + 1)(x + 2) \qquad x^2 - 4 = (x + 2)(x - 2) \qquad \text{(E 11.13)}$$

Second degree equations, which can't be factored by inspection, can be factored with the Quadratic Formula:

$$s_{1,2} = \frac{-B \pm \sqrt{B^2 - 4AC}}{2A} \text{ or, for complex terms} = \frac{-B \pm j\sqrt{4AC - B^2}}{2A}$$

for equations in the form $Ax^2 + Bx + C = 0$ (E 11.14)

MM FOR LT RULE 4B—PARTIAL FRACTION EXPANSION

To successfully use the Inverse Laplace Transform, the s-domain equation must be factored (as above) and partial fraction expanded to the form:

$$F(s) = \frac{\text{Numerator Terms in (s)}}{\text{Denominator Terms in (s)}} = \frac{K_0 (NT_1)(NT_2)(NT_3)\cdots}{(DT_1)(DT_2)(DT_3)\cdots}$$

$$= \frac{K_1}{D_1} + \frac{K_2}{D_2} + \frac{K_3}{D_3}. \qquad \text{(E 11.15)}$$

As an example,

$$F(s) = \quad \text{I} \quad = \quad \text{II} \quad = \quad \text{III}$$

$$F(s) = \frac{s+4}{s^2 + 3s + 2} = \frac{1(s+4)}{(s+1)(s+2)} = \frac{K_1}{s+1} + \frac{K_2}{s+2} \qquad \text{(E 11.16)}$$

The following examples show how to derive the constants K_1, K_2, etc., so that the ILTs can be obtained.

To obtain K_1:

[1] multiply both equations II and III by the DT_1 term (= s + 1) in II and
[2] evaluate both II and III using the "zero" of DT_1 (the term that, substituted for s, will make $DT_1 = 0$) which, in this problem, is -1. Thus,

$$\begin{array}{lccccc}
 & \text{II} & & \text{III} & & \\
[1] & (\cancel{s+1})\dfrac{1\,(s+4)}{(\cancel{s+1})(s+2)}\Big]_{s=-1} & = & (\cancel{s+1})\dfrac{K_1}{\cancel{s+1}}\Big]_{s=-1} & + & (s+1)\dfrac{K_2}{s+2}\Big]_{s=-1} \\
[2] & \dfrac{1\,(s+4)}{(s+2)}\Big]_{s=-1} & = & 1\,K_1\Big]_{s=-1} & + & (s+1)\dfrac{K_2}{s+2}\Big]_{s=-1} \\
 & \dfrac{(-1+4)}{(-1+2)} & = & K_1 & + & 0 \\
 & \dfrac{3}{1} & = & K_1 = 3 & &
\end{array} \qquad \text{(E 11.17)}$$

To obtain K_2:

[1] multiply equation II and III by the DT_2 term in II $(s + 2)$ and
[2] evaluate both II and III using the "zero" of DT_2 (the term that, substituted for s, will make $DI_2 = 0$) which is –2. Thus,

$$\begin{array}{lccccc}
[1] & (\cancel{s+2})\dfrac{1\,(s+4)}{(s+1)(\cancel{s+2})}\Big]_{s=-2} & = & (s+2)\dfrac{K_1}{s+1}\Big]_{s=-2} & + & (\cancel{s+2})\dfrac{K_2}{\cancel{s+2}}\Big]_{s=-2} \\
[2] & \dfrac{(s+4)}{(s+1)}\Big]_{s=-2} & = & (s+2)\,K_1\Big]_{s=-2} & + & (\cancel{s+2})\dfrac{K_2}{\cancel{s+2}}\Big]_{s=-2} \\
 & \dfrac{(-2+4)}{(-2+1)} & = & 0 & + & K_2 \\
 & \dfrac{2}{-1} & = & K_2 \;= -2 \quad \text{thus,} & &
\end{array} \qquad \text{(E 11.18)}$$

$F(s) = V(s) = \dfrac{3}{s+1} - \dfrac{2}{s+2}$ and, using the Functional Transform No. 5,

$$v(s) = 3e^{-t} - 2e^{-2t} \qquad \text{(E 11.19)}$$

RULE for Partial Fraction Expansion:

To obtain ANY K_n:

[1] multiply the factored equation (II) and the Ks expansion (III) by the DT_n term and

[2] evaluate both the factored equation and the Ks equation by the zero of the DT_n term.

TM FOR LT RULE 5—THE INITIAL AND FINAL VALUES

One method of checking the assumptions and the calculations of Laplace Transform problems is to see if the values at the time of switching (t_{0+}) and at the time long after switching (t_∞) make physical sense. The times t_{0+} and t_∞ can be calculated in both the time and complex frequency domains. The mathematical rules for these two time values are the Initial and Final Value Theorems.

MM FOR LT RULE 5—THE INITIAL AND FINAL VALUE THEOREMS

The Initial Value Theorem:

$$\lim_{t\to 0} [\, f(t)\,] = f(t^0) = \lim_{s\to \infty} [\, s\,F(s)\,]$$

As an example, using the problem above,

$$\lim_{s\to \infty} [\, s\,F(s)\,] = \lim_{s\to \infty} \frac{1\,(s+4)}{(s+1)(s+2)} = \lim_{s\to \infty} \frac{\infty}{\infty 2} = \frac{1}{\infty} = 0$$

The Final Value Theorem:

$$\lim_{t\to \infty} [\, f(t)\,] = f(t_\infty) = \lim_{s\to 0} [\, s\,F(s)\,]$$

As an example, using the problem above,

$$\lim_{s\to 0} [\, s\,F(s)\,] = \lim_{s\to 0} \frac{1\,(s+4)}{(s+1)(s+2)} = 2$$

11.5 USING THE LAPLACE TRANSFORM FOR CIRCUITS WITH FIRST DEGREE EQUATIONS

TM LT ANALYSIS OF FIRST ORDER CIRCUITS

First order differential equations, in general, can be solved without the aid of the Laplace Transform. However, their simplicity makes them good practice examples for learning the procedure and rules of the LT method.

GM LT ANALYSIS OF FIRST ORDER CIRCUITS

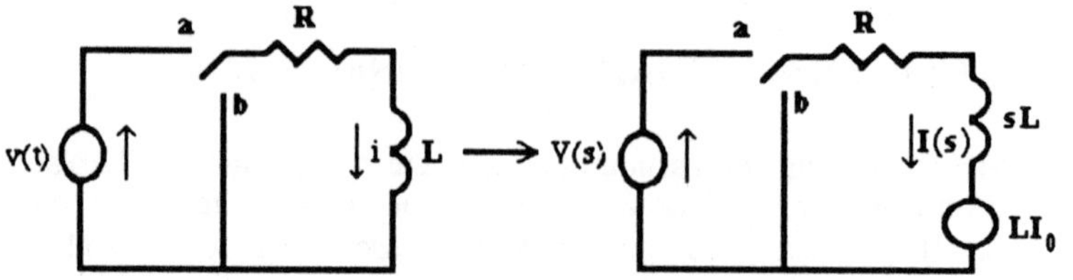

(a) The Time Domain Circuit

(b) The Equivalent s-Domain Circuit

FIGURE 11.1 The Series RL Circuit.

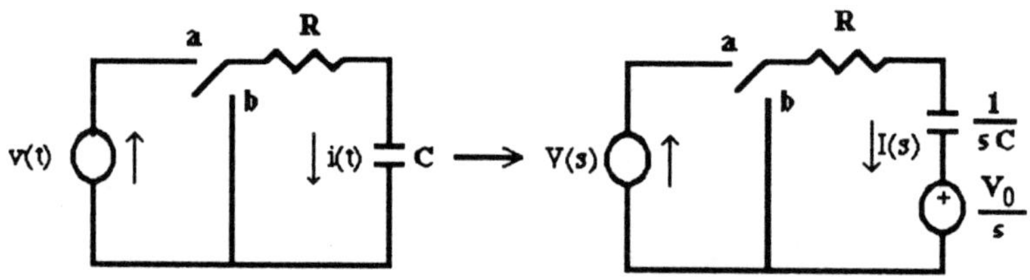

The Time Domain Circuit

The Equivalent s-Domain Circuit

FIGURE 11.2 The Series RC Circuit.

MM LT ANALYSIS OF FIRST ORDER CIRCUITS

For Series RL Circuits:

$v(t) = L\frac{di}{dt} + Ri\,(t)$ is the Kirchhoff Voltage Law (KVL) differential equation (E 11.20)

$V(s) = L\{sI(s) - i(0)\} + RI(s)$ is the equivalent $\mathcal{L}$s-domain equation where i(0) would be any existing inductor current when the switch is closed at t(0). (E 11.21)

$$I(s) = \frac{V(s)}{s(sL + R)} + \frac{LI(0)}{sL + R}$$

$$= \frac{V}{L}\left[\frac{1}{s(s+\alpha)} + \frac{I(0)}{(s+\alpha)}\right], \quad \text{where } \alpha = \frac{R}{L}, \text{ the time constant.} \qquad \text{(E 11.22)}$$

For Series RC Circuits:

$$v(t) = \frac{1}{C}\int_0^t i(t)dt + v_C(0) + Ri\,(t)$$ is the Kirchhoff Voltage Law (KVL) differential equation (E 11.23)

$$V(s) = \frac{1}{sC} I(s) + \frac{v_C(0)}{s} + RI(s)$$ is the equivalent $\mathcal{L}$s-domain equation where $v_C(0)$ would be any existing capacitor voltage when the switch is thrown at t(0).

$$= I(s)\ \{\frac{1}{sC} + R\} + \frac{v_C(0)}{s} \qquad \text{(E 11.24)}$$

$$I(s) = \frac{V(s)}{s}\,[\frac{1}{sC} + R\,] - \frac{V(0)}{s}\,[\frac{1}{sC} + R\,]$$ and next, dividing through by R

$$= \frac{V(s)}{R}\,[\,s + \frac{1}{RC}] - \frac{V(0)}{R}\,[\,s + \frac{1}{RC}]$$ and letting $a = \frac{1}{RC}$

$$= \frac{V(s)}{R}\,[\,s + a] - \frac{V(0)}{R}\,[\,s + a] \qquad \text{(E 11.25)}$$

EXAMPLE 11-1 A SERIES RL CIRCUIT

Find the series current in Figure 11.1 for the forcing functions to follow and by assuming the switch position is a function of time as stated in the different parts of the example. Use the basic equations listed in the series RL circuit **MM**.

{1} Find the current for a UNIT STEP INPUT v(t). Assume the switch has been at terminal b and is switched to terminal a at t = 0. Also, find the poles, the transfer function, and the circuit natural response.

Unfortunately, all the parts of E 11.22 are not correct for using the Inverse Laplace tables. It is now necessary to do a partial-fraction expansion of the first (non-I(0)) part. So,

$$I(s) = \frac{V}{L}\frac{1}{s(s+\alpha)} \qquad = \frac{K_1}{(s)} + \frac{K_2}{(s+\alpha)} \qquad \text{(E 11.26)}$$

$$[1]_{s=0}\ \frac{V}{L}\,s\,\frac{1}{s(s+\alpha)}\,]_{s=0} \qquad = s\,\frac{K_1}{(s)} + s\,\frac{K_2}{(s+\alpha)}\,]_{s=0} \qquad \text{(E 11.27)}$$

$$[2]_{s=0}\ \frac{V}{L}\frac{1}{(s+\alpha)}]_{s=0} \qquad = K_1 + 0\,]_{s=0}\,;\ K_1 = \frac{1}{\alpha} \qquad \text{(E 11.28)}$$

$$[1]_{s=-\alpha}\ \frac{V}{L}(s+\alpha)\frac{1}{s(s+\alpha)} \qquad = (s+\alpha)\frac{K_1}{(s)} + (s+\alpha)\frac{K_2}{(s+\alpha)} \qquad \text{(E 11.29)}$$

$$[2]_{s=-\alpha}\ \frac{V}{L}\frac{1}{s}]_{s=-\alpha} = 0 + K_2\,]_{s} \quad =_{-\alpha}\,;\ K_2 = -\frac{1}{\alpha}, \qquad \text{(E 11.30)}$$

Now, collecting ALL the terms and remembering that $\alpha = L/R$,

$$I(s) = \frac{V}{L}\frac{1}{s(s+\alpha)} = \frac{V}{L}\left[\frac{\frac{1}{\alpha}}{s} + \frac{-\frac{1}{\alpha}}{(s+\alpha)}\right] = \frac{V}{L} * \frac{L}{R} * \frac{1}{s} - \frac{V}{L} * \frac{L}{R} * \frac{1}{s+\alpha} + \left[\frac{I(0)}{s+\alpha}\right]$$

$$= \frac{V}{R}\left[\frac{1}{s} - \frac{1}{s+\alpha}\right] + \left[\frac{I(0)}{s+\alpha}\right]$$ and now , obtaining $\mathcal{L}^{-1}$ from the tables,

(E 11.31)

$$i(t) = \frac{V}{R}\left[\,1 - e^{-\alpha t}\,\right] + I(0)\,e^{-\alpha t} \qquad \text{(E 11.32)}$$

{2} The POLES, from E 11.15, 16, 31, etc., are at $s = 0$ and $s = -\alpha = -R/L$.

{3} The TRANSFER FUNCTION could be defined in several ways, including the output current/the input voltage or the output voltage across a component/the input voltage. Using the first definition,

$$H(s) = \frac{I(s)}{V(s)} = \frac{1}{L}\left[\frac{1}{s(s+\alpha)}\right] \qquad \text{(E 11.33)}$$

{4} The NATURAL RESPONSE of the circuit in Figure 11.1 can be found if it is assumed that the switch has been at terminal a and is switched to terminal b at time t_0.

$0 = L\frac{di}{dt} + Ri\,(t)$ is the Kirchhoff Voltage Low (KVL) differential equation(E 11.34)

$0 = L\{sI(s) - i(0)\} + RI(s)$ is the equivalent $\mathcal{L}$s-domain equation where i(0) would be any existing inductor current when the switch is thrown at t(0). (E 11.35)

$$L\,i(0) = LsI(s) + RI(s) \qquad \text{(E 11.36)}$$

$$I(s) = \frac{L\, I(0)}{sL + R} \text{ and next, dividing through by L,} \qquad \text{(E 11.37)}$$

$$= I(0) \frac{1}{(s + \frac{R}{L})} = I(0) \frac{1}{(s + \alpha)}, \text{ and, using the tables for the ILT,} \qquad \text{(E 11.38)}$$

$$i(t) = i(0)\, e^{-R/L\, t} = i(0)\, e^{-\alpha t} \qquad \text{(E 11.39)}$$

{5} The SINE WAVE RESPONSE of the circuit in Figure 11.1 can be calculated with the LT if the switch is assumed at position a. However, the Laplace Transform is most useful for finding the transient response. Although it is possible to use the LT to find the steady-state (sine/cosine) response, it is so involved and complex (both figuratively and literally), that it is rarely done either in circuit-problem solving or in the literature.

EXAMPLE 11-2 A SERIES RC CIRCUIT

Find the series current in Figure 11.2 for the forcing functions to follow and by assuming the switch position is a function of time as stated in the different parts of the example.

{1} Find the current for a UNIT STEP INPUT v(t). Assume the switch has been at terminal b and is switched to terminal a at t = 0. Also, find the poles, the transfer function, and the circuit natural response.

Using E 11.25, a partial fraction expansion is not necessary, since [s + a] is in the tables so,

$$i(t) = \frac{v(t)}{R} e^{-at} - \frac{v(0)}{R} e^{-at} \text{ and, combining and setting } a = \frac{1}{RC}$$

$$= \frac{v(t) - v(0)}{R} e^{-1/RC\, t} \qquad \text{(E 11.40)}$$

{2} The POLE, from E 11.25, is at $s = -a = -\frac{1}{RC}$.

{3} The TRANSFER FUNCTION is

$$H(s) = \frac{I(s)}{V(s)} = \frac{1}{R} [(s + a)] \qquad \text{(E 11.41)}$$

{4} The NATURAL RESPONSE, using the last part of E 11.40, is

$$i(t) = \frac{v(0)}{R} e^{-1/RC\,t} \qquad \text{(E 11.42)}$$

{2} Using the UNIT INPUT STEP, (a) derive a general equation for the voltage across the capacitor in Figure 11.2 and (b) show the use of this equation with a numerical example.

EXAMPLE 11-3 THE GENERAL EQUATION FOR THE CAPACITOR VOLTAGE IN A SERIES RC CIRCUIT

$$V_C(s) = I(s) * C(s) \qquad \text{(E 11.43)}$$

$$I(s) = \frac{E(s)}{R(s) + C(s)} \qquad \text{(E 11.44)}$$

$$I(s) = \frac{E(s)}{R + \frac{1}{Cs}} \qquad \text{(E 11.45)}$$

$$V_C(s) = I(s) * \frac{1}{Cs} \qquad \text{(E 11.46)}$$

$$= \frac{E(s)}{R + \frac{1}{Cs}} * \frac{1}{Cs} = E(s) \frac{s}{Rs + \frac{1}{C}} * \frac{1}{Cs} \quad \text{is the GENERAL } V_C \text{ equation}$$

for any E(s) input (E 11.47)

$$V_C(s) = \frac{E(s)}{s} * \frac{s}{Rs + \frac{1}{C}} * \frac{1}{Cs} \text{ is the } V_C \text{ equation for a UNIT STEP input}$$

(E 11.48)

$$= \frac{E(s)}{s} * \frac{\frac{s}{R}}{s + \frac{1}{RC}} * \frac{1}{Cs} = \frac{E}{RC} * \frac{1}{s(s + \frac{1}{RC})} = \frac{E}{RC} * \frac{1}{s(s+\alpha)} \qquad \text{(E 11.49)}$$

$$V(s) = \frac{E}{RC} * \frac{1}{s(s+\alpha)} = \frac{K_1}{s} + \frac{K_2}{s+\alpha} \text{ and, using the expansions from before,}$$

(E 11.50)

$$= \frac{E}{RC}\left[\frac{\frac{1}{\alpha}}{s} + \frac{-\frac{1}{\alpha}}{(s+\alpha)}\right] \text{ where } \alpha = \frac{1}{RC} \text{ and } \frac{1}{\alpha} = RC \qquad \text{(E 11.51)}$$

EXAMPLE 11-4

Using the above E 11.51, find $v_C(t)$ for a series RC circuit driven by a step of 10 volts with R = 5 ohms and C = 1 farad.

$$V_C(s) = \frac{10}{5*1}\left[\frac{5*1}{s} - \frac{5*1}{s+5}\right] = 10\left[\frac{1}{s} - \frac{1}{s+5}\right] \qquad \text{(E 11.52)}$$

$$V_C(s) = 10\,(1 - e^{-5t}) \qquad \text{(E 11.53)}$$

11.6 USING THE LAPLACE TRANSFORM FOR CIRCUITS WITH SECOND DEGREE EQUATIONS

TM LT ANALYSIS OF SECOND ORDER CIRCUITS

Second order systems usually are considered RLC resonant circuits, although combinations of simple RL or RC circuits often result in being second degree equations. In using the LT to solve resonant equations, the investigator must remember the concepts of over, critical and underdamping. The form of the Kirchhoff equations (including the possibility of complex terms) that describe these circuits is also important. As with first order circuits, the first requirement is to write the system Kirchhoff differential equations along with producing a diagram of the circuit.

GM LT ANALYSIS OF SECOND ORDER CIRCUITS

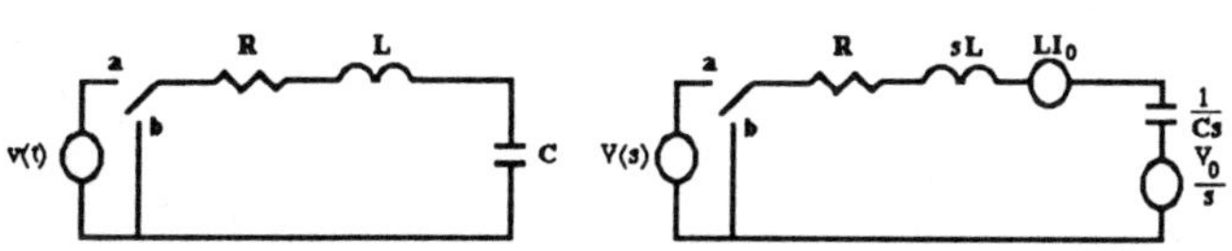

The Time Domain Circuit The Equivalent s-Domain Circuit

FIGURE 11.3 Series RLC circuits.

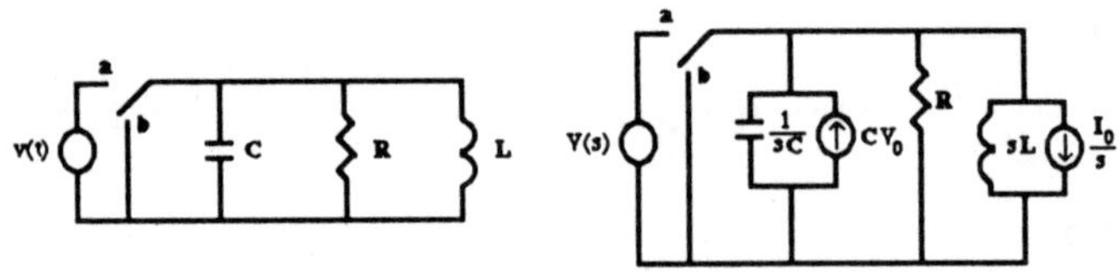

The Time Domain Circuit The Equivalent s-Domain Circuit

FIGURE 11.4 Parallel RLC circuits.

MM LT ANALYSIS OF SECOND ORDER CIRCUITS

For RLC Series Circuits:

$$v(t) = Ri + L\frac{di}{dt} + \frac{1}{C}\int_0^t i\, dt + v_C(0) \text{ is the series time equation} \quad \text{(E 11.54)}$$

$$V(s) = RI(s) + LsI(s) - LI(0) + \frac{1}{sC} I(s) + \frac{v_C(0)}{s} \text{ is the s-domain equation} \quad \text{(E 11.55)}$$

$$I(s) = \frac{1}{R + sL + \frac{1}{sC}} [V(s) + LI(0) - \frac{V_C(0)}{s}] \quad \text{(E 11.56)}$$

$$I(s) = \frac{V(s)}{R + sL + \frac{1}{sC}} \text{ if there is no stored charge or flux} \quad \text{(E 11.57)}$$

For RLC Parallel Circuits:

$$i(t) = i_C + i_R + i_L \quad \text{(E 11.58)}$$

$$i(t) = C\frac{dv}{dt} + \frac{V}{R} + iL \quad \text{(E 11.59)}$$

$$I(s) = sCV - CV(0) + \frac{V}{R} + \frac{V}{sL} + \frac{I(0)}{s} \quad \text{(E 11.60)}$$

$$I(s) = sCV + \frac{V}{R} + \frac{V}{sL} \quad \text{if there is no stored charge or flux} \quad \text{(E 11.61)}$$

$$V(s) = \frac{\frac{I}{C}}{s^2 + \frac{1}{RC}s + \frac{1}{LC}} = \text{the voltage across all components}$$

in the parallel circuit (E 11.62)

$$I_L(s) = \frac{\frac{I}{LC}}{s[s^2 + \frac{1}{RC}s + \frac{1}{LC}]} \quad \text{is the current in the inductor} \qquad \text{(E 11.63)}$$

Special Note: The following examples are used to illustrate the use of the LT in second order circuits. To facilitate understanding, especially when complex numbers are involved, the component values have been simplified and do not necessarily represent practical values.

EXAMPLE 11-5

Using the following circuit, find the s-domain expression for I and the time domain for i for different values of R.

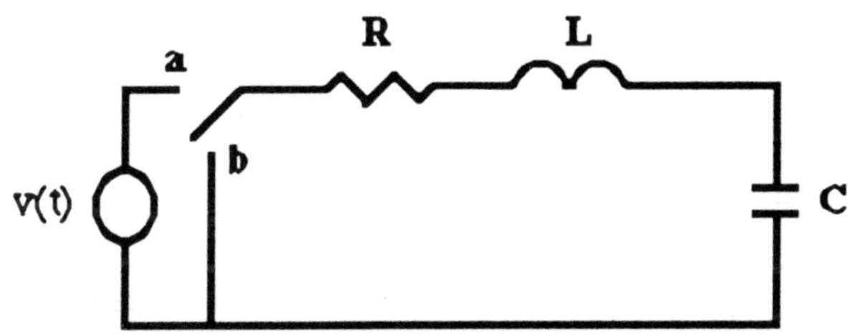

FIGURE 11.5 The circuit for Example 11-5.

{1} The switch has been at b for a long time and is moved to a at t(0). Find I(s), i(t), V_C and v_C for an input pulse of 10 with component values of R = 10 ohms, L = 1 henrys and C = 0.04 farads.

$$I(s) = \frac{V(s)}{R + sL + \frac{1}{sC}} \quad \text{and, for an input of } V(s) = 10/s, \qquad \text{(E 11.64)}$$

$$= \frac{\frac{V}{L}}{s^2 + \frac{R}{L}s + \frac{1}{LC}} \qquad \text{(E 11.65)}$$

$$= \frac{10}{s^2 + 10\,s + 25}$$ this denominator can be factored by inspection (E 11.66)

$$= \frac{10}{(s+5)\,(s+5)}$$ indicating the poles are real and equal for critical damping (E 11.67)

Some tables give F(s) in this form; however, $\frac{10}{(s+5)^2}$ is more common, thus

$i(t) = 10\,t\,e^{-5t}$ which, since the poles are real and equal, is a critically damped circuit. (E 11.67b)

$$V_C = I(s) * sL = \frac{10s}{(s+5)^2} \qquad \text{(E 11.68)}$$

$$v_C = 10\,e^{-5t}\,(1 - 5t) \qquad \text{(E 11.69)}$$

{2} Repeat the above calculations with R = 20 ohms.

$$I(s) = \frac{10}{s^2 + 20\,s + 25} \qquad \text{(E 11.70)}$$

$$= \frac{10}{(s+1.34)\,(s+18.66)}$$ real and distinct roots indicate overdamping (E 11.71)

First, the problem will be solved using partial fraction expansion.

$$I(s) = \frac{10}{(s+1.34)\,(s+18.66)} = \frac{K_1}{(s+1.34)} + \frac{K_2}{(s+18.66)} \qquad \text{(E 11.72)}$$

$$= \frac{(s+1.34)10}{(s+1.34)\,(s+18.66)}\Big|_{s=-1.34} = \frac{(s+1.34)K_1}{(s+1.34)} + \frac{(s+1.34)K_2}{(s+18.66)} \qquad \text{(E 11.73)}$$

$$= \frac{10}{17.32} = K_1 = 0.557 + 0 \quad \text{(E 11.74)}$$

$$= \frac{(s + 18.66)10}{(s + 1.34)\,(s + 18.66)}\Big|_{s = -18.66} = \frac{(s + 18.66)K_1}{(s + 1.34)} + \frac{(s + 18.66)K_2}{(s + 18.66)} \quad \text{(E 11.75)}$$

$$= \frac{10}{-17.32} = 0 + K_2 = -0.577 \quad \text{(E 11.76)}$$

$$I(s) = \frac{10}{(s + 1.34)\,(s + 18.66)} = \frac{0.577}{(s + 1.34)} - \frac{0.577}{(s + 18.66)} \quad \text{(E 11.77)}$$

$$i(t) = 0.577\,[e^{-1.34t} - e^{-18.66t}]$$

Second, the problem will be solved using Laplace tables.

The tables show the form $f(t) = \frac{1}{(b - a)}\,[e^{-at} - e^{-bt}]$; therefore,

$$i(t) = \frac{10}{17.32}\,[e^{-1.34t} - e^{-18.66t}]$$

$$= 0.557\,[e^{-1.34t} - e^{-18.66t}] \quad \text{(E 11.78)}$$

$$V_C = I(s) * sL = \frac{10s}{(s + 1.34)\,(s + 18.66)} \quad \text{(E 11.79)}$$

From the tables,

$$f(t) = \frac{1}{(a - b)}\,[a\,e^{-at} - b\,e^{-bt}] \text{ and, therefore,}$$

$$= \frac{1}{17.32}\,[1.34\,e^{-1.34t} - 18.66\,e^{-18.66t}]$$

$$v_C = 0.0576\,[1.34\,e^{-1.34t} - 18.66\,e^{-18.66t}] \quad \text{(E 11.80)}$$

{3} Repeat the above calculations with R = 6 ohms.

$I(s) = \frac{10}{s^2 + 6s + 25}$ and, factoring by inspection or by using the quadratic formula,

$= \frac{10}{(s+3-j4)(s+3+j4)}$ With complex factors, the system is underdamped. (E 11.81)

Using partial fraction expansion:

$$I(s) = \frac{10}{(s+3-j4)(s+3+j4)} = \frac{K_1}{(s+3-j4)} + \frac{K_2}{(s+3+j4)}$$

$$= \frac{\cancel{(s+3-j4)}\,10}{\cancel{(s+3-j4)}(s+3+j4)}\Big|_{s=-3+j4} = \frac{\cancel{(s+3-j4)}\,K_1}{\cancel{(s+3-j4)}} + \frac{(s+3-j4)\,K_2}{(s+3+j4)}$$

$$\frac{10}{j8} = K_1 = \frac{10}{j8} + 0$$

$$I(s) = \frac{10}{(s+3-j4)(s+3+j4)} = \frac{K_1}{(s+3-j4)} + \frac{K_2}{(s+3+j4)}$$

$$= \frac{\cancel{(s+3+j4)}\,10}{(s+3-j4)\cancel{(s+3+j4)}}\Big|_{s=-3-j4} = \frac{(s+3+j4)\,K_1}{(s+3-j4)} + \frac{\cancel{(s+3+j4)}\,K_2}{\cancel{(s+3+j4)}}$$

$$\frac{10}{-j8} = 0 + K_2 = \frac{10}{-j8}$$

$$I(s) = \frac{10}{(s+3-j4)(s+3+j4)} = \frac{-j1.2}{(s+3-j4)} + \frac{j1.2}{(s+3+j4)}$$

$$I(s) = \frac{10}{(s+3-j4)(s+3+j4)} = \frac{1.2\angle -90^0}{(s+3-j4)} + \frac{1.2\angle 90^0}{(s+3+j4)}$$

Using the partial fraction transform pairs (No. 3),

$$i(t) = 2 * 1.2\, e^{-3t} \cos(4t + 90^0)$$

11.7 THE IMPULSE FUNCTION—ITS THEORETICAL AND PRACTICAL USES

The impulse has theoretical implications in both the time and frequency domains. Although an impulse is defined as a theoretical pulse that has zero width and infinite height, in the real world, switching "spikes" occur that very closely resemble the unit impulse. In fact, an understanding of these very short transients is essential for the design and maintenance of reliable integrated circuits. Likewise, a form of this function, very near to the theoretical ideal, can be used as a very practical, useful tool in circuit analysis and testing.

TM THE IMPULSE FUNCTION IN THE TIME DOMAIN

A simple impulse is defined as a pulse with zero width and infinite height. If the area of the pulse (width times height) is defined as unity or one, it is called a unit impulse or a Dirac/Delta function. The effect of an impulse on a circuit is to "kick" it (in time) but not to disturb any stored energy in any of the circuit elements. The output response to this unit impulse is the *natural response* of the circuit! It is pertinent to note that this is somewhat in conflict with the conventional concept of the natural response, which usually assumes that the natural response does NOT include a forcing function. The unit impulse can exist at t(0) or can be defined at a time (usually) after t(0).

MM THE IMPULSE FUNCTION IN THE TIME DOMAIN

$\delta(t) = 1$ at $t = 0$ and $= 0$ at $t \neq 0$. (E 11.82)

$\delta(t - a) = 1$ at $t = a$ and $= 0$ at $t \neq a$ where δ is delayed a seconds from t(0). (E 11.83)

GM THE IMPULSE FUNCTION IN THE TIME DOMAIN

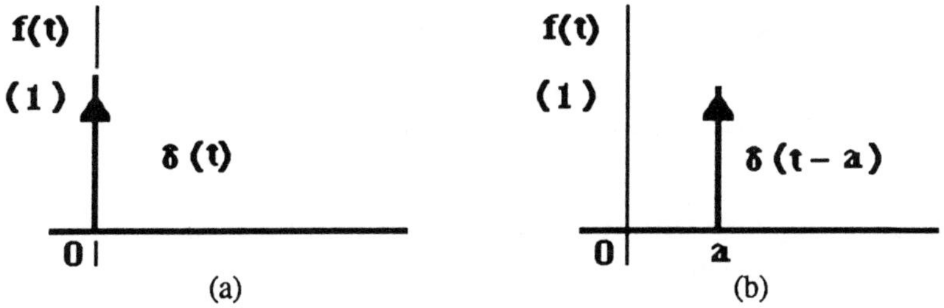

FIGURE 11.6 A unit impulse at t = 0 and t = a.

TM THE IMPULSE FUNCTION IN THE FREQUENCY DOMAIN

Chapter 10 discussed the frequency spectrum of square waves and narrow pulses. As a unit height pulse gets narrower, the number of harmonics—the spectrum—increases. In the limit, if the pulse is infinitely narrow, the spectrum will be infinite. This is another way of looking at the use of a unit pulse as a circuit-analysis tool. Whether it is a theoretically zero width pulse or a very, very narrow pulse, the spectrum is very large. This means that, at least in theory, a unit impulse will provide all possible input frequencies to a circuit and thus can be a test of its frequency (and phase) response.

MM THE IMPULSE FUNCTION IN THE FREQUENCY DOMAIN

From Chapter 10, c_n is the amplitude of the nth harmonic in the spectrum of a pulse. As the pulses get quite narrow, the amplitudes of the first twenty harmonic frequencies (see Table 10.2) are all over 93% of the fundamental. When the pulse width approaches zero, the amplitudes all approach each other.

$$c_n = V_m \frac{\sin n\omega_0\tau/2}{n\omega_0\tau/2} = \text{the form of } 2V_m \frac{\sin nx}{nx} \text{ where n = the harmonic of } \omega_0 \qquad \text{(E 10.33)}$$

EXAMPLE 11-6

Compare the (natural) response of a series RL circuit by {1} switching to terminal b at t(0) in Figure 11.1 (a) and by {2} forcing the unfluxed series RL circuit with a unit impulse.

{1} Figure 6.4, copied below, shows the wave form and the equations for the natural response of a series RC and RL circuit that was obtained by switching. For (b), the (switching) natural response is given as $i(t) = i_0\, e^{-t\, R/L}$.

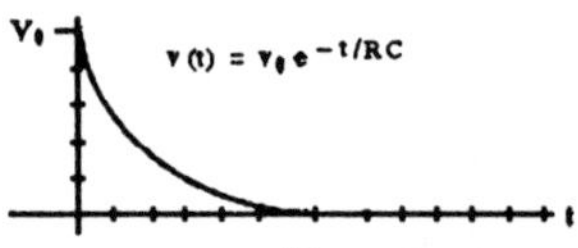

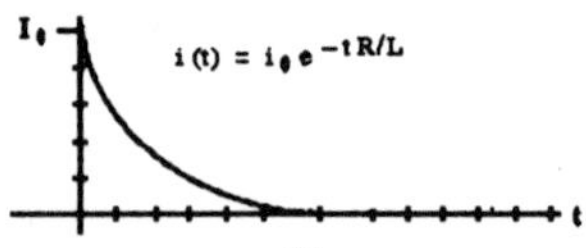

{2} Derive the (natural) response of a series RC circuit, with no stored energy, by forcing it with a unit impulse.

$$L\, i(0) = LsI(s) + RI(s)$$

$I(s) = \frac{L\,I(0)}{sL + R}$ and next, dividing through by L,

$= I(0)\frac{1}{(s + \frac{R}{L})} = I(0)\frac{1}{(s + \alpha)}$, and, using the tables for the ILT,

$I(s) = \frac{V(s)}{R}$ [s + a] See E 11.25 and assume $\delta(s) = V(s) = 1$.

Using the LT tables:

$= I(0)\frac{1}{(s + \frac{R}{L})} = I(0)\frac{1}{(s + \alpha)}$, and, using the tables for the ILT,

$i(t) = i(0)\,e^{-R/L\,t} = i(0)\,e^{-\alpha t}$

Special Note: The unit-impulse response of a system or circuit is often given the special name h(t).

11.8 THE USE OF CONVOLUTION IN LAPLACE TRANSFORM ANALYSIS

TM THE PRINCIPLE OF CONVOLUTION

Convolution, and its convolution integral, is a rather complicated mathematical concept that can be used in a straightforward way to aid in solving a number of circuit analysis problems, including Laplace Transforms. In general, the technique of convolution is based upon the idea that any waveform, whether it be some very jagged and almost indefinable function or a simple pulse or sine wave, can be broken into a series of impulses. All of these impulses can, in turn, be combined (convoluted) to produce a definable function in the s-domain that then can be manipulated and/or transformed to the time-domain. Convolution, like wave-form analysis, can be done graphically as well as mathematically. Although it involves much more than the Laplace Transform, the following rules are guidelines for convolution.

RULE 1: Convolution in the time domain is the equivalent of multiplication in the frequency domain!

RULE 2: The Laplace Transform of the impulse response of a circuit or network is the same as the transfer function of that circuit or network.

RULE 3: The output response R(t)—or $v(t)_{out}$, etc.—for a circuit or network is equal to the (convolution) integral of the forcing function times the impulse response.

MM THE PRINCIPLE OF CONVOLUTION

The Convolution Integral:

$$y(t) = \int_0^t f(\tau)\, h(t-\tau)\, d\tau \qquad \text{(E 11.84)}$$

$$= f(x) * h(t) = f(t) * h(t) \qquad \text{(E 11.85)}$$

where f(x) is the symbol used for the arbitrary input function, h(t) is the integral of the impulses, and the asterisk (∗) is conventionally used as the symbol for the convolution. (The large asterisk [∗] is used in this book for clarity.)

GM THE PRINCIPLE OF CONVOLUTION

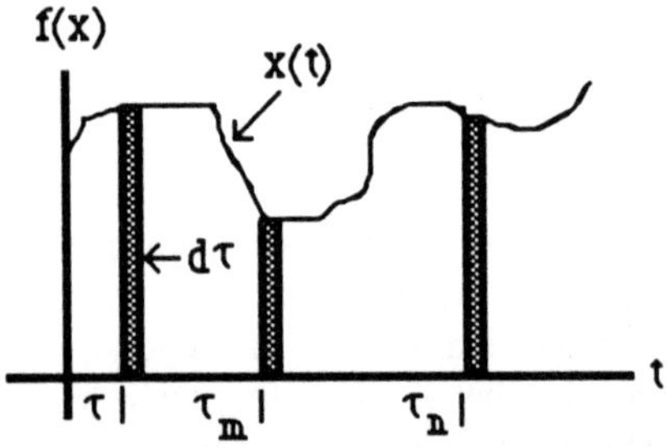

FIGURE 11.7 The Principle of Convolution. Three of an infinite number of impulses are shown. All dτ ⇒ 0.

EXAMPLE 11-7A

Determine the output of a series RC network with an input forcing function u(t) using the convolution integral.

$$H(s) = \frac{V_{out}}{V_{in}} = \frac{1/sC}{R + 1/sC} = \alpha * \frac{1}{s+\alpha} \quad \text{where, as usual, } \alpha = \frac{1}{RC}.$$

From the tables for a u(t) input,

$$h(t) = \alpha\, e^{-\alpha t}$$

By RULE 3, the time response is equal to the convolution of the impulse response and the function. Using E 11.85,

$$v(t) = v_{out} = \int_0^t f(\tau)\, h(t-\tau)\, d\tau$$

$$= \int_0^t u(t-\tau)\, h(\tau)\, d\tau$$

$$= \int_0^t \alpha\, e^{-\alpha\tau}\, d\tau = [-e^{-\alpha t}]_0^t$$

$$v(t) = 1 - e^{-\alpha t}$$

EXAMPLE 11-7B

Determine the output of a series RC network with an input forcing function of cos ωt using the convolution integral.

Special Note: The following derivation is one of the few examples in this book that deviates from the established practice of explaining the problem, line by line. This one example is designed to show the form of using the convolution integral rather than all the calculus details.

$$v(t) = \int_0^t u(t-\tau)\, h(\tau)\, d\tau = \alpha \int_0^t [\cos \omega(t-\tau)]\, e^{-\alpha\tau}\, d\tau \qquad [A]$$

$$= \alpha \int_0^t [\cos \omega(t-\tau)]\, e^{-\alpha\tau}\, d\tau \qquad [B]$$

$$= \alpha \int_0^t [\cos \omega t \cos \omega\tau]\, e^{-\alpha\tau}\, d\tau + \alpha \int_0^t [\sin \omega t \sin \omega\tau]\, e^{-\alpha\tau}\, d\tau \qquad [C]$$

$$= \alpha \cos \omega t \int_0^t [\cos \omega\tau]\, e^{-\alpha\tau}\, d\tau + \alpha \sin \omega t \int_0^t [\sin \omega\tau]\, e^{-\alpha\tau}\, d\tau \qquad [D]$$

From a table of integrals:

$$\int_0^t [\cos \omega\tau]\, e^{-\alpha\tau}\, d\tau = \frac{e^{-\alpha t}[\omega \sin \omega t - \alpha \cos \omega t] + \alpha}{\alpha^2 + \omega^2} \qquad [E]$$

$$\text{and, } \int_0^t [\sin \omega\tau]\, e^{-\alpha\tau}\, d\tau = \frac{e^{-\alpha t}[-\alpha \sin \omega t - \omega \cos \omega t] + \omega}{\alpha^2 + \omega^2} \qquad [F]$$

Using form of equation [D]:

$$v(t) = \frac{\alpha^2}{\alpha^2 + \omega^2} \cos \omega t + \frac{\omega\alpha}{\alpha^2 + \omega^2} \sin \omega t + \frac{\alpha}{\alpha^2 + \omega^2} e^{-\alpha t}$$
$$[\cos \omega t\, (\omega \sin \omega t - \alpha \cos \omega t) - \sin \omega t\, (\alpha \sin \omega t + \omega \cos \omega t)\,] \qquad [G]$$

Now, simplifying and combining,

$$\frac{\alpha}{\alpha^2 + \omega^2} \cos \omega t + \frac{\omega}{\alpha^2 + \omega^2} \sin \omega t = \frac{1}{\sqrt{\alpha^2 + \omega^2}} (\cos \omega t - \tan^{-1} \frac{\alpha}{\omega}) \qquad [H]$$

and,

$$\cos \omega t\, (\omega \sin \omega t - \alpha \cos \omega t) - \sin \omega t\, (\alpha \sin \omega t + \omega \cos \omega t\,) = -\alpha \text{, so} \qquad [I]$$

$$v(t) = \frac{\alpha}{\sqrt{\alpha^2 + \omega^2}} (\cos \omega t - \tan^{-1} \frac{\alpha}{\omega}) - \frac{\alpha^2}{\alpha^2 + \omega^2} e^{-\alpha t} \qquad [J]$$

In equation [J], the first term is the steady-state term and the second term is the transient term.

12

Spectrum Analysis: The Fourier Series and Transforms

12.1 AN INTRODUCTION TO THE CONTINUOUS FOURIER TRANSFORM

This chapter extends the methods of circuit analysis based upon the work of Jean Baptiste Joseph Fourier (1768–1830) from the analog world into the digital (sampled data) world. The following is not a complete outline of the chapter but a general list of the Fourier topics that are reviewed or introduced.

1. THE CLASSICAL (CONTINUOUS) FOURIER SERIES—The spectral components that make up a continuous (analog), periodic signal.
2. THE CLASSICAL (CONTINUOUS) FOURIER INTEGRAL—The spectral components that make up a continuous (analog) but aperiodic signal.
3. THE DISCRETE-TIME (SAMPLED) FOURIER SERIES—The spectrum of a periodic, sampled signal.
4. THE DISCRETE-TIME (SAMPLED) FOURIER TRANSFORM (DFT)—The spectrum of an aperiodic, sampled signal.
5. THE FAST FOURIER TRANSFORM (FFT)—Methods of rapidly doing the calculations required for the DFT.

12.2 THE CLASSICAL (CONTINUOUS) FOURIER SERIES IN REVIEW—WHAT MAKES UP A CONTINUOUS (ANALOG), PERIODIC SIGNAL

The classical Fourier Series, introduced in Chapter 10, can represent almost any periodic wave form with a sum of sine and cosine functions. The trigonometric Fourier series is composed of sin and cos terms, that represent the sinusoid at the fundamental repetition rate (period) and the related harmonics—usually presented as functions of $n\omega_0$. The complex form is presented with Euler equivalents as a function of $e^{jn\omega}$. In the trigonometric form, the spectrum of a periodic wave form will be: $f(t) = a_0 + a_1 \cos \omega_0 t + a_2 \cos 2\omega_0 t + \cdots \cdots + b_1 \sin \omega_0 t + b_2 \sin 2\omega_0 t + \cdots$. Most computer analysis programs are capable of presenting the Fourier components as real and imaginary or as amplitude and phase.

12.3 THE CLASSICAL (CONTINUOUS) FOURIER INTEGRAL EXTENSION OF THE FOURIER SERIES—THE SPECTRAL COMPONENTS THAT MAKE UP A CONTINUOUS (ANALOG) BUT APERIODIC SIGNAL

The Fourier Integral spectrum of a continuous but aperiodic (analog) signal is a unique extension of the classical Fourier Series. Whereas the Fourier Series is limited to decomposing periodic wave forms into sinusoidal components, it can be extended, using a "trick" of conceptual mathematics, to include non-periodic or single wave forms. This extension of the Fourier Series is called the continuous Fourier Transform or, simply, the Fourier Transform (FT). The key image used here for the FT, as it was for the Fourier Series, is the spectrum.

The Fourier Transform is the technique, using the Fourier Integral, for decomposing a non-periodic pulse, in the time domain, to its spectral components in the frequency domain. The Fourier Transform and the Fourier Integral refer to the same general concept and technique. The Fourier Integral is based upon the idea, introduced in Chapter 10, of continually reducing the width of a pulse. In the limit, this width can go to zero. This zero-width pulse will, again in the limit, contain an infinite spectrum with all frequencies at equal amplitudes. The Fourier Transform uses a form of the Fourier Integral to obtain the spectrum of a non-periodic signal by changing—transforming—an equation from the time domain to the frequency domain.

TM THE CLASSICAL FOURIER TRANSFORM (FT)

The FT is derived by conceptually extending the time (T), between successive waves in a periodic function, to infinity. Thus, a non-periodic wave becomes a periodic wave where the "next" wave form never happens. The FT and its inverse (IFT), like the Laplace Transform, can be implemented by deriving functional transforms or by table look up. The frequencies in a FT spectrum of a wave form can be specified as real and imaginary or by amplitude and phase.

The Fourier Transform and the Fourier Integral are often used interchangeably to define the same principle and technique.

MM THE FOURIER TRANSFORM

(a) In review, the Trigonometric Fourier Series:

For a background, from Chapter 10, a review of the definitions of the (continuous) trigonometric Fourier Series:

$$f(t) = \frac{1}{2} a_0 + \sum_{n=1}^{\infty} a_n \cos n\omega_0 t + \sum_{n=1}^{\infty} b_n \sin n\omega_0 t$$ adapted from (E 10.10)

$$a_0 = \frac{1}{T} \int_0^{T=2\pi} f(t)\, dt$$ = the average value of the wave form

adapted from (E 10.13)

$$a_n = \frac{2}{T} \int_0^{T=2\pi} f(t) \cos n\omega_0 t\, dt \quad n = 0, 1, 2, 3, \cdots$$ adapted from (E 10.14)

$$b_n = \frac{2}{T} \int_0^{T=2\pi} f(t) \sin n\omega_0 t\, dt \quad n = 1, 2, 3, \cdots$$ adapted from (E 10.15)

Note that the above integrals start from zero.

(b) The Exponential or Complex Form of this (Continuous) Fourier Series:

$$f(t) = a_0 + \sum_{n=1}^{\infty} \frac{a_n}{2} [(e^{jn\omega_0 t} + e^{-jn\omega_0 t}) + \frac{b_n}{j2} (e^{jn\omega_0 t} - e^{-jn\omega_0 t})]$$

$$= c_0 + \sum_{n=-\infty}^{\infty} c_n e^{jn\omega_0 t}$$ adapted from (E 10.30)

$$c_n = \frac{1}{T}\int_{t_0}^{t_0+T} f(t)\, e^{-jn\omega_0 t}\, dt$$

$$= V_m \frac{\sin n\omega_0\tau/2}{n\omega_0\tau/2} = \text{the form of } 2V_m \frac{\sin nx}{nx} \quad n = \text{the harmonic of } \omega_0$$

from (E 10.31) and (E 10.33)

$$c_0 = \frac{1}{T}\int_{t_0}^{t_0+T} f(t)\, dt = a_0$$ adapted from (E 10.32)

(c) The Fourier Integral (f) in its Trigonometric Form:

$$f\{f(t)\} = \int_{-\infty}^{\infty} A(\omega) \cos \omega t\, dt + \int_{-\infty}^{\infty} B(\omega) \sin \omega t\, dt$$

(this is also somtimes termed X(ω)) (E 12.1)

Note that the above integrals start from $-\infty$.

And, the Fourier Integral in its exponential form:

$$f\{f(t)\} = \sum_{n=-\infty}^{\infty} C_n e^{jn\omega_0 t} \qquad \text{(E 12.2)}$$

$$C_n = \frac{1}{T}\int_{t_0}^{t_0+T} f(t)\, e^{jn\omega_0 t}\, dt \qquad \text{(E 12.3)}$$

$$f\{f(t)\} = X(\omega) = \int_{-\infty}^{\infty} x(t)\, e^{-j\omega t}\, dt \text{ is the FT definition} \qquad \text{(E 12.4)}$$

$$f^{-1} X(\omega) = \frac{1}{2\pi}\int_{-\infty}^{\infty} F(\omega)\, e^{j\omega t}\, dt \text{ is the IFT definition} \qquad \text{(E 12.5)}$$

GM THE FOURIER TRANSFORM

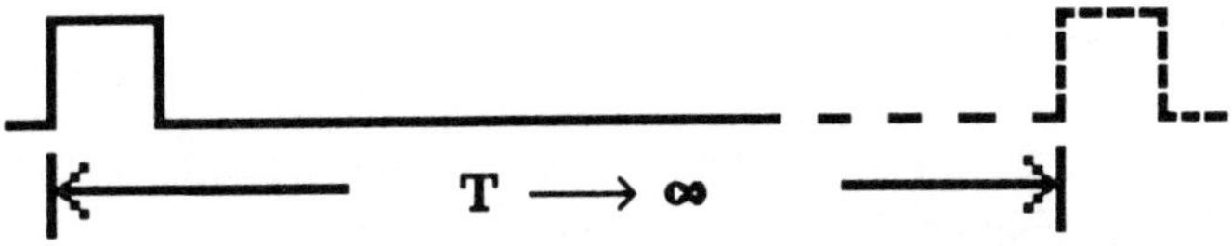

FIGURE 12.1 The idea of a periodic pulse as the repetition time goes to infinity.

12.4 THE ATTRIBUTES OF THE FOURIER INTEGRAL/FOURIER TRANSFORM

The Fourier Integral is an extremely powerful mathematical technique for working with circuit transient problems as well as spectrum analysis. It has been used very successfully for the study of many electrical communication problems, including television and radar. THE landmark paper on this subject was T. Murakami and M. S. Corrington, "Applications of the Fourier Integral in The Analysis of Color TV Systems," *IRE Trans. on Circuit Theory*, September 1955, Vol. CT-2, No. 3. Unfortunately, the integrals are usually quite involved and require a good understanding of both the philosophy and working techniques of higher mathematics. The Fourier Integral was introduced here, not necessarily as a working tool for everyday circuit analysis, but as background for the Discrete and the Fast Fourier Transform.

12.5 AN INTRODUCTION TO SAMPLED DATA

For the appropriate background, it is now necessary to introduce some of the concepts of sampled data.

Just as it is important to find ways to investigate the transient response and the spectrum of continuous signals, it is also important to study these features of sampled signals. Sampled signals exist in analog circuits as well as digital circuits. In common amplitude modulation, as used in broadcast radio, the carrier is CONTINUALLY sampling the input speech or music. In digital communication, the sampling is discrete; i.e., it is NOT CONTINUOUS but takes place only at separate and distinct units of time. This, of course, requires somewhat different analysis techniques than continuous sampling.

TM SAMPLING AND QUANTIZING ANALOG SIGNALS

ANALOG TO DIGITAL

Digital sampling is the process of "taking a quick look" and capturing a short segment of the signal (amplitude) for processing. A similar process does happen in the analog domain, but in this chapter it will be assumed that the *sampling* is a digital method or process. Although the sampling process is now assumed to be exclusively digital, the signal that is sampled may be analog or digital. Most often, when the sampled input signal is analog, the sampling process will include the conversion of the analog amplitude to a digital number (it is *quantized*). Thus, after the sampling and usually an almost instantaneous conversion process, the continuous analog signal is represented by a series of discrete digital numbers.

Any sampling process, be it analog or digital, must obey the famous rules of Shannon and Nyquist. These rules state that, to get a faithful, distortionless (sampling distortion is called *aliasing*) sample of a continuous, periodic signal, the sampling rate (frequency) must be AT LEAST TWICE the rate (frequency) of the highest frequency in the signal. Most often, the sampling rate is much higher than the minimum value. This sampling theorem is one of the most important concepts in electrical communication.

MM SAMPLING AND QUANTIZING

$R_{sampling} \geq 2\ R_{max\ sig}$ where R_{sample} is the sampling rate and $R_{max\ sig}$ is the maximum rate of any signal to be sampled. (E 12.6)

GM SAMPLING AND DIGITIZING

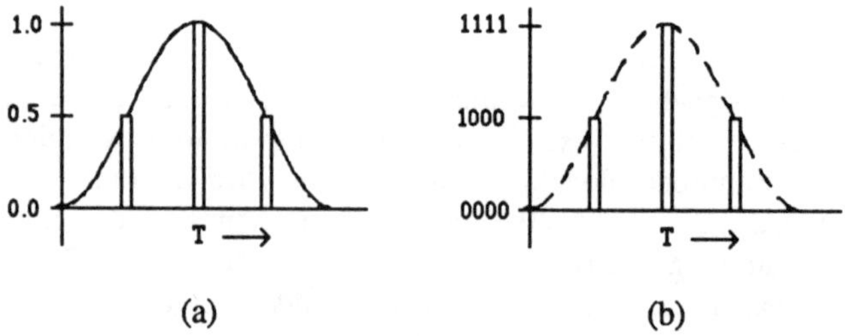

FIGURE 12.2 An example of digital sampling: (a) A half-wave analog signal that is being sampled. Note that if the peak amplitude value is 1.0, the wave shape is sampled where the amplitudes are 0.0, 0.5, 1.0, 0.5, and 0.0. (There would be eight samples for a complete full wave.); (b) In the digital domain, the corresponding digital numbers, using a 4-bit code, would be 0000, 1000, 1111, 1000, and 0000.

12.6 THE ANALYSIS OF DISCRETE-TIME (DT) SIGNALS AND PROCESSING SYSTEMS

TM THE FORMAT OF DISCRETE SIGNALS AND PROCESSED DATA

By definition, data in discrete-sampled signals come in specific clusters or units. In Figure 12.2(b), these discrete clusters (digital numbers) resulted from sampling an analog signal. There are also signals that originate as discrete-time clusters. For example, the hourly temperature or barometric pressure or the hourly or daily price of a stock represent discrete-signal data. Thus, the pressure might be read at 1 a.m., 2 a.m., 3 a.m., etc., throughout the day. When this is put in discrete format, the first reading would be the 0th item, the next the 1th item, the next the 2th item, etc. Thus, each item is delayed one unit of time (T), which means that the first reading would be p_0 or p(n), the second would be p(n–1), the next p(n–2), . . .

Often, there is a need to get an average of the discrete-data series, such as the average temperature or price for the day or month. As with any average calculation, sum the series and divide by the total number (N) of items in the series.

Sometimes discrete-data are captured or calculated on the basis of the variation from a "bogey" or standard number. Thus, each discrete term will have a "k" (the symbol is different in different books) that changes the amplitude from the standard or from the "1" of a discrete unit pulse. Some references call the k the *weighting factor* and the delay number (n-?) the *shifting factor* or shifting property.

The phrase "*Difference Equation*" is used to describe the series of weighted and shifted discrete terms that mathematically define a discrete signal or a processor output.

Special Note: Horizontal axis polarity rules for digital-signal figures.

1. If the digital figure has an abscissa scale that is only used to illustrate "n" digital samples, that scale is positive to the right of any zero or other reference point and negative to the left of the reference point (see Figure 12.4).
2. If the digital figure has an abscissa scale that is used to illustrate time or time delay, the scale is negative to the right of any zero or other reference point and positive to the left of the reference point (see Figure 12.3).

These two rules are consistent with most texts; for instance, see the Lynn and Fuerst and the Strum and Kirk references.

MM THE FORMAT OF DISCRETE-SIGNAL DATA OR, SIMPLY, THE DISCRETE-SIGNAL

The Discrete-Signal (Data) Format using the [n-?] term:

$$y[n] = x[n] + x[n-1] + x[n-2] + x[n-3] + \cdots \quad \text{(Note that } x[n-0] = x[n]\text{)} \qquad \text{(E 12.7)}$$

The Average of Discrete Signal:

$$y[n]_{Ave} = \frac{1}{N} x[n] + x[n-1] + x[n-2] + x[n-3] + \cdots \quad \text{with N terms in the series} \qquad \text{(E 12.8)}$$

The use of "k" weighting with a unit impulse = x[k]δ[n]:

$$y[n] = x[5]\delta[n] + x[4]\delta[n-1] + x[2]\delta[n-2] + x[0]\delta[n-3] + x[-2]\delta[n-2] \cdots \qquad \text{(E 12.9)}$$

GM THE FORMAT OF DISCRETE-SIGNAL DATA

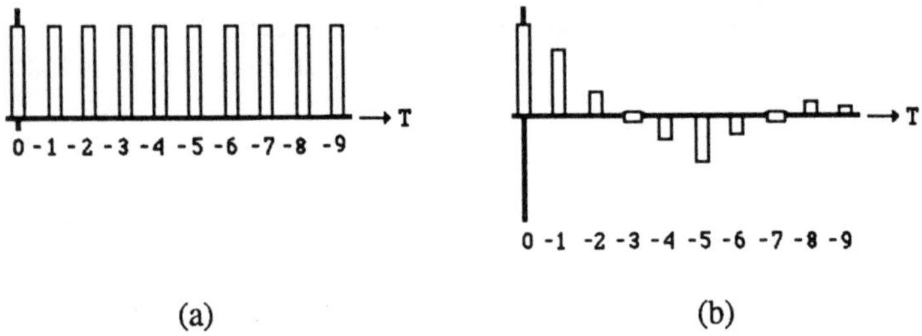

FIGURE 12.3 (a) Uniform discrete digital data, in the form of E 12.7, that might represent a step function; (b) Weighted discrete data, in the form of E 12.9, that might represent a decaying sinusoid. It should be remembered that each rectangle is a symbol for a coded digital number.

12.7 THE DISCRETE TIME-DOMAIN ANALYSIS OF INPUT SIGNALS

The two most important analysis signals in the discrete-time domain are the step function and the unit impulse. As with continuous signals, the step and impulse function permit analysis in both the (discrete) time and frequency domains.

Likewise, many other discrete input signals can be structured by adding together a series of unit impulse signals.

TM THE DISCRETE-TIME (DT) IMPULSE AND THE DISCRETE-TIME STEP

The DT unit impulse is similar, but not exactly the same, as its continuous counterpart. The digital-domain unit impulse, δ[n], is a digital number, usually with a value corresponding to the binary equivalent of unity or one, lasting for one delay time (usually one clock-pulse width).

The DT Unit Step is defined as a series of DT unit impulses. As in the continuous domain, the DT unit step is assumed to go on for "a very long time" relative to other signals or processes that are involved and, in the limit, goes to infinity.

MM THE DISCRETE-TIME IMPULSE AND THE DISCRETE-TIME STEP

The DT unit impulse:

δ[n] = 1 when t = 0 and δ[n] = 0 at all other times.
Here the "1" is a digital 1. (E 12.10)

The DT step:

$$u[n] = \sum_{0}^{\infty} \delta[n] \qquad \text{(E 12.11)}$$

GM THE DISCRETE-TIME IMPULSE AND THE DISCRETE-TIME STEP

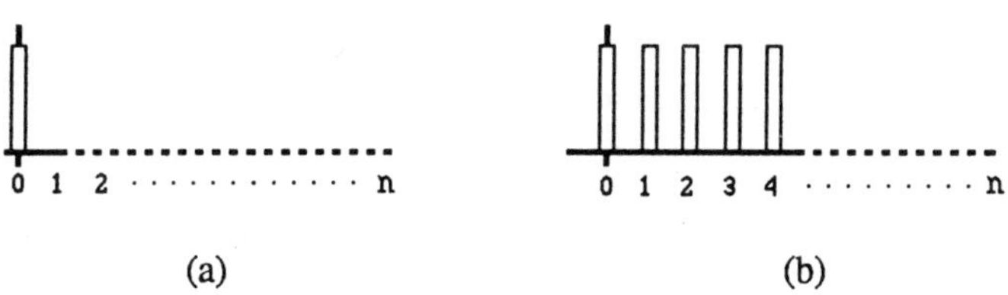

FIGURE 12.4 (a) A discrete-time impulse; (b) A discrete-time step.

THE DISCRETE TIME-DOMAIN ANALYSIS OF SYSTEMS OR PROCESSORS

If a DT unit pulse or a DT step is used to drive a digital system, their effect on the output response is somewhat similar to the comparable effect on continuous systems using the equivalent continuous-time forcing functions. However, it is well to be aware that the DT "system" can be a (digital) circuit or a (micro) processor, which means that the flow of current or voltage may be replaced with the flow of a program. A program can have delay and memory just as components, in a circuit, can have delay and storage. The output of a program is a response just as is the output of a circuit.

TM THE DT TRANSFER FUNCTION AND RESPONSE OF A SYSTEM

The transfer function of a digital process or system relates its output response to its input similar to an analog system. Likewise, there exist such parameters as natural and steady-state DT responses. Like its analog counterpart, the response of a DT system is characterized by its (digital) impulse response.

MM THE DT TRANSFER FUNCTION AND RESPONSE OF A SYSTEM

If the system equation is written in the form,

$$y[n] = x[n] + x[n-1] + x[n-2] + x[n-3] + \cdots \qquad \text{a repeat of (E 12.7)}$$

then the impulse response, and thus the transfer function can be in the form,

$$h[n\} = \delta[n] + \delta[n-1] + \delta[n-2] + \delta[n-3] + \cdots \text{ which is also the system natural response} \qquad \text{(E 12.12)}$$

Since a DT step is a series of impulses, the system step response s[n] takes the form,

$$s[n] = \sum_{n=-\infty}^{\infty} h[n] \text{ or } \sum_{n=0}^{\infty} h[n] \text{ (if the step starts at } n = 0) \qquad \text{(E 12.13)}$$

Considering each sample of an input as an impulse x[k], the General System Response, to any DT input function, can now be written:

$$y[n] = \sum_{k=-\infty}^{\infty} x[k]\, h[n-k]$$ this is known as the *Convolution Sum* of a DT system and its input. (E 12.14)

12.8 DIGITAL CONVOLUTION—COMBINING DIGITAL SIGNALS IN THE DIGITAL DOMAIN

TM DIGITAL CONVOLUTION

Time-domain convolution can be used with (discrete) digital signals just as it can be used with (continuous) analog signals. The normal form of convolution is to combine (convolve) an input-signal function with a system or process transfer function to obtain an output response. Although discrete-signal convolution is much simpler, both in concept and execution, than analog convolution, it is not a commonly used technique in overall digital signal processing. In general, the Fast Fourier Transform (covered later in this chapter) has replaced convolution.

MM DIGITAL CONVOLUTION

For review, the (continuous) Convolution Integral:

$$y(t) = \int_0^t f(\tau)\, h(t-\tau)\, d\tau$$ from (E 11.84)

$= f(x) * h(t) = f(t) * h(t)$ where * is used as the symbol of convolution in this book from (E 11.85)

Digital Convolution:

$$y[n] = \sum_{k=-\infty}^{\infty} x[k]\, h[n-k]$$ (E 12.15)

12.9 THE DISCRETE-TIME (SAMPLED) FOURIER SERIES—THE SPECTRUM OF A PERIODIC, SAMPLED SIGNAL

TM THE SAMPLED FOURIER SERIES (PERIODIC)

A sampled, digital periodic signal can be broken down into sine and cosine component parts, just as a continuous periodic signal.

MM THE SAMPLED FOURIER SERIES (PERIODIC)

$$x[n] = \sum_{k=0}^{k=N-1} a_k e^{j2\pi kn/N}$$

where k = the kth spectral component (harmonic), N = number of samples per period, and n = the nth sample (E 12.16)

$$a_k = \frac{1}{N} \sum_{k=0}^{k=N-1} x[n]\, e^{-j2\pi kn/N} \qquad a_k = \text{the magnitude of the } k^{th} \text{ harmonic}$$

$$= \frac{1}{N} \sum_{k=0}^{k=N-1} x[n]\left(\cos 2\pi k \frac{n}{N} - j \sin 2\pi k \frac{n}{N}\right) \qquad \text{(E 12.17)}$$

GM

See models at the end of this chapter.

12.10 THE DISCRETE-TIME (SAMPLED) FOURIER TRANSFORM (DFT) AND ITS FAST FOURIER TRANSFORM (FFT)—THE SPECTRUM OF AN APERIODIC, SAMPLED SIGNAL

Just as with the continuous Fourier Series, it would seem that after the Discrete (sampled) Fourier Series for periodic signals, there should be an extended form for aperiodic signals called the Discrete Fourier Transform. Well, there is, but . .

As digital computer hardware was developing in the 1950s and 1960s, so did the need for software that, among many other things, would solve problems much more rapidly. This was especially true for methods to calculate the DFT. About 1965, James Cooley and John Tukey developed a method (an algorithm) for materially reducing the time required to obtain the DFT spectrum of a signal.

Their research showed that the conventional methods of doing a DFT require many calculation steps that were redundant and thus unnecessary. Their "fast" method, which eliminated these calculations, and other approaches that followed, truly revolutionized both the philosophy and technique of digital signal processing! The FFT is a very special, fast form of the DFT.

For an example of the utility of this "fast" method, remember that the convolution of two signals in the digital domain is the same as multiplying their equivalents in the frequency domain, and these frequency-domain multiplications are much faster than the time-domain convolution. Thus, to "convolve" two time signals, it is usually much faster (sometimes by a factor of 1000) to follow this procedure:

1. Take the FFTs of the two original time signals—this puts them in the frequency domain.
2. Multiply the two results in the frequency domain.
3. Take the Inverse Fast Fourier Transform (IFFT) of this subsequent frequency-domain signal to again obtain a time-domain signal. This complete process is the equivalent of the convolution of the original signals.

Thus, although digital convolution and other similar digital processing techniques are still used, the FFT and IFFT are the dominant techniques used for the equivalent of convolving or for just finding the spectrum of time signals.

TM THE SAMPLED FOURIER TRANSFORM AND THE FFT (APERIODIC)

As with the Fourier Series, the Fourier Transform can be presented in a discrete form. Again, the DFT is based upon the principle of a repetition rate going to zero—the next one never happens. This generic DFT is rarely used any more, since there are now techniques, called Fast Fourier Transforms, to materially speed up computer calculations.

MM THE SAMPLED, DISCRETE-TIME FOURIER TRANSFORM (APERIODIC)

$$X[k] = X[\Omega] = \sum_{-\infty}^{\infty} x[n]\, e^{-j2\pi kn/N} = \sum_{-\infty}^{\infty} x[n]\, e^{-j\Omega n} \text{ (where } \Omega = 2\pi k/N)$$

Note: different books use different symbols. (E 12.18)

$$x[n] = \sum_{k=0}^{k=N-1} \frac{X_k(2\pi/N)}{N} e^{jk2\pi n/N} \qquad \text{(E 12.19)}$$

12.11 GRAPHIC MODELS FOR THE DISCRETE FOURIER SERIES AND THE DISCRETE FAST FOURIER TRANSFORM (APERIODIC)

A small amount of paper and pencil calculations using E12.16, .17, .18, or .19 will indicate that such calculations become quite involved and "messy." There are few better illustrations of the observation that it requires computers to perform computer calculations! Even simple examples, let alone serious design or analysis in digital signal processing (DSP), require the use of both hardware and software tailored to DSP needs. More and more textbooks are now including program listings or computer disks to allow the student or the practicing engineer to investigate the principles and refinements of topics in DSP.

The following graphics illustrate the use of simple but powerful programs for study and demonstration. Figure 12.5 illustrates the use of E12.17 to find the REAL components of a discrete periodic signal that has seven samples per period with the values (magnitudes) of x[n] equal to x[0]=2, x[1]=1, x[2]=0, x[3]=2, x[4]=1, x[5]=0 and x[6]=2. All the calculations were done with the spreadsheet.

Figures 12.6 through 12.8 use a DSP program written by Dr. Banu Onaral and Marianne R. Cain of Drexel University and published as *Discrete Signals and Systems II*, 1989 and made available through Kinko's Academic Courseware Exchange, P.O. Box 8000, 255 West Stanley Ave., Ventura, California.

THE SAMPLED, DISCRETE-TIME FOURIER SERIES

The a[k] harmonics of a wave form.

$$a_k = \frac{1}{N} \sum_{k=0}^{k=N-1} x[n] \left(\cos 2\pi k \frac{n}{N} - j \sin 2\pi k \frac{n}{N}\right) \qquad \text{copy of (E 12.17)}$$

where a[k] = the total part of this k^{th} harmonic and a[n] is the contribution from each x[n] sample.

As an example, see the first calculation in Figure 12.5. For N = 7 as given above and n = 0:

$$a[0] = \frac{1}{7} * 2 \cos (2\pi * 1 * 0) = \frac{1}{7} * 2 \cos 0 = \frac{1}{7} * 2 * 1 = 0.28571428$$

THE REAL SPECTRAL COMPONENTS OF A PERIODIC DIGITAL SIGNAL

x[n]	k	n	a[n]
2	0	0	0.28571428
1	0	1	0.14285714
0	0	2	0
2	0	3	0.28571428
1	0	4	0.14285714
0	0	5	0
2	0	6	0.28571428
		a[0]=	1.14285712

x[n]	k	n	a[n]
2	3	0	0.28571428
1	3	1	-0.1287098
0	3	2	0
2	3	3	-0.0635774
1	3	4	-0.0317887
0	3	5	0
2	3	6	-0.2574197
		a[3]=	-0.1957813

x[n]	k	n	a[n]
2	6	0	0.28571428
1	6	1	0.08906997
0	6	2	0
2	6	3	-0.2574197
1	6	4	-0.1287098
0	6	5	0
2	6	6	0.17813994
		a[6]=	0.16679468

x[n]	k	n	a[n]
2	1	0	0.28571428
1	1	1	0.08906997
0	1	2	0
2	1	3	-0.2574197
1	1	4	-0.1287098
0	1	5	0
2	1	6	0.17813994
		a[1]=	0.16679468

x[n]	k	n	a[n]
2	4	0	0.28571428
1	4	1	-0.1287098
0	4	2	0
2	4	3	-0.0635774
1	4	4	-0.0317887
0	4	5	0
2	4	6	-0.2574197
		a[4]=	-0.1957813

x[n]	k	n	a[n]
2	2	0	0.28571428
1	2	1	-0.0317887
0	2	2	0
2	2	3	0.17813994
1	2	4	0.08906997
0	2	5	0
2	2	6	-0.0635774
		a[2]=	0.45755808

x[n]	k	n	a[n]
2	5	0	0.28571428
1	5	1	-0.0317887
0	5	2	0
2	5	3	0.17813994
1	5	4	0.08906997
0	5	5	0
2	5	6	-0.0635774
		a[5]=	0.45755808

FIGURE 12.5 The REAL components of a discrete periodic signal.

GM EXAMPLES OF THE FFT OF A (SAMPLED) SINE WAVE, IMPULSE, AND STEP FUNCTION

The figures arbitrarily show the data as real and imaginary; a magnitude and phase option is also available in this and many similar DSP programs.

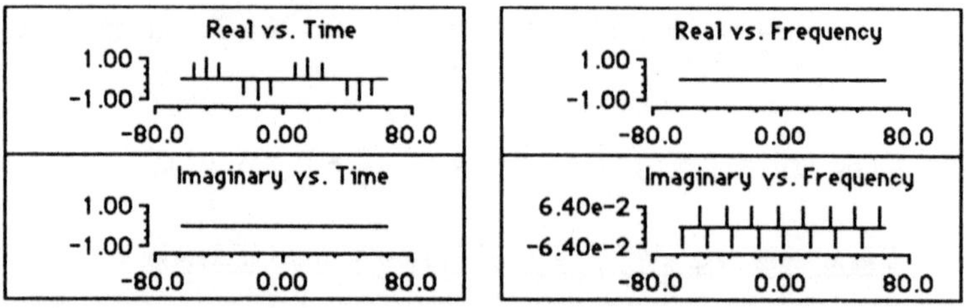

(a) (b)

FIGURE 12.6 (a) The time domain components of a discrete sine wave; (b) The frequency domain components of a discrete sine wave obtained with an FFT.

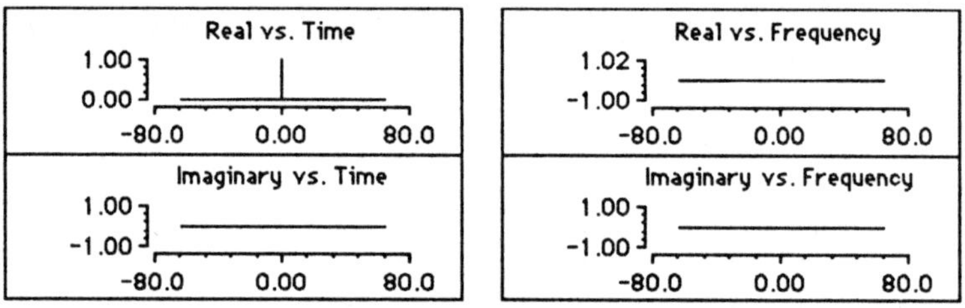

(a) (b)

FIGURE 12.7 (a) The time domain components of a discrete impulse; (b) The frequency domain components of a discrete impulse obtained with an FFT.

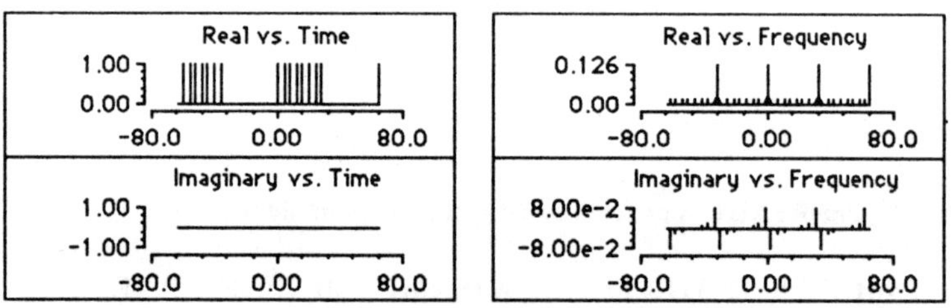

(a) (b)

FIGURE 12.8 (a) The time domain components of a discrete square wave; (b) The frequency domain components of a discrete square wave obtained with an FFT.

13

DSP, Digital Filters, and the Z-Transform

13.1 AN INTRODUCTION

One of the most interesting—and very useful—constituents of digital signal processing (DSP) is the technique of digital filtering. In analog circuits, the concept of a simple circuit, such as a resistor and capacitor, used to low-pass a signal comes rather easily. However, the premise that a relatively simple collection of digital components or a small computer program can be used to "filter" a series of numbers is often difficult to comprehend. Before the analysis and design of digital filters can be described, it is necessary to review or reinforce the background in digital signals and digital processing. Thus, the digital filter will be the ultimate goal, the icing on the cake, for this chapter. The following is not a complete outline of the chapter but a general list of the topics that are reviewed or introduced.

1. DIGITAL SIGNALS IN THE TIME DOMAIN
2. DIGITAL SIGNALS IN THE FREQUENCY DOMAIN
3. DIGITAL PROCESSES IN THE TIME DOMAIN
4. DIGITAL PROCESSES IN THE FREQUENCY DOMAIN AND THE Z-TRANSFORM
5. DIGITAL FILTERS

13.2 SYMBOLS AND EQUATIONS FOR ELEMENTARY DIGITAL SIGNALS

TM THE DIGITAL SIGNAL AND ITS EQUATION

A DIGITAL SIGNAL will be defined as a series of related digitally coded numbers (usually, but not necessarily, binary) that combine to represent some function or event. Digital signals originate in two forms. Often, an analog or continuous time signal is sampled and converted to a series of related digital (discrete) numbers. Likewise, many signals, such as stock market sales, hourly or daily temperature or pressure readings, etc., start in discrete number form. If, for now, it is assumed that the number series is an input, it is common to call that input "x." Since there are "n" number codes in this input signal, the form is usually stated as x(n). This x(n) can likewise be broken into its component "n" parts.

It is common to relate a number series to some time reference, usually a time zero. This means that it is possible to have discrete numbers in the series occurring before the established time zero (+) and also to have numbers occurring after the time zero (-). An example of time zero is the first monthly payment in a year. Thus, the January payment would (usually) be the "now" or x(n) payment and February would be the later or delayed payment x(n −1). The December payment, of the previous year, would be the earlier x(n + 1) payment.

The fundamental building block used to represent a digital series is the discrete unit impulse or δ. When all the signal pulses in a signal, such as a digital step function, have a value of unity, the digital equation can be written with or without including the δ in the terms. However, if the signal equation includes weighting, the δ must be included.

MM THE DIGITAL SAMPLED SIGNAL AND ITS EQUATION

$$x(n) = \cdots x(n+2) + x(n+1) + x(n) + x(n-1) + x(n-2) + \cdots \quad \text{and}$$

$$= \cdots x\,\delta(n+2) + x\,\delta(n+1) + x\,\delta(n) + x\,\delta(n-1) + x\,\delta(n-2) + \cdots$$

are the same (E 13.1)

GM THE DIGITAL SIGNAL AND ITS EQUATION

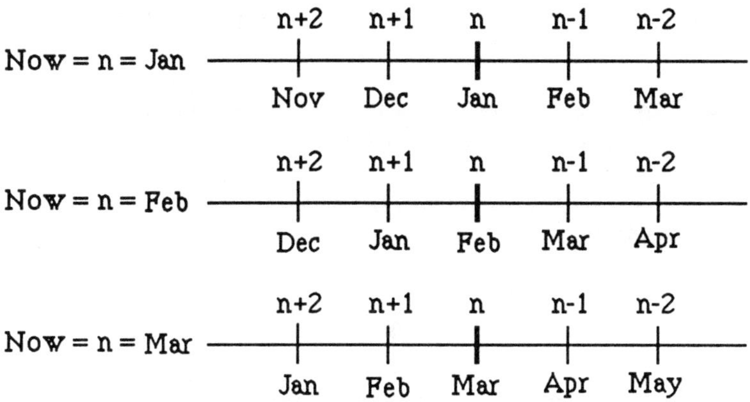

FIGURE 13.1 The "n" of a digital number series illustrated by monthly payments. In the lower equations, the months are shifted to the left. See "Horizontal Axis Polarity Rules for Digital-Signal Figures" in Chapter 12.

TM SHIFTING AND WEIGHTING DIGITAL (UNIT IMPULSE) SIGNALS

Digital signals can be manipulated (processed) in many ways. Two of the most rudimentary are the shift and weighting functions. Shifting will move the signal (or the X axis) left or right relative to its present location (usually in time) or relative to an "n" position (usually $x(n) = n_0$). Weighting is multiplying each digital number in the signal series by some constant value. DSP programs will often have instructions such as "X-Axis Shift" to shift or move the time axis (similar to moving the signal) and "Y-Axis Scale," which multiplies each signal number by a particular constant.

MM SHIFTING AND WEIGHTING DIGITAL (UNIT IMPULSE) SIGNALS

Referring to Figure 13.2 (a):

$x(n) = x[6]\,\delta(n) + x[4]\,\delta(n-1) + x[2]\,\delta(n-2)$ is the top signal equation (E 13.2)

$= x[6]\,\delta(n-2) + x[4]\,\delta(n-3) + x[2]\,\delta(n-4)$ is the equation of the bottom SHIFTED signal (E 13.3)

Referring to Figure 13.2 (b):

$x(n) = x[6]\,\delta(n) + x[4]\,\delta(n-1) + x[2]\,\delta(n-2)$ is the top signal equation (E 13.4)

$= x[3]\,\delta(n-2) + x[2]\,\delta(n-3) + x[1]\,\delta(n-4)$ is the equation of the bottom SHIFTED and WEIGHTED signal (E 13.5)

GM SHIFTING AND WEIGHTING DIGITAL (UNIT IMPULSE) SIGNALS

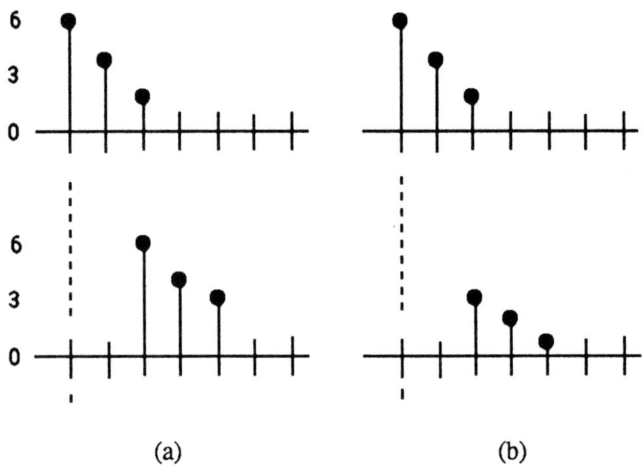

FIGURE 13.2 (a) An illustration of signal shifting. The bottom signal is shifted (delayed) relative to the top signal; (b) An illustration of signal shifting AND weighted. The bottom signal is shifted like (a) and also weighted (multiplied) by a factor of 0.5.

13.3 DIGITAL SIGNALS IN THE TIME DOMAIN

TM THE DISCRETE UNIT STEP TIME DOMAIN FUNCTION

The Discrete Step Function is one of the three major discrete signals used in elementary DSP. As in continuous signal processing, the Discrete Step is one of the tools used to find the transient response of discrete systems. It can be shifted and weighted to become a discrete (drop the unit) step function. The Discrete Step is formulated by using a series of Discrete Impulse Functions.

MM THE DISCRETE UNIT STEP TIME DOMAIN FUNCTION

$$u(n) = 0, n < 0$$

$$= 1, n \geq 0 \quad \text{(E 13.6)}$$

GM THE DISCRETE UNIT STEP TIME DOMAIN FUNCTION

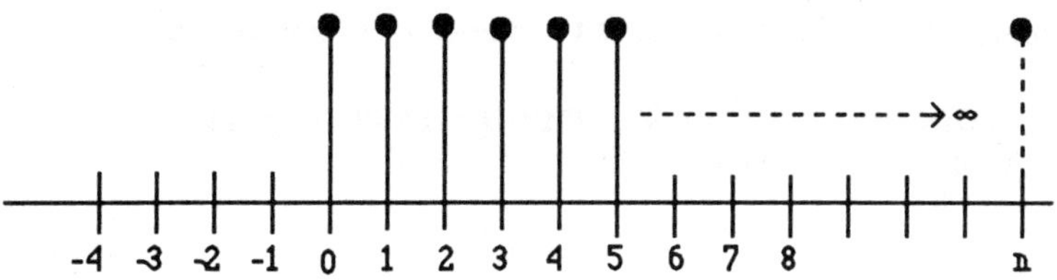

FIGURE 13.3 The Discrete Step Function.

TM THE DISCRETE IMPULSE TIME DOMAIN FUNCTION

The Discrete Unit Impulse is THE primary building block of DSP. It represents the idea that most digital input or output signals can be represented as a series of digital numbers or digital codes. It will be shown later that, like continuous signal processing, the impulse response of a processor will give its total response, including the transient and steady state.

MM THE DISCRETE IMPULSE TIME DOMAIN FUNCTION

$$\delta = 0, n \neq 0$$

$$= 1, n = 0 \quad \text{(E 13.7)}$$

GM THE DISCRETE IMPULSE TIME DOMAIN FUNCTION

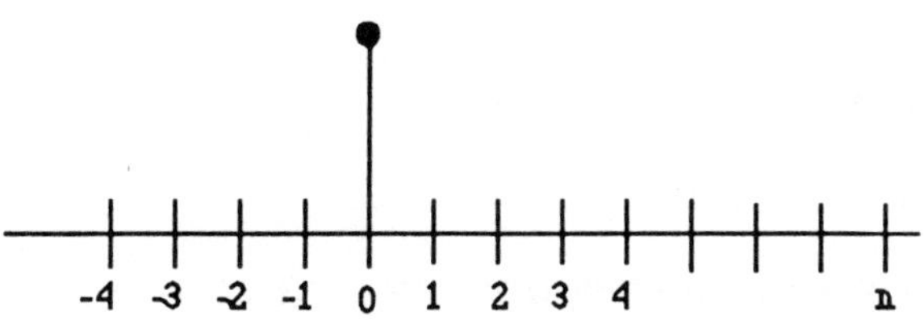

FIGURE 13.4 The Discrete Impulse Function.

TM THE DISCRETE IMPULSE TIME DOMAIN FUNCTION AS A SAMPLING FUNCTION

The Discrete Unit Impulse also represents the ideal sampler. Using this concept, a continuous signal v(t) can be sampled every T seconds and will produce a pulse train, representing the original signal of $\delta(t - nT)$ pulses, which occur every nT seconds.

Most discrete terminology drops the T in the nT part of the equation so that $\delta(t - nT)$ becomes $\delta(t - n)$. The T is ALWAYS inferred—just not written.

MM THE DISCRETE IMPULSE TIME DOMAIN FUNCTION AS A SAMPLING FUNCTION

$$\delta_T(t) = \sum_{n=0}^{\infty} \delta(t - nT) = \sum_{n=0}^{\infty} \delta(t - n) \quad \text{(E 13.8)}$$

$$v^*(t) = \sum_{n=0}^{\infty} v(nT)\, \delta(t - nT) = \sum_{n=0}^{\infty} v(n)\, \delta(t - n) \quad \text{(E 13.9)}$$

Where: $v^*(t)$ = the output of a $\delta_T(t)$ sampler
$v(nT)$ or $n(n)$ = the input signal
$\delta(t - nT)$ or $\delta(t - n)$ = the sampling function

GM THE DISCRETE IMPULSE TIME DOMAIN FUNCTION AS A SAMPLING FUNCTION

The usual way of visualizing a sampler is to draw a switch which closes briefly at each sample time.

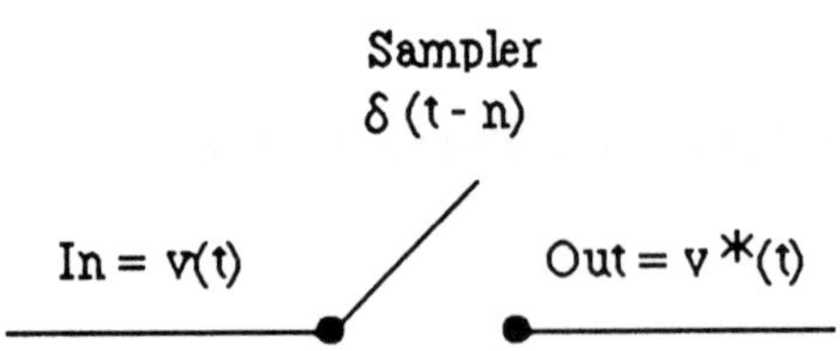

FIGURE 13.5 The impulse as a sampler.

TM THE DISCRETE SINUSOID TIME DOMAIN FUNCTIONS

The Discrete Sine or Cosine can be a sampled continuous signal or a truncated group of impulses representing these wave forms. As with a continuous analog signal, there is a peak amplitude A, a radian frequency $2\pi fn$ (where n is "this" particular sample), and a phase angle ϕ which advances or retards the wave form from n_0 (t_0). The sinusoid can be stated in trigonometric or exponential form—see the next section.

MM **THE DISCRETE SINUSOID TIME DOMAIN FUNCTIONS**

$$x(n) = A \sin(2\pi f T n \pm \phi) \qquad \text{(E 13.10)}$$

$$x(n) = A \cos(2\pi f T n \pm \phi) \qquad \text{(E 13.11)}$$

A = the peak amplitude
ϕ = the phase angle ($\pm$) at $n = 0$
$T = 1/f_{sample}$

GM **THE DISCRETE SINUSOID TIME DOMAIN FUNCTIONS**

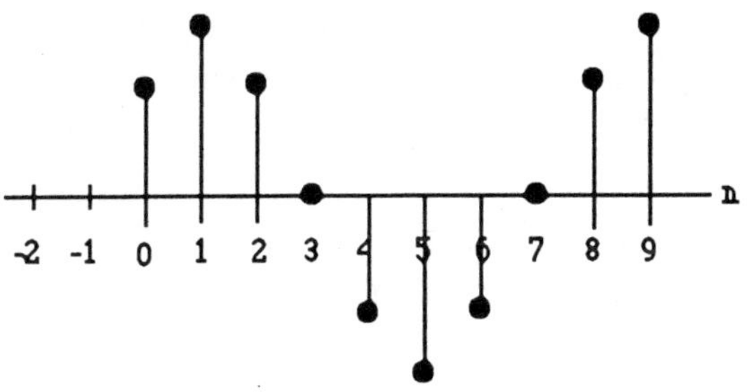

FIGURE 13.6 A Discrete Cosine Function: $A = 1$, $F = \frac{1}{N} = \frac{1}{8}$, $\phi = -45°$.

13.4 DIGITAL SIGNALS IN THE FREQUENCY DOMAIN

Signals in the discrete frequency domain can be divided into periodic and aperiodic. As with continuous analog signals, the spectrum components of discrete, periodic digital signals can be found with a form of the Fourier Series and the spectrum components of aperiodic digital signals can be found with a form of the Fourier Transform.

The spectrum of a periodic x[n] digital signal can be represented by the Discrete Fourier Series, which can be transformed to provide the spectral coefficients. If a signal is periodic, be it a continuous or discrete signal, its spectrum (spectral line or lines) is aperiodic—it only occurs at one general location in frequency.

The spectrum of an aperiodic x[n] digital signal can be represented by the Discrete Fourier Transform, which can also be transformed to provide the spectral coefficients. If a signal is aperiodic, its spectrum is periodic—the spectral line (or lines) representing its spectral coefficients continually repeat in frequency.

TM THE GENERAL FORM OF PERIODIC DIGITAL SIGNALS IN THE FREQUENCY DOMAIN

A Periodic Digital Signal can be represented as a series algebraic equation, such as E 13.1, or as a complex exponential series. The periodicity of the signal can be observed by graphing the terms in the algebraic equation or by observing the exponential terms as a function of n. Each exponential term represents a vector that is rotating in increments of $2\pi/N * 1, 2\pi/N * 2, 2\pi/N * 3, \cdots 2\pi/N * n$, for the cosine or sine components that make up the signal's Discrete Fourier Series. Likewise, each term of the series contributes to each item in the exponential series. Figure 12.5 provides a good example of each series term (x(n)) contributing to the harmonic terms a[0], a[1], a[2], etc.

A continuous digital signal can be represented by a discrete Fourier Series. Any periodic digital signal with N samples per period can be represented in the frequency domain by N harmonics and this spectrum repeats every 2π—a consequence of sampling.

MM THE GENERAL FORM OF PERIODIC DIGITAL SIGNALS IN THE FREQUENCY DOMAIN EXPONENTIALS AND SINUSOIDS

For $x[n] = A\,e^{\kappa n}$

If $\kappa > 0$, the exponential will rise as in a charging capacitor

If $\kappa = 0$, the exponential has a value of $A*1$ which is a constant

If $\kappa < 0$, the exponential will fall as in a discharging capacitor (E 13.12)

If $x_1[n] = A\,e^{jn\Omega} = A\cos(n\Omega) + j\,A\sin(n\Omega)$ and

$x_2[n] = A\,e^{-jn\Omega} = A\cos(n\Omega) + j\,A\sin(-n\Omega)$ then,

$x_1 + x_2 = 2A\cos(n\Omega)$ and,

$$A\cos(n\Omega) = A\cos(\omega nT) = A\cos(2\pi f\,nT) = \frac{1}{2}(x_1[n] + x_2[n])$$

$$= \frac{A}{2}e^{jn\Omega} + \frac{A}{2}e^{-jn\Omega} \text{ and,} \quad \text{(E13.13)}$$

$$A\sin(n\Omega) = \frac{A}{2j}e^{jn\Omega} - \frac{A}{2j}e^{-jn\Omega} \quad \text{(E 13.14)}$$

THE SPECTRUM OF PERIODIC DIGITAL SIGNALS

$$x[n] = \sum_{k=0}^{k=N-1} a_k \, e^{j2\pi kn/N}$$

where k = the kth spectral component (harmonic), N = number of sampler per period, and n = the nth sample repeat of (E 12.16)

$$a_k = \frac{1}{N} \sum_{k=0}^{k=N-1} x[n] \, e^{-j2\pi kn/N}$$ a_k = the magnitude of the kth harmonic

$$= \frac{1}{N} \sum_{k=0}^{k=N-1} x[n] \, (\cos 2\pi k \frac{n}{N} - j \sin 2\pi k \frac{n}{N})$$ repeat of (E 12.17)

The form of calculating the Real spectral coefficients of the harmonic amplitudes a_k for the discrete Fourier Series:

$$a_{k=0} = \frac{1}{N} \text{Cos} \{ 2 \pi * (k=0) * ((n = 0 \text{ to } N-1)/N) \}$$

gives N values to sum for a_0

$$a_{k=1} = \frac{1}{N} \text{Cos} \{ 2 \pi * (k=1) * ((n = 0 \text{ to } N-1)/N) \}$$

gives N values to sum for a_1

$$a_{k=N-1} = \frac{1}{N} \text{Cos} \{ 2 \pi * (k=0 \text{ to } N-1) * ((n = 0 \text{ to } N-1)/N) \}$$

gives N values to sum for a_{N-1}

The form of calculating the Imaginary spectral coefficients of the harmonic amplitudes a_k for the discrete Fourier Series:

$$a_{k=0 \text{ to } N-1} = \frac{1}{N} \text{Sin} \{ 2 \pi * (k= 0 \text{ to } N-1) * ((n = 0 \text{ to } N-1)/N) \}$$

gives N * N values for all a_k Imaginary coefficients (E 13.17)

GM THE GENERAL FORM OF PERIODIC DIGITAL SIGNALS IN THE FREQUENCY DOMAIN

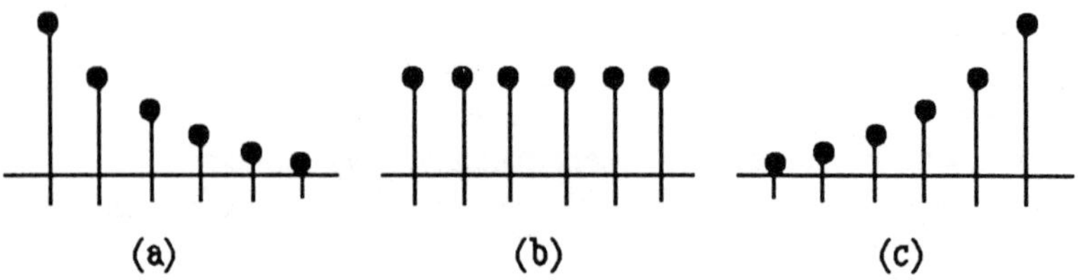

FIGURE 13.7 Falling, constant, and rising exponential digital signals of the general form $x[n] = A\,e^{\kappa n}$: (a) $\kappa < 0$; (b) $\kappa = 0$; (c) $\kappa > 0$.

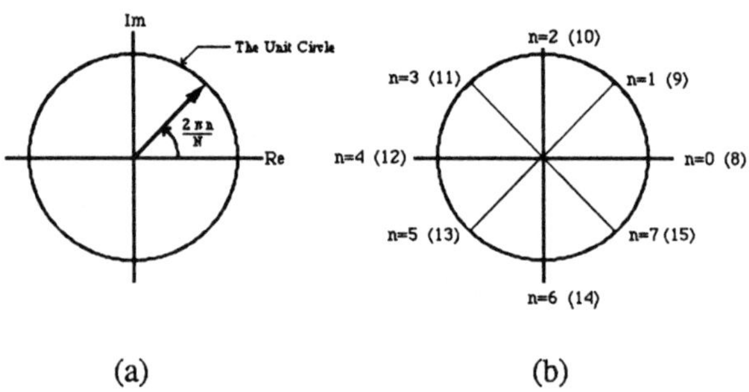

FIGURE 13.8 (a) The general rotating vector representing the discrete harmonic $x[n] = A\left(\cos 2\pi k \frac{n}{N} - j \sin 2\pi k \frac{n}{N}\right)$. The vector does not rotate in a smooth, continuous manner but "jumps" in discrete $\frac{2\pi n}{N}$ steps. (b) The positions of a vector when N is 8. Note that the position of the vector at n=8 is the same as n=0 and the position is the same for n=9 and n=1, etc. This shows that the spectrum, which is represented by the vector positions, is repeating on and on ad infinitum.

TM THE GENERAL FORM OF APERIODIC DIGITAL SIGNALS IN THE FREQUENCY DOMAIN

Aperiodic digital signals can be represented by the Discrete Fourier Transform. The general form of the DFT resembles the DFS, but there are some major differences. In the DFS, the signal is assumed to be periodic and thus repeating. The Fourier Series representation is similar to a continuous-signal Fourier Series. However, the DFT represents only one isolated signal and thus the total of the harmonics and their amplitudes (sometime called the density of the spectrum) is much smaller than the total of the harmonics and their amplitudes in the DFS. In the Discrete Fourier Transform, the averaging term 1/N is not included. In fact, the "a_k" of the DFS has been replaced by "Na_k" = X in the DFT. This X, or usually X(Ω), uses the N factor to keep the spectral energy finite since, using the DFT, the a_k coefficients can, in the limit, vanish to zero. See Chapter 12 for the fundamental DFT equations.

MM THE GENERAL FORM OF APERIODIC DIGITAL SIGNALS IN THE FREQUENCY DOMAIN

THE SPECTRUM OF APERIODIC DIGITAL SIGNALS

$$X(\Omega) = Na_k = \sum_{-\infty}^{\infty} x[n]\, e^{-j\Omega n}$$

Example 13.1: Find the Fourier Transform (spectrum) of a simple aperiodic signal shown in Figure 13.9.

1. The signal equation is $x[n] = \delta[n+1] + \delta[n] + \delta[n-1]$ (note that no weighting factors are included)
2. The Fourier Transform (spectrum) is

   $$X(\Omega) = \sum_{-\infty}^{\infty} x[n]\, e^{-j\Omega n} \text{ (where } \Omega = 2\pi k/N\text{) and,}$$

3. $X(\Omega) = 1 * e^{-j\Omega n} + 1*1 + 1* e^{j\Omega n}$

(See Figure 13.9, page 302.)

13.5 DIGITAL PROCESSES IN THE TIME DOMAIN

The most important characteristic of a digital processor is its transfer characteristic, termed h[n] in the digital time domain or H(Ω) in the digital frequency domain. It is this h or H that modifies the input signal to give it the desired output amplitude, frequency response, wave shape, etc. It should be remembered that digital processing can take place in hardware or in a program (software).

In the time domain, this transfer processing is usually digital convolution. There are two important ideas to be remembered about digital convolution. First, both the concept and the implementation of digital convolution are much simpler than continuous signal or analog convolution. Second, many times, digital convolution is a slow process and it is replaced by the FFT (see Chapter 12).

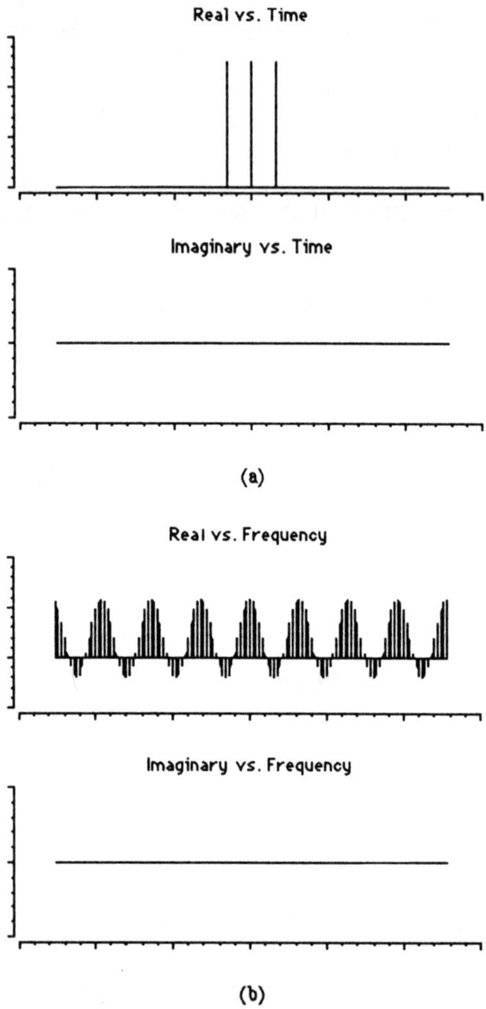

FIGURE 13.9 (a) The Example 13.1 Discrete Signal; (b) The Example 13.1 Frequency Spectrum. (These displays were constructed using "Discrete Signals and Systems II." See the Onaral and Cain reference.)

Even so, later it will be shown that the basic processing technique used in all digital filters is, in fact, digital-domain convolution.

TM TIME-DOMAIN DIGITAL PROCESSING AND THE PROCESSOR TRANSFER FUNCTION

A digital transfer function, like an analog transfer function, is defined as the ratio of the output to the input. The input and output signals can be generated by hardware or software, or some combination of the two. The transfer function or the "characteristic" of a processor can be obtained with an applicable hardware or software unit-impulse input.

MM THE DIGITAL PROCESSOR TRANSFER FUNCTION

THE FORM OF THE TRANSFER FUNCTION IN THE TIME DOMAIN

The form of the transfer function:

$$h[n] = \frac{y[n]}{x[n]} \text{ or } y[n] = x[n] * h[n] \text{ where here the "*" denotes time-domain convolution} \qquad \text{(E 13.18)}$$

Using the Impulse Function:

$$y[n] = \delta[n] * h[n] = 1 * h[n] = h[n] \qquad \text{(E 13.19)}$$

Special Note: This last equation represents more than just a nice, neat mathematical statement. When testing processors like digital filters, the use of a temporary input "unit impulse," produced by one clock pulse, will often indicate whether the processor is producing approximately its proper output (function).

Using the Step Function:

$$y[n] = u[n] * h[n] \qquad \text{(E 13.20)}$$

GM THE DIGITAL PROCESSOR TRANSFER FUNCTION

THE TIME-DOMAIN DIGITAL PROCESSOR

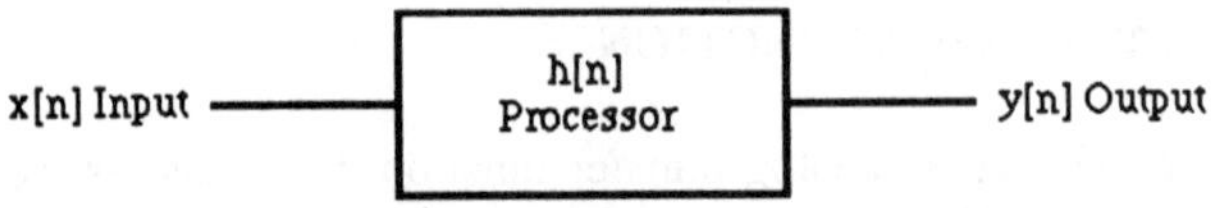

FIGURE 13.10 The relationship between the digital processor and its input and output in the time domain.

13.6 DIGITAL PROCESSES IN THE FREQUENCY DOMAIN

The processing and analysis of digital signals in the frequency domain have appeal for at least two reasons. First, very often the input or output signal is either specified or visualized as a function of frequency. Second, the FFT has tended to create frequency-domain devotees of us all.

TM FREQUENCY-DOMAIN PROCESSING AND THE PROCESSOR TRANSFER FUNCTION

The key concept in digital frequency-domain processing is the *frequency response* just as the *spectrum* is the key word in frequency-domain signal specification. The frequency response is usually specified by amplitude and phase. The major signals that are used for frequency-domain analysis are sinusoids; the impulse, since it theoretically contains an infinite number of sinusoids; and the step.

The transfer function of the signal processor in the frequency domain is characterized by its effect on the amplitude and phase of the input signal. If a processor is normal and "well behaved," it is termed a *linear, time-invariant* (LTI) system. An LTI system acts on all signals, regardless of amplitude (over its operating range), in the same way—it obeys superposition. It also operates the same way each time it is used—it is time invariant. The transfer function may change a signal's magnitude (amplitude) or its phase, but NOT its period or repetition rate.

MM THE DIGITAL PROCESSOR TRANSFER FUNCTION

The Form of the Frequency-Domain Transfer Function:

$$H[n] = \frac{Y[n]}{X[n]} \qquad \text{(E 13.21)}$$

$Y[n] = X[n] * H[n]$ (multiplication in the frequency domain = convolution in the time domain) (E 13.22)

The Frequency-Domain Transfer Function:

$$H(e^{j\theta}) = \sum_{m=-\infty}^{m=+\infty} h(m)\, e^{-j\theta m} \qquad \text{(E 13.23)}$$

The Frequency-Domain Processor Output for a Sinusoidal Input:

$$Y[n] = H(e^{j\theta}) * X[n] \qquad \text{(E 13.24)}$$

$$Y[n] = \sum_{m=-\infty}^{m=+\infty} h(m)\, e^{-j\theta m} * e^{j\theta n} \text{ or } = \sum_{m=-\infty}^{m=+\infty} h(m) * e^{j\theta(n-m)} \qquad \text{(E 13.25)}$$

Special Note: The new variable, "m," is used to denote that the sampled phase of the transfer-function MAY be different than that of the input function. Also, since the term $e^{j\theta n}$ represents the input function, it may be brought to the left, leaving only terms related to the transfer function.

Example 13.2: Show the relative phase change of the input X[n], the transfer function H[m], and the ouput Y[n]. Assume a system with a unit-sample response (the transfer function) of $H[n] = \delta[n] + \delta[n-1]$ and an input of $X[n] = e^{j(\pi/2)n}$. Find its output = Y[n].

1. Put the transfer function in the form dictated by the general equation E 13.23 and the specific h[n] given in this problem:

$$H(e^{j\theta}) = \sum_{m=-\infty}^{m=+\infty} h(m)\, e^{-j\theta m}$$

$$= \sum_{m=-\infty}^{m=+\infty} [\delta(m) + \delta(m-1)]\, e^{-j\theta m}$$

$$= 1\, e^{j0} + 1\, e^{-j\theta} = 1 + e^{-j\theta}$$

2. Now put the transfer function in a form dictated from the input:

$$H(e^{j\pi/2}) = 1 + e^{-j\pi/2} = 1 - j = 1.41\, e^{-j\pi/4}$$

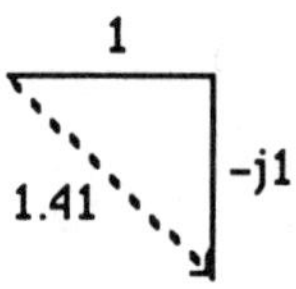

FIGURE 13.11 The π/4 or 45° right triangle.

3. Find the final Y[n]:

$$Y[n] = e^{j\theta n} * H(e^{j\theta})$$

$$= H(e^{j\pi/2})^n * 1.41\, e^{-j\pi/4} = 1.41\, e^{j(\pi n/2 - \pi/4)}$$

Note the input π/2 angle is changed by a –π/4 and the input amplitude of 1 is changed to an output amplitude of 1.41.

GM THE DIGITAL PROCESSOR TRANSFER FUNCTION

THE FREQUENCY-DOMAIN DIGITAL PROCESSOR

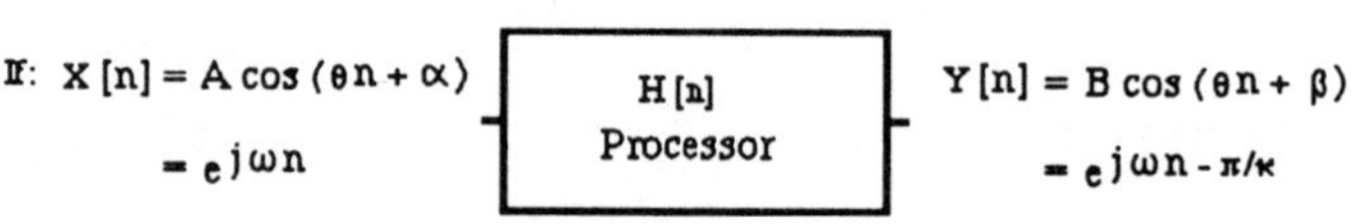

FIGURE 13.12 The relationship between the digital processor and its input and output in the frequency domain. Note that the processor may change the amplitude and/or the phase of the input signal but NOT its period!

13.7 THE Z-TRANSFORM

TM THE ELEMENTARY Z-TRANSFORM

The Z-Transform provides a method of working with discrete signal difference equations just as the Laplace Transform provides a method of working with continuous signal differential equations.

The idea and techniques of the Z-Transform include the z-plane, with poles and zeros, and also the inverse Z-Transform. Although the Z-Transform somewhat parallels the Laplace Transform in principle, its implementation is quite different. The z-plane gives the user an insight into the frequency characteristics (amplitude and phase) of processing functions such as digital filters, something like the s-plane gives an insight into transient characteristics. In fact, just as most textbooks specify the Laplace Transform as a general analog transient technique, they often place the Z-Transform under digital steady state or frequency techniques. The z-plane display is especially helpful for sampled signals since, if the s-plane were used, some or all of the poles would continually repeat to infinity.

The Z-Transform of a discrete signal can be placed in the form of a power series in $(z^{-1})^n$. It is possible to use this power series form to inverse transform back to the original, or processed signal, X(n). There are also other methods of using the inverse Z-Transform including look up tables, partial-fraction expansion (used in Chapter 11) and a technique, beyond the scope of this book, called contour integration. (Note that some books use x(n) for all discrete signals but, for clarity, this book uses X(n) for the frequency-domain signal equations and x(n) for the time-domain equations.)

Since input and output (processed) discrete signal equations can be written as X(n) and Y(n), using the appropriate rules and techniques, these can be (z) transformed into X(z) and Y(z). Likewise, the transfer function, in the z-domain, is $H(z) = Y(z)/X(z)$.

The functions of the Z-Transform can be displayed in a complex z-plane just as the Laplace functions are displayed in the s-plane. The z-plane has real and imaginary axes similar to the s-plane plus an important boundary area called the *unit circle*. The use of this unit circle can be defined as:

1. Poles of difference equations of most "real-world" functions map on or inside the z-plane unit circle just as most "real-world" poles of differential equations map on or to the left side of the s-plane imaginary axis.
2. A pole or a series of repeating poles that would lie on the s-plane $j\omega$ axis will map to the unit circle.

3. A pole or a series of repeating poles that would lie to the left of the s-plane jω axis, including complex poles, will map inside the unit circle on lesser circles.

MM THE ELEMENTARY Z-TRANSFORM

$$X(z) = \sum_{n=0}^{n=\infty} x[n]\, z^{-n}$$ where the z^{-n} is a unit delay—often one clock pulse

(E 13.26)

Example 13.2: If as difference equation series is

$$X[n] = a_1 x[n] + a_2 x[n-1] + a_3 x[n-2] + \cdots,$$ then

$$Z[n] = a_1 x[z] + a_2 x[z] z^{-1} + a_3 x[z] z^{-2} + \cdots$$

THE Z-TRANSFORM OF THE IMPULSE

If $X(n) = A\,\delta(n)$

$$X(z)_{Imp} = \sum_{n=0}^{n=\infty} A\,\delta(n)\, z^{-n} = Az^0 = A \qquad \text{(E 13.27)}$$

THE Z-TRANSFORM OF THE STEP

If $X(n) = 0$ for $n < 0$ and A for $n \geq 0$,

$$X(z)_{step} = \sum_{n=0}^{n=\infty} A^*\, z^{-n} = A\,\{1 + z + z^{-1} + z^{-2} + \cdots\} \qquad \text{(E 13.28)}$$

or, using the closed power (geometric) series form where, if the $(S)_{series} = a + ar + ar^2 + \cdots$, then

$$(S)_{series} = \frac{a}{1-r}$$ and, using this closed form of E13.28,

$$X(z)_{step} = \frac{A}{1 - z^{-1}} = \frac{Az}{z-1} \qquad \text{(E 13.29)}$$

THE Z-TRANSFORM OF DECREASING EXPONENTIAL

If $X(n) = A\, a^n\, u(n)$,

$$X(z)_{Exp} = \sum_{n=0}^{n=\infty} (A * z^{-1})^n = \frac{A}{1 - az^{-1}} = \frac{Az}{z - a} \tag{E 13.30}$$

THE Z-TRANSFORM OF A SINE FUNCTION

If the continuous sine wave is $x = \sin \omega_0 t$ and the sampled sequence is $X(n) = \sin n\omega_0 T$,

$$X(z)_{sin} = \sum_{n=0}^{n=\infty} (\sin n\omega_0 T)\, z^{-n}.$$

Now, using the Euler identity $\sin n\omega_0 T = \dfrac{e^{j\omega_0 T} - e^{-j\omega_0 T}}{2j}$

$$= \sum_{n=0}^{n=\infty} \frac{e^{j\omega_0 T} - e^{-j\omega_0 T}}{2j} z^{-n} = \frac{1}{2j} [\sum_{n=0}^{n=\infty} (e^{j\omega_0 T}) - \sum_{n=0}^{n=\infty} (e^{-j\omega_0 T} -)]\, z^{-n}$$

$$= \frac{1}{2j} [\frac{z}{z - e^{j\omega_0 T}} - \frac{z}{z - e^{-j\omega_0 T}}]$$

$$= \frac{1}{2j} \frac{z(e^{j\omega_0 T} - e^{-j\omega_0 T})}{z^2 - (e^{j\omega_0 T} - e^{-j\omega_0 T})z + 1}$$

$$= \frac{z \sin \omega_0 T}{z^2 - 2z \cos \omega_0 T + 1}$$, or in factored form,

$$= \frac{z \sin \omega_0 T}{(z - e^{j\omega_0 T})(z - e^{-j\omega_0 T})} \tag{E 13.31}$$

Special Note: The details of the derivation of E 13.31 are beyond the scope of this book. It was presented in this form to indicate the general procedure and to indicate how the second order came about. The equations are used to illustrate the use of the z-plane and Z-Transform pairs later in this chapter. Z-Transform pairs, like Laplace Transform pairs, are also listed in many textbooks.

THE EXPONENTIAL FORM OF THE Z-TRANSFORM

$$Z = e^{st} = e^{(\alpha + j\omega)T}, \text{ where } |Z|_{mag} = e^{\alpha T} \text{ and } \angle Z_{ang} = \omega T \qquad \text{(E 13.32)}$$

GM THE ELEMENTARY Z-PLANE

A COMPARISON WITH THE LAPLACE S-PLANE

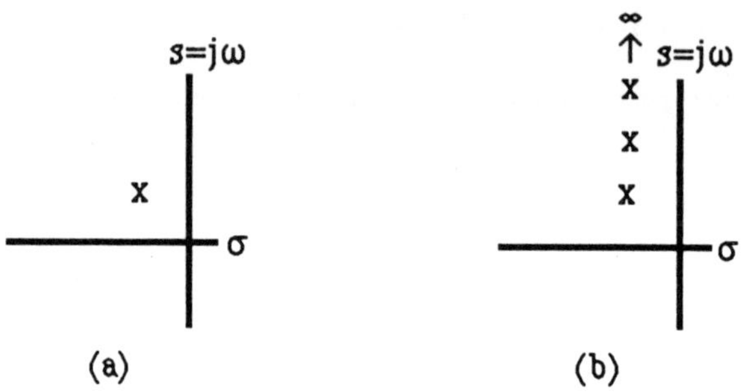

FIGURE 13.13 THE ELEMENTARY S-PLANE WITH DISCRETE POLES: (a) A representative pole of a CONTINUOUS function in the s-plane; (b) The same s-plane representation except now the function is SAMPLED. These poles continually repeat in the s-plane, which prevents its use for sampled signals.

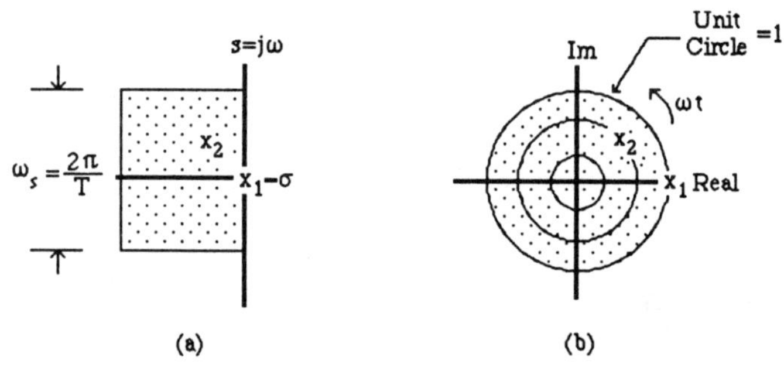

FIGURE 13.14 MAPPING FROM THE S- TO THE Z-PLANE: (a) Poles in the s-plane (for clarity, not shown repeating) (see Figure 13.13 (a) and (b)); (b) Poles in the z-plane.

1. **The imaginary (jω) axis in the s-plane (a) maps (transforms) into the unit circle in the z-plane (b). A pole on the jω axis will map onto the unit circle. A pole x_1 at the intersection of σ and jω will map to the unit circle on the z-plane real axis.**

2. **A single pole or one of two complex poles x_2, to the left of the $j\omega$ axis, maps to a lesser circle in the z-plane. A stable signal or system must have poles on or to the left of the $j\omega$ axis and likewise on or inside the unit circle.**

A BRIEF TABLE OF Z-TRANSFORM PAIRS

WAVEFORM	SIGNAL x[n]	SPECTRUM X(Z)	Z-PLANE POLES AND ZEROS
Unit Impulse	$\delta[n]$	1	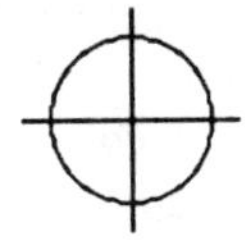
Unit Step	$u[n]$	$\frac{z}{z-1}$	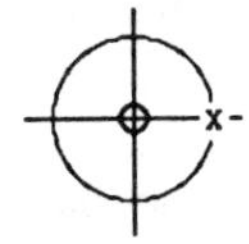
Decreasing Exponential	$a^n u[n]$	$\frac{z}{z-a}$	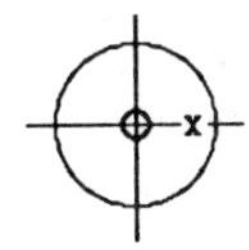
Sine Wave	$\sin \omega t$ $\sin n\Omega_0 u[n]$	$\frac{z \sin \omega T}{z^2 - 2z\cos\omega T + 1}$	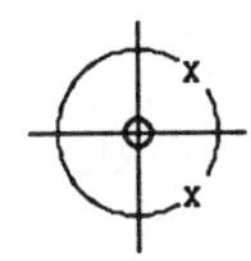
Cosine Wave	$\cos \omega t$ $\cos n\Omega_0 u[n]$	$\frac{z(z - \cos \omega T)}{z^2 - 2z\cos\omega T + 1}$	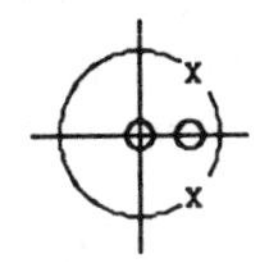

FIGURE 13.15 Z-Transform pairs.

13.8 ELEMENTARY DIGITAL FILTER CONCEPTS— SIMPLE DIGITAL FILTER MODELS

TM HOW DO DIGITAL (AND ANALOG) FILTERS REALLY WORK?

There are two models that apply to the physical action of simple analog filters. In the TIME domain, the charging of a capacitor or the production of magnetic flux in an inductor cannot happen instantaneously. Circuits that use these elements, combined with resistors, take time to charge or discharge, flux or decay, thus their slow or rapid step response. In the FREQUENCY domain, the ability to charge or discharge, flux or decay, is a function of the rate of change of the input signal. By definition, a low-pass analog filter can follow the changes of a slowly moving signal, but not the changes of a rapidly (high-frequency) moving signal.

There are likewise at least two models that apply to digital filters. The simplest digital filter is the AVERAGER. As with any averaging technique, the digital averager tends to discard the rapid (high-frequency) extremes. A second digital filter method uses the techniques of delay (shift) and add. If a symmetrical, continuing signal is delayed (time shifted) an amount equal to one-half of its period and then added to its original, continuing self, the result will be zero. (The shifted signal will be the exact inverse of the original and the two will completely cancel when added.) Likewise, if a signal is delayed (shifted) only a very, very small amount, the addition will be about the same wave shape of the original signal but, in the limit, twice the amplitude. Thus, this shift and add algorithm will produce a resultant that varies from an amplitude of twice (2A) the original signal, when the delay was equal to zero, to no output at all when the signal is delayed one-half a period. The delay should not be more than one-half of the signal period, since this would violate the rules of sampling and produce aliasing.

Digital filters are configured in two general forms—without feedback and with feedback. The digital filter without feedback is somewhat reminiscent of a simple RC or RL analog filter and the feedback digital filter is somewhat like the active-feedback analog filter. The output y(n) of the non-feedback digital filter is only dependent upon the input signal x(n) (and the filter transfer function), whereas the output of the feedback digital filter is, like the active analog filter, also dependent upon its (fed back) output.

MM HOW DIGITAL FILTERS REALLY WORK—ELEMENTARY EQUATIONS

For a very simple non-feedback (*non-recursive*) averaging digital filter:

$$y(n) = \frac{1}{N}\,[x(n) + x(n-1) + x(n-2) + \cdots]$$

where N is the number of terms (delays) (E 13.33)

For a very simple feedback (*recursive*) digital filter:

$$y(n) = x(n) + y(n-1) \qquad \text{(E 13.34)}$$

GM HOW DIGITAL FILTERS REALLY WORK—ELEMENTARY GRAPHIC MODELS

The graphical representation of a non-feedback digital filter is an elementary delay (shift) and add circuit:

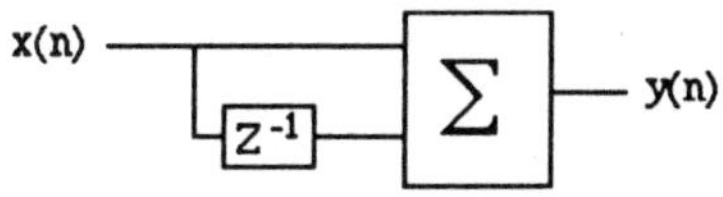

FIGURE 13.16 A very simple delay (shift) and add configuration; the Z^{-1} represents one unit delay, which is often the delay time of one clock pulse.

An elementary, non-feedback (non-recursive) averaging digital filter:

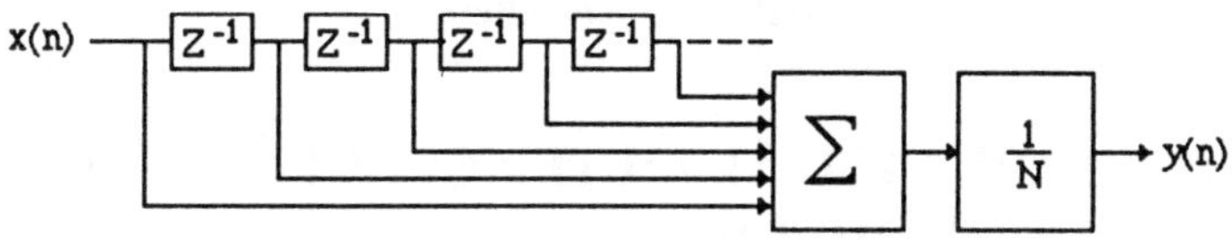

FIGURE 13.17 A simple non-recursive digital filter (see E 13.33).

Simplified wave forms illustrating the use of delay (shift) and add:

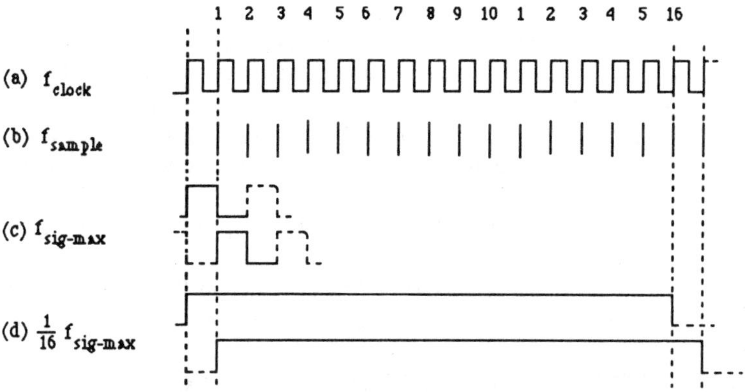

FIGURE 13.18 (a) A wave form representing the clock frequency for a simple digital filter system; (b) The impulses that represent the sampling time or the sampling pulse. As illustrated, the f_{sample} pulse would be timed to the leading or rising edge of the clock pulses; (c) Wave forms indicating the maximum signal frequency—with two samples per period—that can be sampled without aliasing; (d) Lower-frequency wave forms with 32 samples per period.

When the lower (d) wave form is delayed and added to the upper (d) wave form, the resulting amplitude will be quite close to twice that of the original wave form. If the lower (c) wave form is delayed and added to the upper (c) wave form, resulting amplitude will be zero. Thus, low-pass digital filters produce output amplitudes that are close to 2A at very low frequencies and then drop off to zero as the frequency of the input signal increases to a value of one-half that of the sample frequency.

For a very simple feedback (recursive) digital filter:

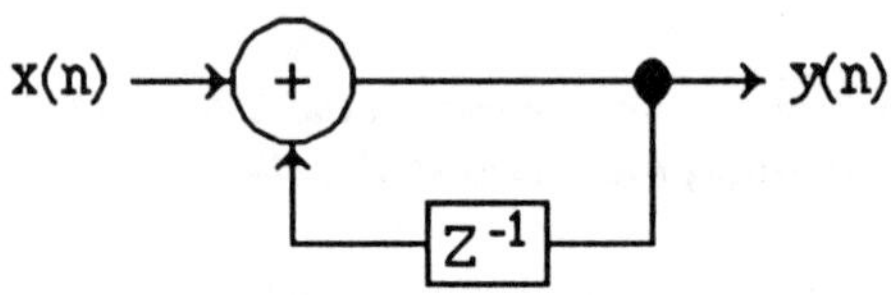

FIGURE 13.19 A simple recursive (feedback) digital filter.

13.9 ELEMENTARY DIGITAL FILTER CONCEPTS—THE SAVINGS ACCOUNT MODEL

TM THE SAVINGS ACCOUNT AND DIGITAL FILTERS

Assume that a savings account is opened on January 2 with a deposit of ten dollars and it is the intention to deposit ten dollars at the first of each month. For now, disregard the interest that is earned. The idea of a regular monthly deposit can be used to model [1] a digital accumulator, [2] a digital averager, and [3] a moving averager. Assume that six deposits—January through June—have been faithfully made and now some calculations are in order. (Note that the ideas of an averager and a moving averager are analogous to the concepts of simple digital filters.)

A digital accumulator will sum the total payments (samples) to date. A digital averager will find the average payment and a *moving* averager will find the average payment for the first six payments (January through June), the next six months (February through July), and the next, and the next, etc.

MM THE SAVINGS ACCOUNT AND DIGITAL FILTERS

1. For a non-feedback digital payment accumulator:

The Output Total = Jan + Feb + Mar + · · · · · June (E 13.35)

$$y[n] = x[n] + x[n-1] + x[n-2] + \cdots x[n-5] \qquad \text{(E 13.36)}$$

$$y[n] = \sum_{k=0}^{k=M-1} x[n-k]$$

where M is the number of monthly payments (samples) (E 13.37)

2. For a non-feedback digital payment averager:

$$\text{The Average Output} = \frac{1}{M}\{\text{Jan} + \text{Feb} + \text{Mar} + \cdots\cdots \text{June}\} \quad \text{(E 13.38)}$$

$$y[n] = \frac{1}{M}\{x[n] + x[n-1] + x[n-2] + \cdots\cdots x[n-5]\} \quad \text{(E 13.39)}$$

$$y[n] = \frac{1}{M}\sum_{k=0}^{k=M-1} x[n-k] \quad \text{(E 13.40)}$$

3. For a feedback (recursive) digital payment accumulator:

The Output = The Next Input + The Last TOTAL

$$y[n] = x[n] + y[n-1] \quad \text{(E 13.41)}$$

GM THE SAVINGS ACCOUNT AND DIGITAL FILTERS

See Figure 13.19.

13.10 DIGITAL FILTERS WITH WEIGHTING COEFFICIENTS

TM DIGITAL FILTERS WITH WEIGHTING COEFFICIENTS

The elementary moving average or "savings account" filters have very limited applications. Their frequency cutoff is very broad and imprecise. In fact, if even a general sinc or sin x/x low-pass (or high-pass) characteristic is desired, many delays are required in a non-feedback configuration. The words "filter coefficient" should bring to mind the ideas of the Butterworth or Chebyshev polynomials and their resulting analog filter coefficients. Practical digital filters with sharp cutoffs require circuits that will allow weighting (filter) coefficients, such as these, to be incorporated in the stages of both non-feedback and feedback design. In fact, it is the use of these very versatile weighting coefficients that make digital filters so desirable. Practical, stable filters, with frequency cutoffs of more than 40 dB per octave, are almost impossible with analog designs but are a "piece of cake" in the digital world.

The addition of weighting coefficients to the elementary filter circuits, or their equivalent software routines, typically requires a two-input multiply function. For reference, one can think of the elementary moving average filter delay elements as incorporating IMPLIED multiply-by-one coefficients.

MM DIGITAL FILTERS WITH WEIGHTING COEFFICIENTS

The complete processor (filter) difference equation including recursive and non-recursive terms:

y[n]	=	Recursive Terms Including Coefficients	Non-Recursive Terms Including Coefficients
	=	$a_1y(n-1) + a_2y(n-2) + a_3y(n-3) + \cdots$	$b_0x[n] + b_1x[n] + b_1x[n-1] + b_2x[n-2] + \cdots$

(E 13.42)

Equation 13.42 can be broken into y[n] = recursive filter terms only or y[n] = non-recursive terms only or used, as it is presented, to include a processor/filter with both recursive and non-recursive terms.

GM DIGITAL FILTERS WITH WEIGHTING COEFFICIENTS

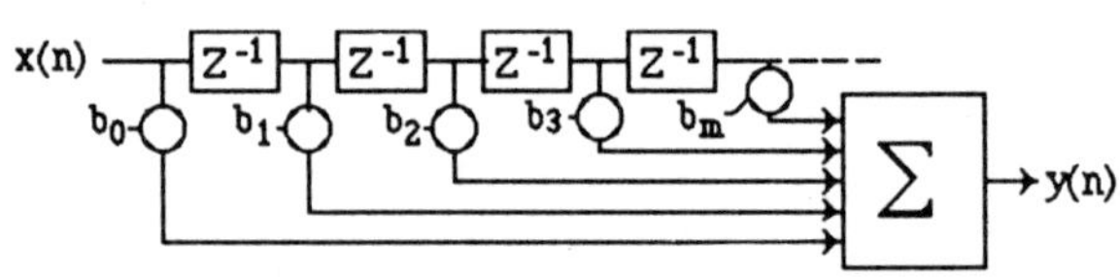

FIGURE 13.20 An elementary non-recursive filter, including coefficient multipliers.

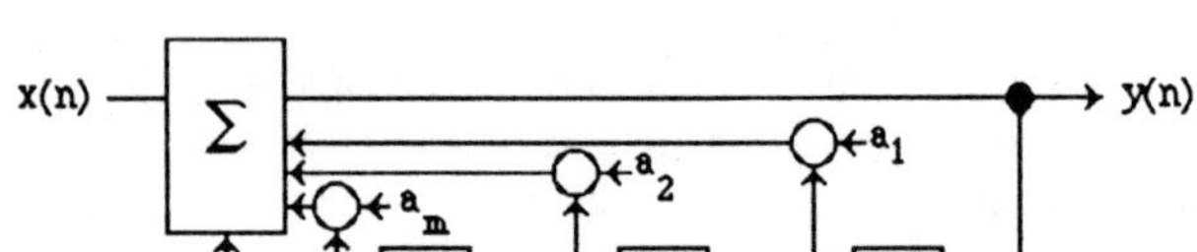

FIGURE 13.21 An elementary recursive filter, including coefficient multipliers.

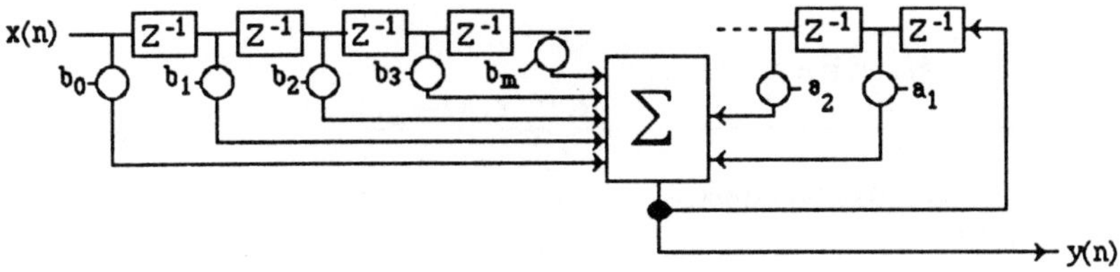

FIGURE 13.22 An elementary digital filter model with both non-recursive and recursive sections that include processing coefficients (see E 13.42).

13.11 DIFFERENT FORMS OF DIGITAL FILTER EQUATIONS—RELATIONSHIPS

TM DIGITAL FILTER EQUATIONS AND RELATIONSHIPS

The defining equation for the digital filter/processor is its transfer function. This relates the parameters of the input, such as amplitude, phase, spectrum, impulse or step content and timing, etc. The "work" of the filter/processor, as described in the appropriate equations, is to *convolve* the input signal with the transfer function in the time domain or to *multiply* the input signal by the transfer function in the frequency domain.

If the equation for the transfer function is known—as a function of Z (the Z-Transform)—the equation for the output in time or frequency can be derived. Likewise, it is possible to work "backward or forward" from most forms of the filter equations to derive the other forms.

MM DIGITAL FILTER EQUATIONS AND RELATIONSHIPS

The forms of the transfer equation:

$$H(z) = \frac{\sum_{k=0}^{k=M} b_k z^{-k}}{\sum_{k=0}^{k=N} a_k z^{-k}}$$

$$= \frac{\sum_{k=0}^{k=M} \text{non-recursive part (zeros)}}{\sum_{k=0}^{k=N} \text{recursive part (poles)}} = \frac{K\,(z-z_1)(z-z_2)(z-z_3)\cdots}{(z-p_1)(z-p_2)(z-p_3)\cdots} \qquad \text{(E 13.43)}$$

$$H(\Omega) = \frac{K\,\exp(j\Omega-z_1)\,\exp(j\Omega-z_2)\,\exp(j\Omega-z_3)\cdots}{\exp(j\Omega-p_1)\,\exp(j\Omega-p_2)\,\exp(j\Omega-p_3)\cdots} \qquad \text{(E 13.44)}$$

$$H(z) = \frac{Y(z)}{X(z)} \qquad \text{(E 13.45)}$$

$$Y(z) = H(z) * X(z) \qquad \text{(E 13.46)}$$

Equation 13.46 can be put in the form of an output difference equation with the rule that states $z^{+\kappa}$ is the equivalent of going forward in time κ intervals $(n+\kappa)$ in the transfer-function equation and $z^{-\kappa}$ is the equivalent of going back in time κ intervals $(n-\kappa)$.

Example 13.3: Convert the following arbitrary transfer-function equation to an output difference equation.

$$H(z) = \frac{Y(z)}{X(z)} = \frac{1}{4z^2+1}$$

$$Y(z)\,(4z^2+1) = X(z)$$

$4z^2\,Y(z) + Y(z) = X(z)$ now, using the rule for z^κ time movements,

$4y\,[n+2] + y[(n+1] = x[n]$ and remembering that a difference equation can be manipulated just as any other algebraic equation, by subtracting 2 from the delay terms,

$4y\,[n] + y[n-1] = x[n-2]$ and manipulating and dividing both sides of the equations by 4,

$$y[n] = -\,0.25\,y[n-1] + 0.25\,x[n-2]$$

GM DIGITAL FILTER EQUATIONS AND RELATIONSHIPS

	INPUT SIGNAL	TRANSFER FUNCTION	OUTPUT SIGNAL
In Time:	x[n]	h[n]	y[n] = x[n] * h[n]
In Frequency:	X(Ω)	H(Ω)	Y(Ω) = X(Ω) * H(Ω)
In z:	X(z)	H(z)	Y(z) = X(z) * H(z)

FIGURE 13.23 The relationships between time, frequency and z.

13.12 DIGITAL FILTER REPRESENTATIONS IN THE Z-PLANE

TM ZEROS, POLES AND COEFFICIENTS

The H(z) transfer function and its z-plane graphical representation, just as the analog transfer function and the s-plane representation, give an excellent understanding of the character and limits of a digital filter. Digital filters must operate in a frequency range dictated by the sampling frequency. The resulting frequency axis is usually designated from zero to π or from (the equivalent) zero to $f_{sample}/2$.

There are four general pole and zero locations for digital filters. Low-pass filters are usually considered as the prototype standard and high-pass, band-pass, and band-reject filters are often adapted from their fundamental designs. The number of zeros or poles depends upon the particular design but, in general, the more zeros in a non-recursive (non-feedback) filter and the more poles in a recursive (feedback) filter, the steeper the cutoff. It is possible to have all zero non-recursive filters and all pole-recursive filters, but this is not common. Poles in non-recursive and zeros in recursive filters are used to provide proper or desired delay or to assure stable operation.

One common way to design an upper band digital filter is first to design an equivalent low-pass filter and then shift this prototype design up the frequency axis. One common shift method is to multiply the time-domain impulse-difference equation h[n] by a factor cos (nπ/P), where n is each of the nth sample and 1/P is the position on the frequency (0 to π) axis.

Coefficients are the weighting factors for the particular zeros or poles. The non-recursive savings account examples had assumed coefficients of unity because a "unit" payment (say, ten dollars) was made each month. A different weighting factor would be included if the payments varied. Unit coefficients position the zeros or poles on the unit circle of the z-plane. Coefficients of less than unity are used for three general reasons.

1. The design, especially of recursive filters, most often calls for less-than-unity coefficients.
2. Rounding and other arithmetic errors often result in the equivalent of placing a pole outside the unit circle which, by the definition of the z-plane and its unit circle, makes the system unstable. A coefficient of 0.9 or 0.95 will usually insure stable operation. It is often good practice not to use unity pole coefficients.
3. As a special case of this last item, the coefficients of filters adapted from analog designs, such as Butterworth, Chebyshev, etc., usually do not contain unity coefficients.

MM ZEROS, POLES AND COEFFICIENTS

One method of shifting a low-pass prototype filter to form its equivalent band-pass or high-pass counterpart:

$$h[n]_{\text{new equivalent}} = \frac{1}{C} \cos n \pi \frac{1}{P} \tag{E 13.47}$$

where there are C coefficients and $\frac{1}{P}$ is the frequency position relative to π such as $\frac{1}{2}\pi$, $\frac{1}{3}\pi$, etc.

Example 13.4: Assume a five-term moving average filter with an impulse response shown in Figure 13.24 (a).

1. Shift the impulse response to make the low-pass filter *causal* (a causal impulse response always starts at h[n] = 0). See Figure 13.24 (b).
2. Convert this new casual low-pass filter to a band-pass filter centered at $\pi/2$.

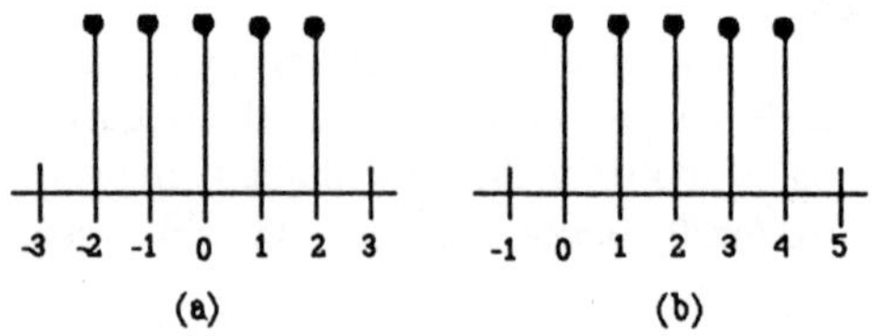

FIGURE 13.24 (a) The non-causal filter impulse response; (b) The causal impulse response.

1. Since there are five terms in the filter, the causal zero terms will be spaced on the unit circle every $\pi/5$ or 72° and be symmetric about the left real axis. The frequency-domain equation will be:

$$H(\Omega) = K(\cos 36^\circ + \cos 108^\circ + \cos 180^\circ + \cos 252^\circ + \cos 324^\circ)$$

K is a gain term based upon the method of calculation.

2. Using E 13.47 to shift each zero term $\pi/2$ or 90° to convert to a band-pass filter.

$$H(\Omega) = K'(\cos 126^\circ + \cos 198^\circ + \cos 270^\circ + \cos 342^\circ + \cos 54^\circ)$$

The K' term would include the $\frac{1}{C}$ plus any additional calculation method adjustments.

Note that the zeros are symmetric about the 180° point for the low-pass filter and symmetric about the 270° point for the band-pass filter.

GM ZEROS, POLES AND COEFFICIENTS

1. The general zero and pole positions in the z-plane:

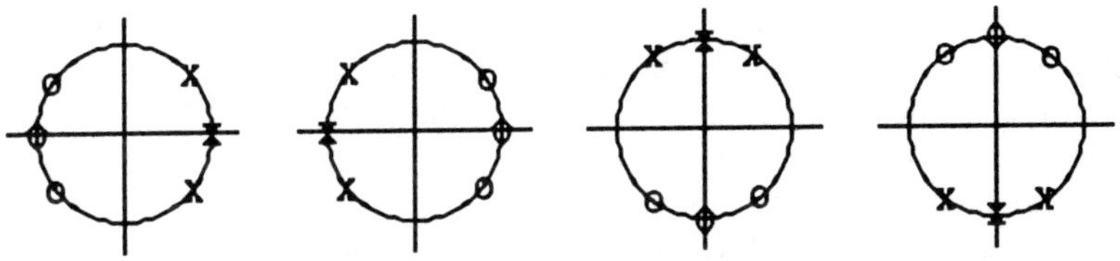

(a) (b) (c) (d)

FIGURE 13.25 The general zero and pole positions for digital filters. Each filter may contain all zeros, all poles or a combination of the two. Likewise, the zeros and/or the poles may extend around the circle, depending upon their position and number relative to the sampling frequency. (a) For a low-pass filter, the zeros tend to gravitate to the left real-axis side and poles to the right real-axis side; (b) For a high-pass filter, the zero-pole positions are reversed; (c) The zero-pole positions for a band-pass filter; (d) The zero-pole positions for a band-rejection filter.

2. The general positions of low-pass, band-pass, and high-pass filters relative to the sampling frequency:

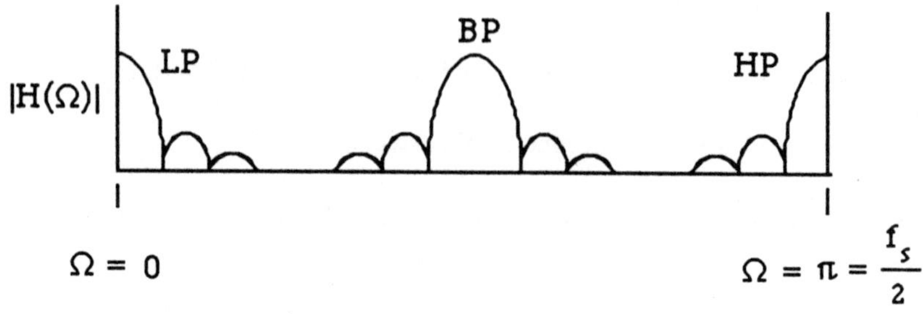

FIGURE 13.26 The relative frequency positions of digital filters.

13.13 THE DSP LEARNING, DESIGN AND ANALYSIS LABORATORY

THE DSP LEARNING LABORATORY

To master the ideas and techniques of digital signal processing, one must use digital signal processing! As in any discipline of engineering, learning DSP requires the proper mix of exposure to the conceptual and to the applied. The regimen of the laboratory, with its frustration and defeats, as well as its victories in truly learning how to apply an analysis or design concept, is mandatory. To obtain a usable knowledge of the workings and the potential of DSP, the very necessary formal or informal classroom-type of learning must be augmented with hands-on laboratory experiences.

The prerequisites for a DSP learning laboratory include the rudimentary knowledge of personal computer or work station operation, at least one computer language and a rudimentary ability to use word processing and spreadsheet programs. These must be the "givens" in the learning equation. They have replaced the pencil, ruler, compass, slide rule, and, many times, the soldering iron of the past.

The requisites for the learning DSP laboratory depend upon the level of experience and sophistication desired. For instance, rudimentary experience can be obtained if the investigator, with the previously mentioned prerequisites, uses only a series of computer programs and techniques similar to those listed in [1], which follow. More practical experience, and thus more valuable skills and polish, can be obtained if, in addition, the laboratory experience includes the actual building and testing of DSP elements, such as digital filters, using integrated logic circuits on a wire-wrap board. This is outlined in [2]. A third laboratory experience should include the building, testing, and analyzing of DSP building blocks using the complete, specialized digital processing integrated circuits available from most large IC manufacturers.

More and more textbooks are including software (sometimes even a disk is included) that is designed to provide the student with the facility to create and analyze DSP building blocks. As one excellent example, see the Lynn and Fuerst reference.

1. THE REQUISITE COMPUTER PROGRAMS FOR A BASIC DSP LEARNING LABORATORY

The individual who wishes to learn or upgrade to a knowledge of DSP should have, or acquire, as a minimum, the following:

(a) Programs that will provide the facility to create and manipulate digital signals. Such operations should include signal scaling, shifting, folding, arithmetic operations on one or more signals, convolution, correlation, and forward and reverse Fast Fourier Transforms.

(b) Programs that will provide the facility to create and manipulate digital processes. These processes should include the above rudimentary operations plus the ability to create and analyze both elementary FIR and IIR digital filters and a facility to create and analyze digital filters based upon the major analog prototype designs (Butterworth, Chebyshev, Bessel, etc.).

2. THE REQUISITE HARDWARE AND SOFTWARE FOR THE BASIC BUILDING, TESTING AND ANALYZING OF FUNDAMENTAL DSP ELEMENTS

(a) One of the easiest and most cost-effective techniques to learn the principles of DSP is the use of computer-generated software simulations. Many personal computers now include the facilities to output synthesized music routines. The general area of computer music and computer synthesizers is one of the better ways to create and understand the workings of DSP. Likewise, there is a wealth of literature and design information available in this area.

(b) Building DSP hardware elements, such as digital filters, requires somewhat sophisticated prototyping plug boards, access to prototype IC boards, or some form of wire-wrapping equipment. Of these three, wire wrapping is the most versatile.

(c) DSP hardware module boards, of course, require appropriate test and analysis hardware. It is often more instructive to use self-made input and clock signals than to use commercial test or clock generators. Likewise, although it is sometime laborious, the use of sychronized oscilloscopes is usually more instructive than using dedicated logic analyzers. Once a plan for a DSP hardware laboratory is carefully formulated and executed, the facilities can be used over and over for ongoing classes.

3. THE REQUISITE HARDWARE AND SOFTWARE FOR USING, TESTING AND ANALYZING DEDICATED SIGNAL PROCESSING INTEGRATED CIRCUITS

(a) A large sector of the "real-world" DSP is produced with dedicated signal processor integrated circuits such as the Texas Instruments TMS320 family, the Motorola DSP56000 family and several others (see the Terrell reference). These not only provide the basic DSP operations but can and are being used as processors in the fields of robotics, voice and speech recognition and synthesis, medical signal processing, telephone processing, etc. In most cases, the manufacturers supply abundant application and interfacing information and references to their applications in industry and research.

(b) Many of these DSP circuit manufacturers also provide complete hardware development, test, and analysis packages so that the investigators can easily interface their basic unit with other systems, such as a personal computer. Usually, much of the beginning software, as well as the hardware, is included in these development packages.

14

Calculus for Circuit Analysis

14.1 AN INTRODUCTION—WHAT IS CALCULUS?

The contributions to the invention of calculus by Newton, Leibniz, Cauchy, and others (circa 1700) provided a method of solving a group of problems that scholars had pondered for many centuries—at least since the time of the ancient Greek civilization. Calculus is the study of change, often irregular change, and the rate at which a given change, or a collection of changes, takes place. The philosophy and techniques of calculus enable us also to determine results or consequences of change. Since electronic signals are usually changing, both in the transient and steady states, it is almost imperative that calculus must be used to study and analyze all but the simplest of them.

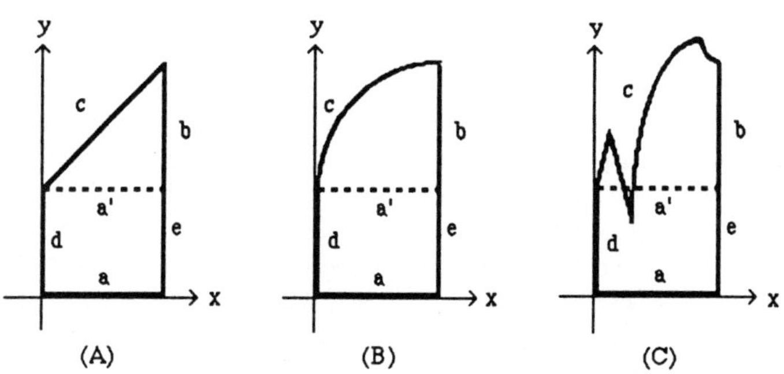

FIGURE 14.1 Geometric figures to illustrate the functions of calculus.

One of the first elementary concepts of calculus is concerned with the changes of a first quantity "y" as a result of the change in a second quantity "x". A line or curve can represent the movement of a point from the line's stated beginning to its stated end. Using the x and y axes for references, in Figure 14.1 (A), the line c represents the movement of a point up in the y direction as it also progresses from left to right in the x direction. There is a change in the y direction as it moves in the x direction. This form of the change in one parameter as a second is changing is familiar in circuit analysis as the change of current in an inductor with changing time or di/dt. In (A), the line c's "y" (parameter) is changing in a very regular way as "x" changes. However, in (B), the change, in the y direction with respect to the x direction, is not a simple straight line—it might be defined as part of a circle or maybe a parabola. The change of c in (C) is quite irregular and would require careful analysis before it might be possible to describe its y-direction change as a function (or functions) of its x-direction change.

Calculus is also concerned with finding maximum and minimum values or points. Again, using line c in (A), finding the minimum and maximum x and y values is very simple. The minimum and maximum value of line c can also be easily obtained in (B), but it would be a wee bit more difficult to write them in the form of f(x). For line c in (C), it would be more difficult still to write the minimum and maximum as f(x).

Calculus is also concerned with finding the "area under the curve." This "area" may indeed be a physical area or it may also be the charge on a capacitor or the flux in a magnetic circuit. In Figure 14.1 (A), the area of either a'bc or aebcd is quite easy to find by simple geometry. However, these areas in (B) and (C) become more difficult and require the methods of integral calculus.

14.2 THE GEOMETRIC FOUNDATIONS FOR THE CALCULUS OF CIRCUIT ANALYSIS

TM PLANES AND LINES

The most fundamental graphical building block to model a signal or the change in a signal is the line. Most signals are represented with lines, triangles, sine waves, etc., in a two-dimensional x-y or *Cartesian* plane. (Other planes such as the impedance plane, the s-plane or the z-plane are extensions of this basic two-dimensional form.) Most often the horizontal reference axis—the plane's abscissa—is labeled "x" and the vertical axis—the ordinate—is labeled "y". Straight (signal) lines, drawn in this x-y plane, can be defined as horizontal, vertical, or angle lines—where the angle is most often referenced to horizontal. Each line can be characterized with its own equation and its own *slope*. The slope of the line (usually termed "m") is the tangent of the angle it makes with the horizontal. If the angle is zero to 90 degrees (termed the first quadrant), its tangent, and thus its slope, is positive. If the angle is 90 to 180 degrees (the second quadrant), the tangent and the slope are negative. The slope of a line is positive in the third quadrant and negative in the fourth quadrant.

MM PLANES AND LINES

The slope of a line:

$$m = \mathrm{Tan}\ \theta = \frac{y_2 - y_1}{x_2 - x_1} \tag{E 14.1}$$

The slope of a horizontal line is 0 and of a vertical line is ∞

Some forms of the equation of a line:

$y = mx + b$ where m is the slope and b is the y–intercept (crossing) (E 14.2)

$$y_2 - y_1 = m(x_2 - x_1) \tag{E 14.3}$$

$\frac{x}{a} + \frac{y}{b} = 1$ where a and b are the intercepts on the x and y axes (E 14.4)

GM PLANES AND LINES

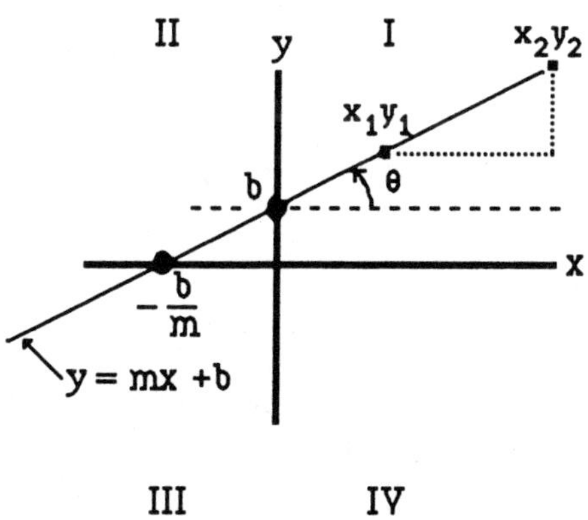

FIGURE 14.2 A sloping line in the x-y plane. The Roman numerals show the quadrants. The slope $m = \tan\theta = \frac{y_2 - y_1}{x_2 - x_1}$.

TM RIGHT TRIANGLES

The right triangle, which is a particular combination of lines, is the next logical extension from the elementary single line in the x-y plane. The impedance plane, the s-plane and the z-plane are all derived from the concept of the right triangle. Likewise, the right triangle, plus the square and the circle, are used as the fundamental units for understanding irregular shapes in calculus.

MM RIGHT TRIANGLES

The sides of an ABC right triangle:

$$c^2 = a^2 + b^2 \text{ where c is the hypotenuse} \quad \text{(E 14.5)}$$

The length of the line opposite the angle θ = length of the hypotenuse * sin θ or $b = c * \sin\theta$ (E 14.6)

The length of the line adjacent to the angle θ = length of the hypotenuse * cos θ or $a = c * \cos\theta$ (E 14.7)

$$\text{Tan } \theta = \frac{\text{The length of the line opposite the angle } \theta}{\text{The length of the line adjacent to the angle } \theta} = \frac{b}{a} \quad \text{(E 14.8)}$$

The area of a right triangle:

$$\text{Area } = \frac{1}{2}(a * b) \quad \text{(E 14.9)}$$

GM RIGHT TRIANGLES

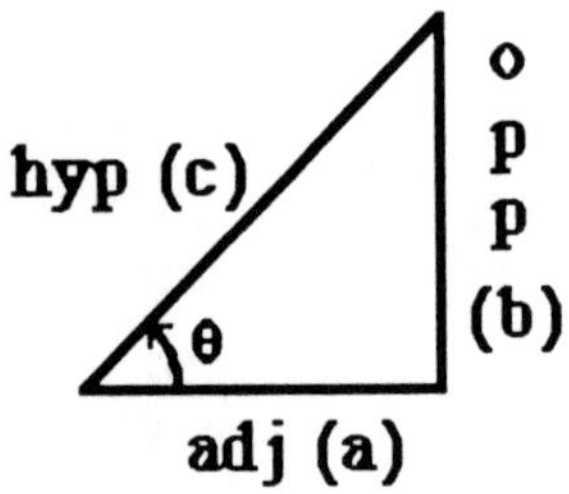

FIGURE 14.3 The prototype right triangle.

TM CIRCLES, PARABOLAS, AND ELLIPSES

The circle is one of the general forms important as the basis of many signals analyzed using calculus. The circle is the basic form of vector representation for analog signals and for the z-plane.

A parabola is a special curved line where all specified x,y points on the line are exactly the same distance from a fixed point and a special fixed line. The fixed point is termed the focus, the special straight line the directrix, and the midpoint on the parabola the vertex. Many second order signals and circuits graph into parabolas. The center line of the parabola is termed the axis. The sound intensity pattern of many microphones is parabolic.

For completeness, the ellipse is included. An ellipse is a closed curve with two foci. Since it is egg-shaped, it is defined with a long or major axis and also with a perpendicular (at a right angle), shorter or minor axis. The orbit of most satellites is some form of an ellipse.

MM CIRCLES, PARABOLAS, AND ELLIPSES

Let the radius of a circle be r, let the center point of the circle be in the x-y plane at x_1, y_1, and let any point on the circumference be x_2, y_2. (Many books use x_1, y_1 = x, y and x_2, y_2 = h,k.)

1a. For a circle with a center at 0 (xy):

$$r^2 = x_2^2 + y_2^2 \text{ or}$$

$$r = \sqrt{x_2^2 + y_2^2} \qquad \text{(E 14.10)}$$

1b. For a circle with a center somewhere in the plane—not at 0(xy):

$$r^2 = (x_2 - x_1)^2 + (y_2 - y_1)^2 \text{ or}$$

$$r = \sqrt{(x_2 - x_1)^2 + (y_2 - y_1)^2} \qquad \text{(E 14.11)}$$

2a. For a parabola with its center horizontal line on the x axis and the vertex at the origin:

$$y^2 = 2px$$

where p/2 is the distance from the focus to any point on the parabola and p is the distance from the focus to the directrix. (E 14.12)

(Note: Some books use p as the distance from the focus to a point on the parabola and 2p as the distance from the focus to the directrix. This makes $y^2 = 4px$.)

2b. For a parabola with its center line vertical on the y axis and the vertex at the origin:

$x^2 = 2py$ where p is the distance from the focus to any point on the parabola. (E 14.13)

3a. For an ellipse that is symmetric about the x and y axes and with the two foci, –F and F, on the major x axis, any point x,y on the ellipse can be described by:

$\frac{x^2}{a^2} + \frac{y^2}{b^2} = 1$ where ±a are the x-coordinate boundary points and ±b are the y-coordinate boundary points. (E 14.14)

3b. For an ellipse that is symmetric about the x and y axes and with the two foci, –F and F, on the major y axis, any point x,y on the ellipse can be described by:

$$\frac{x^2}{b^2} + \frac{y^2}{a^2} = 1 \qquad \text{(E 14.15)}$$

GM CIRCLES, PARABOLAS, AND ELLIPSES

1. For a circle with a radius "r":

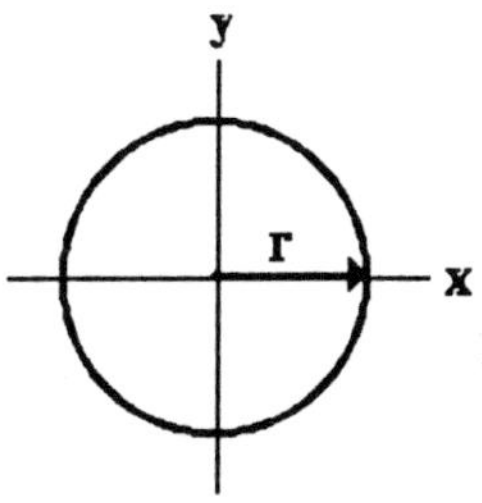

FIGURE 14.4 The prototype circle.

2. For the parabola:

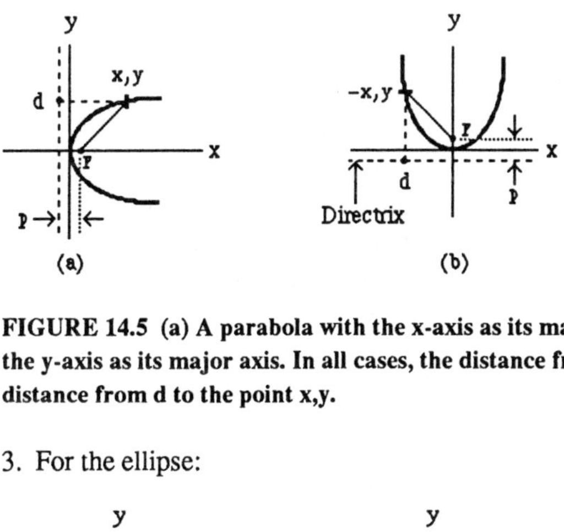

FIGURE 14.5 (a) A parabola with the x-axis as its major axis; (b) A parabola with the y-axis as its major axis. In all cases, the distance from F to a point x,y equals the distance from d to the point x,y.

3. For the ellipse:

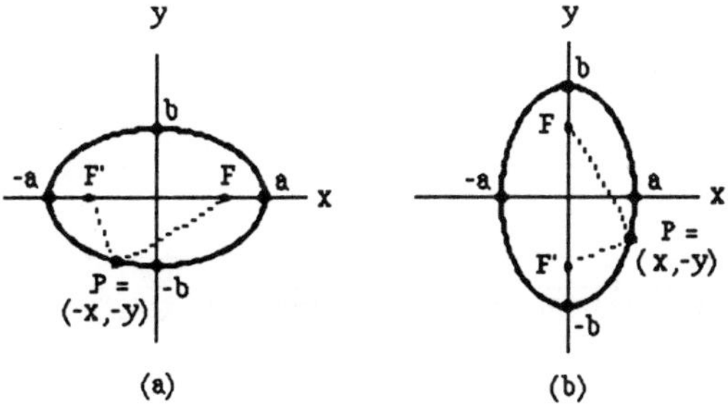

FIGURE 14.6 (a) An ellipse with the x-axis as its major axis; (b) An ellipse with the y-axis as its major axis.

TM SINUSOIDS AND EXPONENTIALS

Sinusoids and exponentials are among the most important geometric forms represented in circuit analysis. However, they are well documented and illustrated elsewhere in this book.

14.3 TRENDS AND LIMITS

TM TRENDS AND LIMITS

One of the most interesting and challenging concepts in calculus is the idea of the limit! What happens when such things as a series of numbers or an equation (or an idea) is taken down a trend path to its very end—to its limit? This is one of the first fun things about calculus because it lets you—really, it challenges you to—think about an idea that is completely foreign to elementary "two plus two" mathematics.

Example 14.1: Using the common circuit analysis equation $\frac{\sin x}{x}$, what happens when x goes to a value of zero?

First, it is important to remember that dividing by zero is NOT allowed because it will give an indefinable! One way to approach a problem like this is to try to establish a trend and see where the problem is going. What is the value just a very tiny distance or amount before the x goes to zero?

MM TRENDS AND LIMITS

Some forms of a limit problem:

$$\lim_{x \to 0} \frac{x^2}{x+1} = 0 \quad \text{(E 14.16)}$$

$$\lim_{x \to 0} \frac{\sin x}{x} = 1 \quad \text{(E 14.17)}$$

x	1	0.5	0.25	0.1	0.01	0.001
sin x / x	0.84147098	0.95885108	0.98961584	0.99833417	0.99998333	0.99999983

FIGURE 14.7 A table showing the calculations that establish a trend for $\lim_{x \to 0} \frac{\sin x}{x}$.

GM TRENDS AND LIMITS

The graphic of $\frac{\sin x}{x}$ showing the "y" value as "x" approaches (tends toward) zero.

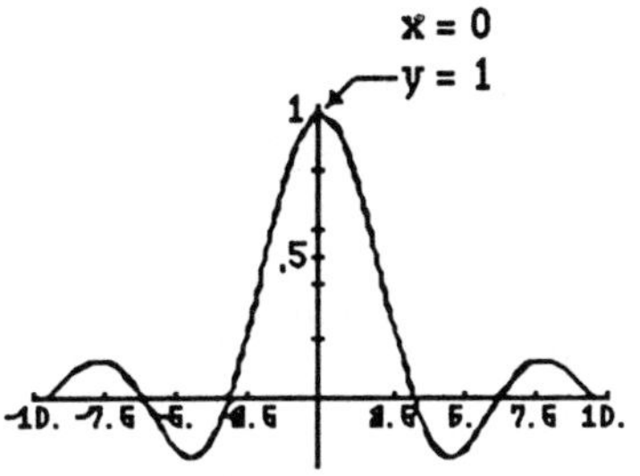

FIGURE 14.8 The familiar sin x/x curve with y=1 at x=0.

14.4 LINES, TANGENTS, AND CURVE FUNCTIONS

Since calculus is the mathematics of change, it can define the changes in curves and curved surfaces or their rates of change. Using this approach, it is possible to develop a background understanding for the branch of calculus known as differentiation.

TM LINES, TANGENTS, AND CURVES

The form of the slope equation $m = \frac{y_2 - y_1}{x_2 - x_1}$ can be used to help explain and define the curvature—and the change of curvature—at a given point on functions such as circles, parabolas, or sinusoids. One way to define the curvature or slope at a point "P" is to visualize or construct a *tangent* at that point P.

A tangent to a curved surface, such as a circle or a sinusoid, is a straight line that just touches the curve at a given point P and has the identical slope as the curve (the function) at that point P. Since any curve can be defined as the series of infinitely small points, then there must be a tangent for every point on a curve and thus, in the limit, there are an infinite number of tangents possible for every curve.

Assume two points, P1 and P2, on a curve are selected for reference. At each of these there is a certain curvature or slope. The question may be asked, "If the function of the curve is known, can the change (rate-of-change) of the curve's vertical or y value be calculated with respect to the change in its horizontal or x value?"

1. For a curved line, the change of the vertical or y distance from P2 back to P1 is $y_2 - y_1$ or Δy. (The Greek letter [delta] Δ means the change in or the change from.) Likewise, the change in the horizontal or x distance from P2 to P1 is $x_2 - x_1$ or Δx.
2. If the distance from P2 to P1 is made smaller and smaller, the ratio of $y_2 - y_1/x_2 - x_1$ or $\Delta y/\Delta x$ changes from simply the average slope of a line through P1 and P2 (termed a chord or c) to, in the limit, the *function* of, or point on, the curve.
3. Since at the exact point P1, y also equals f(x), then going just a wee bit farther on the function (curve) will equal the original point P1 or f(x) plus Δx. This small increment traversed will then equal the original point value f(x), plus the additional increment Δx, less the original value of f(x).
4. Using [3] above, in the limit, the ratio $\Delta y/\Delta x$ can be written as the quantity f(x), at P, plus any additional increment of x, minus the original value of f(x), all divided by Δx or $\Delta y/\Delta x = \{[f(x) + \Delta x) - f(x)] / \Delta x\}$.

MM LINES, TANGENTS, AND CURVES

[1] $\frac{y_2 - y_1}{x_2 - x_1} = \frac{\Delta y}{\Delta x}$ for two points on a curve (function) (E 14.18)

[2] $\lim_{\Delta x \to 0} \frac{\Delta y}{\Delta x}$ f(x) (if Δx is truly zero, we have not moved, so in the limit, $P1 = f(x_1)$) (E 14.19)

[3] $\Delta y = f(x + \Delta x) - f(x)$ (E 14.20)

[4] $\frac{\Delta y}{\Delta x} = \lim_{\Delta x \to 0} = \frac{f(x+\Delta x) - f(x)}{\Delta x}$ (E 14.21)

GM LINES, TANGENTS, AND CURVES

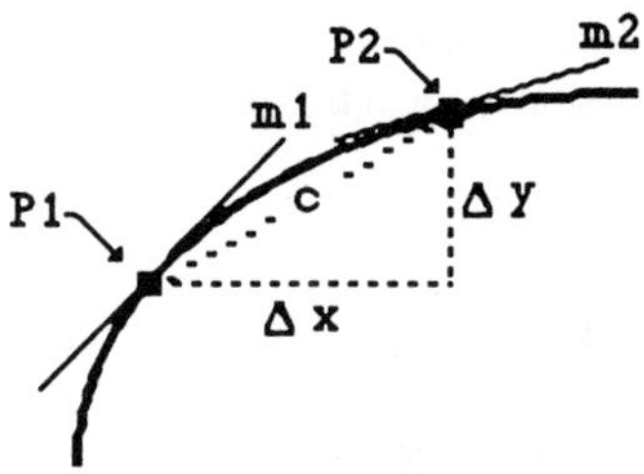

FIGURE 14.9 Two points on a curve. In the limit, P2 approaches P1. m1 and m2 are the original slopes or tangents at P1 and P2.

14.5 THE DERIVATIVE AND DIFFERENTIATION

TM THE DERIVATIVE

Based upon the idea of tangents, slopes and the limit, in the limit: The derivative of a function is the ratio of the increment of a function to the increment of an independent variable. Often the symbol f ' or *f* ' is used as the symbol for the derivative.

MM THE DERIVATIVE

$$f' = \lim_{\Delta x \to 0} \frac{\Delta y}{\Delta x} = \frac{f(x+\Delta x) - f(x)}{\Delta x} \qquad \text{(E 14.22)}$$

Special Note: This E 14.22 is one of the two or three most important equations in calculus! It is the basis for the differentiation of all functions.

To find the derivative of a function $y = f(x)$:

1. Put the equation $y = f(x)$ in the form $y + \Delta y = f(x + \Delta x)$
2. Manipulate the form to $\Delta y = f(x + \Delta x) - y = f(x + \Delta x) - f(x)$
3. Divide by Δx for $f' = \lim_{\Delta x \to 0} \frac{\Delta y}{\Delta x} = \frac{f(x+\Delta x) - f(x)}{\Delta x}$

(E 14.23)

It is necessary to use this equation, and the technique it represents, to truly understand the underlining process used in the differentiation of circuit-analysis functions. However, often a function, or a differentiation model similar to a function, can be found in a look up table. Likewise, many mathematics programs, using personal computers or work stations, can do much of the work of differentiation. The following section first indicates the general technique of differentiation and then lists a number of differentiation models that are common in circuit analysis.

14.6 MODELS FOR DIFFERENTIATION

MM and GM DIFFERENTIATE A FIRST-ORDER FUNCTION

Special Note: In the following graphics, the symbol "f(x)" = y.

Differentiate $f(x) = y = x$:

$$y + \Delta y = f(x + \Delta x) = (x + \Delta x)$$

$$\Delta y = (x + \Delta x) - x$$

$$\lim_{\Delta x \to 0} \frac{\Delta y}{\Delta x} = \frac{(x + \Delta x) - x}{\Delta x} = \frac{\Delta x}{\Delta x}$$

$$f' = \lim_{\Delta x \to 0} = 1$$

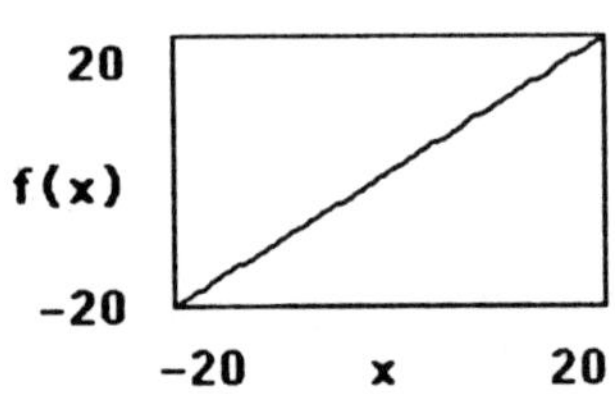

FIGURE 14.10 y = x.

MM and GM DIFFERENTIATE A SECOND-ORDER FUNCTION

Differentiate $f(x) = y = x^2$

$$y + \Delta y = f(x + \Delta x) = (x + \Delta x)^2$$

$$\Delta y = (x + \Delta x)^2 - x^2$$

$$\lim_{\Delta x \to 0} \frac{\Delta y}{\Delta x} = \frac{(x+\Delta x)^2 - x^2}{\Delta x}$$

$$= \frac{x^2 + 2x\Delta x + (\Delta x)^2 - x^2}{\Delta x}$$

$$= \frac{2x\Delta x + \Delta x^2}{\Delta x} = \frac{\Delta x(2x+\Delta x)}{\Delta x}$$

$$= 2x + \Delta x$$

$$\lim_{\Delta x \to 0} = 2x$$

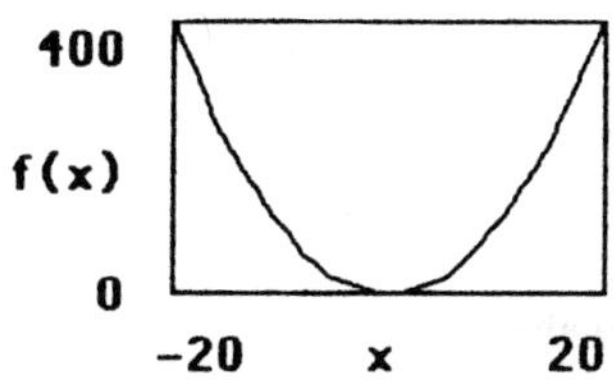

FIGURE 14.11 $y = x^2$.

MM and GM DIFFERENTIATE A CONSTANT FUNCTION

Special Note: The complete derivation will not be included in the remainder of this section.

Differentiate y = a (a constant or number)

$$f' = \frac{d}{dx}(a) = 0 \text{ (this says there is no } \Delta \text{ slope)} \qquad \text{(E 14.25)}$$

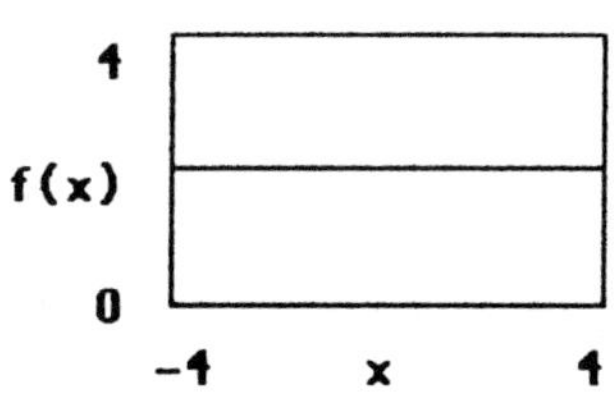

FIGURE 14.12 y = a (= 2).

MM and GM DIFFERENTIATE AN N-ORDER FUNCTION

Differentiate $y = x^a$

$$f' = \frac{d}{dx}(x^a) = \frac{d}{dx}(u^n) = nu^{n-1}\frac{du}{dx}$$

An example: differentiate $y = x^3$

$$\frac{d}{dx}(x^3) = 3x^2 \qquad \text{(E 14.26)}$$

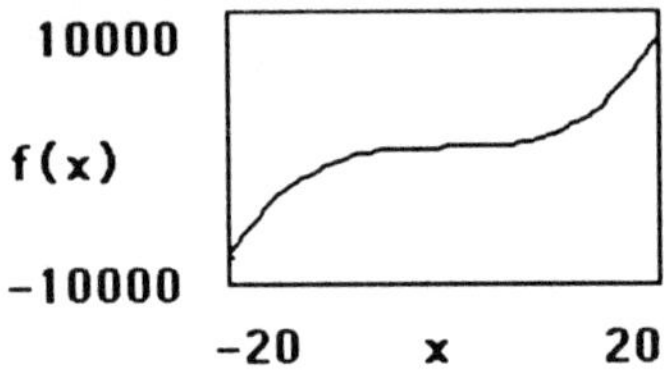

FIGURE 14.13 $y = x^3$ (See $y = x$ and $y = x^2$ illustrated above.)

Special Note: The remainder of this section follows the format of most books on calculus by using the letters "u" and "v" for functions and "a", "n" and "e" to represent constants.

MM and GM DIFFERENTIATE THE SUM OF TWO FUNCTIONS

$$f' = \frac{d}{dx}(u + v) = \frac{du}{dx} + \frac{dv}{dx} \qquad \text{(E 14.27)}$$

An example: Assume $y_1 = x^2$ and $y_2 = 20x$

$$y_1 + y_2 = y = x^2 + 20x$$

$$\frac{d}{dx}(x^2 + 20x) = 2x + 20$$

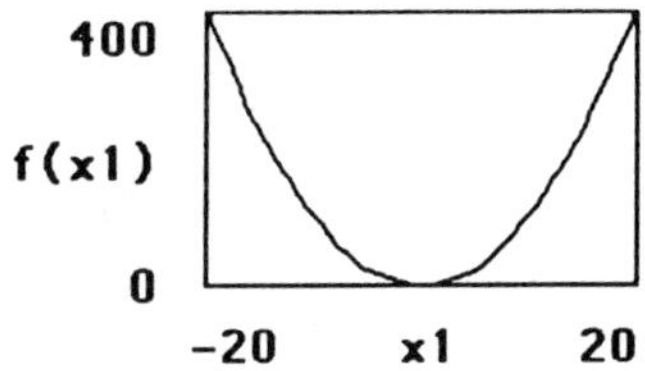

FIGURE 14.14 $y = x^2$.

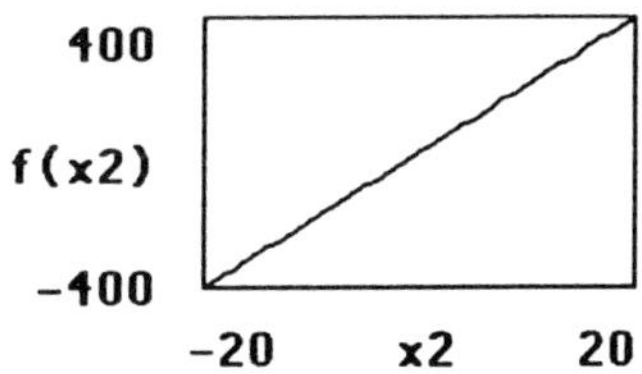

FIGURE 14.15 $y = 20x$.

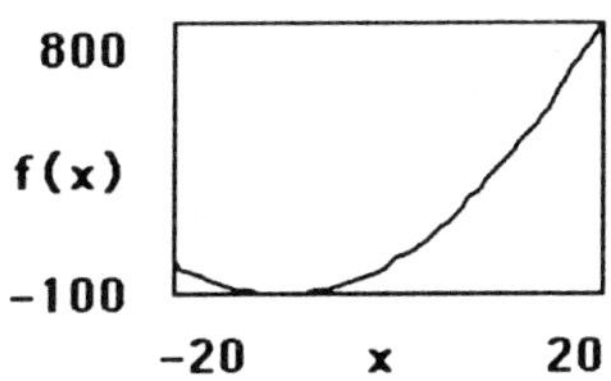

FIGURE 14.16 $y=x^2 + 20x$.

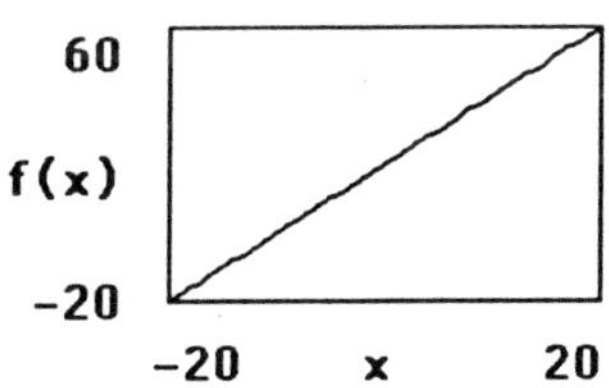

FIGURE 14.17 $y = 2x + 20$.

MM and GM DIFFERENTIATE THE PRODUCT OF TWO FUNCTIONS

$$f' = \frac{d}{dx}\,[\text{(Function one) times (function two)}]$$

$$f' = \frac{d}{dx}(u * v) = \frac{d(uv)}{dx} = u\frac{dv}{dx} + v\frac{du}{dx}$$

An example: Assume $u = y_1 = x^2$ and $v = y_2 = 20x$

$$u\frac{dv}{dx} + v\frac{du}{dx} = (x^2 * 20) + (20x * 2x) = 20x^2 + 40x^2 = 60x^2 \qquad \text{(E 14.28)}$$

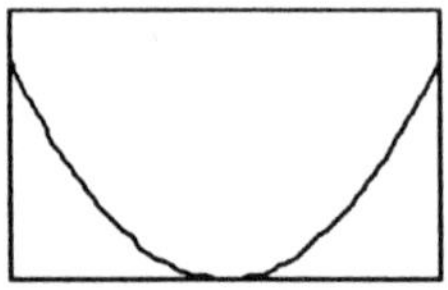

FIGURE 14.18 $y = 60x^2$.

MM and GM DIFFERENTIATE THE QUOTIENT OF TWO FUNCTIONS

$$f' = \frac{d}{dx}\text{ [(Function one) divided by (Function two)]}$$

$$f' = \frac{d}{dx}\left(\frac{u}{v}\right) = \frac{v\frac{du}{dx} - u\frac{dv}{dx}}{v^2} \qquad \text{(E 14.29)}$$

An example: Assume $u = y_1 = x^2$ and $v = y_2 = 20x$

$$\frac{v\frac{du}{dx} - u\frac{dv}{dx}}{v^2} = \frac{(20x * 2x) - (x^2 * 20)}{400x^2}$$

$$= \frac{(40x^2) - (20x^2)}{400x^2} = \frac{1}{20} = \text{a constant}$$

(See its graphic FORM in Figure 14.12.)

TM DIFFERENTIATE LOGARITHMIC FUNCTIONS

The logarithm of a number is the power to which a specified base must be raised to equal that number. The logarithm of 100 to base 10 is 2. The logarithm of 6 to base "e" is 1.79 · · · . The logarithm of 8 to base 2 is 3, etc.

MM and GM DIFFERENTIATE LOGARITHMIC FUNCTIONS

$$f' = \frac{d}{dx}(\log_a u) = \frac{1}{u}(\log_a e)\frac{du}{dx} \text{ This is the general equation.} \qquad \text{(E 14.30)}$$

$$f' = \frac{d}{dx}(\log_e u) = \frac{1}{u}(\ln u) = \frac{1}{u}\frac{du}{dx} \qquad \text{(E 14.31)}$$

$$f' = \frac{d}{dx}(\log_{10} u) = \frac{1}{u}(\log_{10} e)\frac{du}{dx} \qquad \text{(E 14.32)}$$

An example: Differentiate $y = \ln 25\, x^3$

$$f' = \frac{1}{u}\frac{du}{dx} = \frac{1}{25\, x^3} * 75\, x^2 = \frac{3}{x}$$

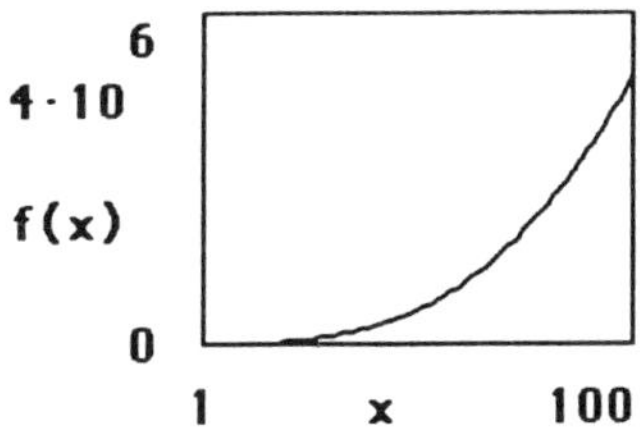

FIGURE 14.19 $y = \ln 25\, x^3$.

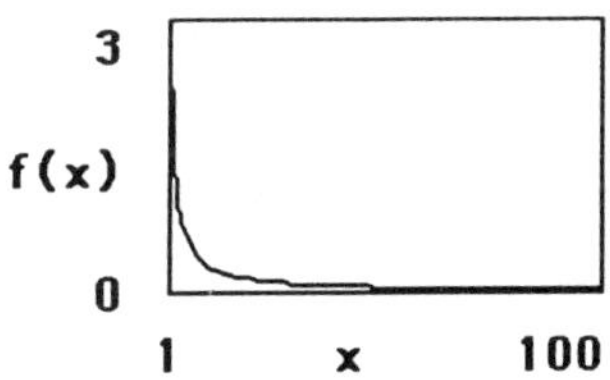

FIGURE 14.20 $y = \frac{1}{25\, x^3} * 75\, x^2 = \frac{3}{x}$.

An example: Differentiate $y = \log_{10} 25\, x^3$

$$f' = \frac{1}{u}(\log_{10} e)\frac{du}{dx} = \frac{1}{25 * x^3} * 0.434 * 75\, x^2 = \frac{1.302}{x}$$

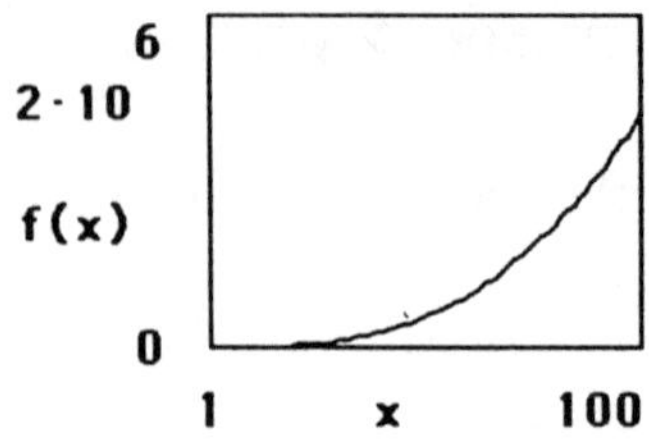

FIGURE 14.21 $y = \log_{10} 25\, x^3$.

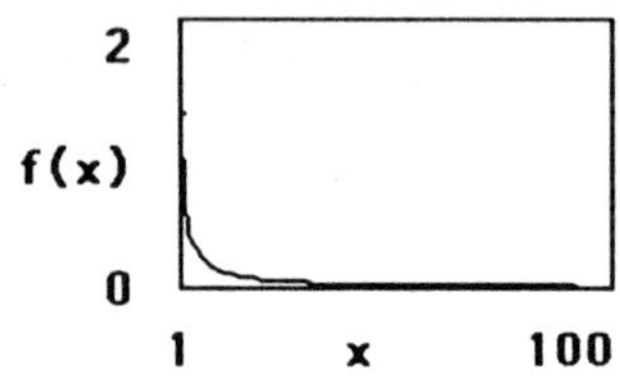

FIGURE 14.22 $y = \dfrac{1.302}{x}$.

TM DIFFERENTIATE FUNCTIONS TO A POWER

When differentiating a function to a power, the power can be positive, which can be termed an exponent, or negative, which can be termed a root (such as a square or cube root).

MM and GM DIFFERENTIATE FUNCTIONS TO A POWER

$$f' = \frac{d}{dx}(u^n) = nu^{n-1}\frac{du}{dx} \qquad \text{repeat of (E 14.26)}$$

$$f' = \frac{d}{dx}(a^u) = a^u * \ln a * \frac{du}{dx} \qquad \text{(E 14.33)}$$

An example: Assume $y = e^{x^{1/2}}$

$$f' = e^{x^{1/2}} * \ln e * \frac{d}{dx} x^{1/2}$$

$$= e^{x^{1/2}} * 1 * \frac{1}{2} x^{-1/2}$$

$$= e^{x^{1/2}} * 1 * \frac{1}{2\sqrt{x}}$$

$$= \frac{1}{2\sqrt{x}} e^{x^{1/2}}$$

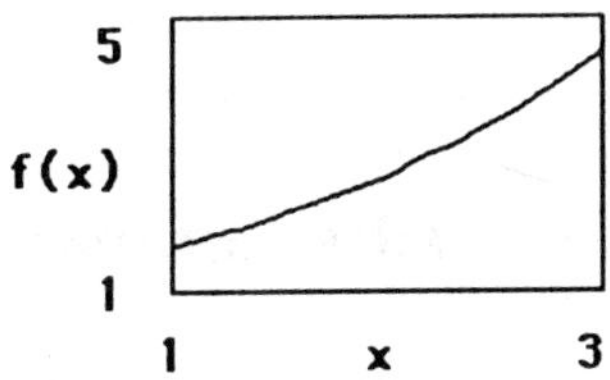

FIGURE 14.23 $y = e^{x^{1/2}}$.

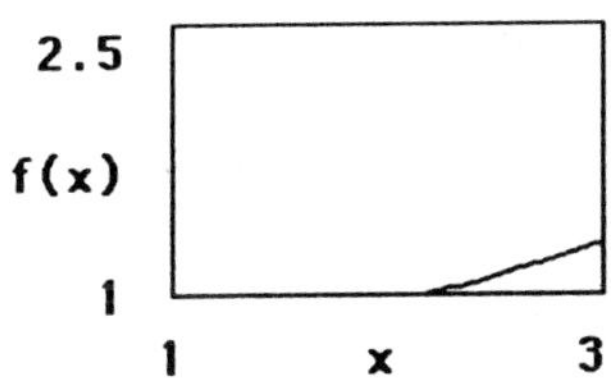

FIGURE 14.24 $y = = \frac{1}{2\sqrt{x}} e^{x^{1/2}}$.

An example: Assume $y = 12^{x^{1/2}}$

$$f' = 12^{x^{1/2}} * \ln 12 * \frac{d}{dx} x^{1/2}$$

$$= 12^{x^{1/2}} * 2.485 * \frac{1}{2} x^{-1/2}$$

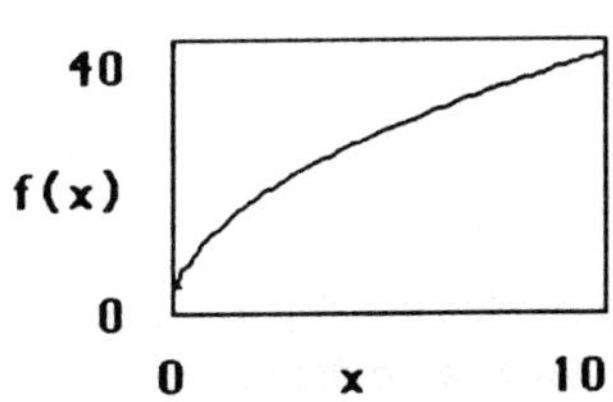

FIGURE 14.25 $y = 12^{x^{1/2}}$.

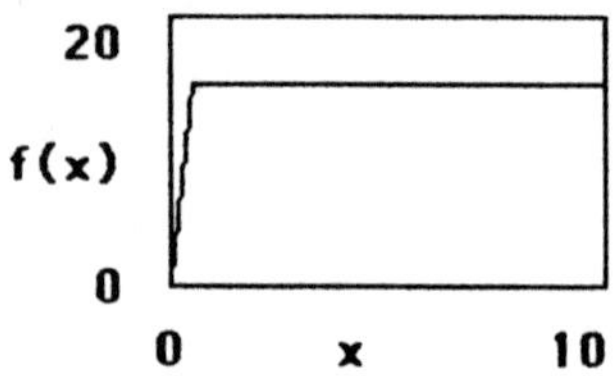

FIGURE 14.26 $y = 12^{x^{1/2}} * 2.485 * \frac{1}{2} x^{-1/2}$.

MM and GM DIFFERENTIATE TRIGONOMETRIC FUNCTIONS

$$f' = \frac{d}{dx}(\sin u) = \cos u \frac{du}{dx} \quad \text{(E 14.34)}$$

$$f' = \frac{d}{dx}(\cos u) = -\sin u \frac{du}{dx} \quad \text{(E 14.35)}$$

$$f' = \frac{d}{dx}(\tan u) = \sec^2 u \frac{du}{dx} \quad \text{(E 14.36)}$$

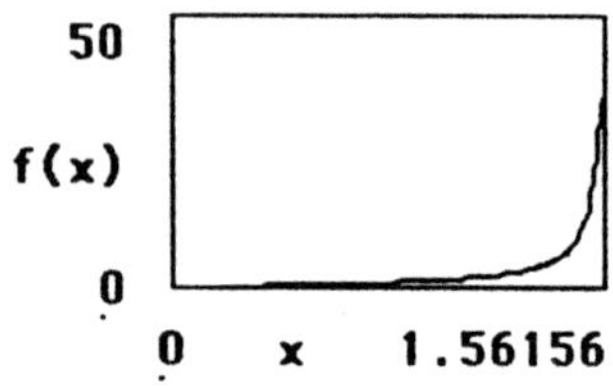

FIGURE 14.27 $y \approx \tan(u)$ (from 0 to $\approx 0.499\,\pi$).

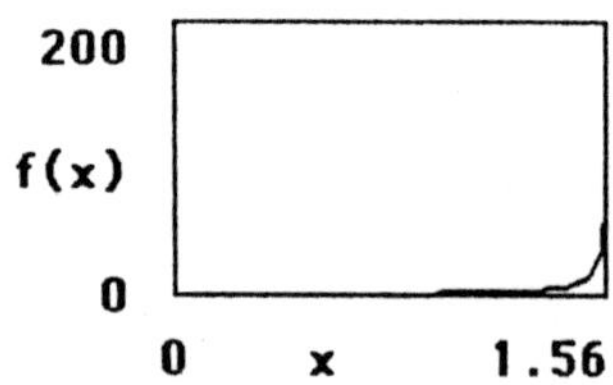

FIGURE 14.28 $y \approx \frac{d}{dx}(\tan u)$ (from 0 to $\approx 0.499\,\pi$).

Special Note: In the above two figures, the equations include an "approximately equal" (≈) sign. When the calculations for these figures were double checked with a hand calculator, some rounding errors in the second and third place were found due to the (one of many) methods of calculation and graphing that were chosen. These very slight operator-technique errors were left to illustrate that once again, even with these wonderful, inherently very accurate computer and graphing programs, IT IS THE RESPONSIBILITY OF THE USER TO UNDERSTAND THEIR BASIC ASSUMPTIONS AND THEIR TECHNIQUE OPTIONS.

TM DIFFERENTIATE A FUNCTION OF A FUNCTION

The preceding differentiating models have, in general, taken the form of "find the derivative of $f(x)$" or find the derivative of $y = f(x)$. There is also a type of differentiation problem where it is required to find the derivative of a function of a function as (f of g of x) or $f'(g(x))$.

MM DIFFERENTIATE A FUNCTION OF A FUNCTION—THE CHAIN RULE

If $y = f(u)$ and $u = f(x)$:

$$f'(x) = \frac{dy}{dx} = \frac{dy}{du} * \frac{du}{dx} \qquad \text{(E 14.37)}$$

An example: Assume $y = u^2 + 4$ and $u = 4x + 4$. Two methods will be used to find the derivative of this function of a function.

By substitution and by the Chain Rule:

$$y = (4x + 4)^2 + 4 = 16x^2 + 32x + 16 + 4$$

$$= 16x^2 + 32x + 20$$

$$f'(x) = 32x + 32$$

$$y = u^2 + 4 \text{ and } u = 4x + 4$$

$$\frac{dy}{du} = 2u \text{ and } \frac{du}{dx} = 4$$

$$\frac{dy}{du} * \frac{du}{dx} = (2u) * (4) = (8x + 8) * 4$$

$$\frac{dy}{du} * \frac{du}{dx} = 32x + 32$$

14.7 ANTIDERIVATIVES, THE INTEGRAL, AND INTEGRATION

TM ANTIDERIVATIVES, THE INTEGRAL AND INTEGRATION

The preceding section dealt with finding a new function, a derivative, based upon the change, or really the rate of change, of the first function. For instance, a cosine function f_2 is the derivative of a sine function f_1. A few moments' reflection will indicate that most (original) functions are themselves derivatives. Just as the cosine function f_2 was the derivative of a sine function f_1, so might that original f_1 be a derivative of SOME other function—say an f_0. The function f_0 can be derived by "going backward" from the original function f_1 just as f_1 was derived by going backward from f_2. This "going backward" process is termed finding the antiderivative or, more commonly, *integrating* or *integration.*

Integrating or integration can also be broadly defined as a process for "finding the area under the curve." Although this concept of finding the area is a somewhat limited definition of integration, its understanding is an excellent first step in appreciating the ideas and techniques of integration.

Integration or finding the antiderivative can be visualized using these three ideas:

1. Assume there is a function f_2 that has been created by finding the derivative of a first function f_1. Since f_1 is defined as a series of points, the derivative f_2 is actually the slope or rate of change (dy/dx), at a series of points, of f_1. The words "at a series of points" are the key to this part of the process. It is assumed that the function f_1 was some definable curve (say a sin x or sin ω) and that the derivative also defines another series of points such as cos x or cos ω. The derivative can be stated as a general equation such as y = cos ω (a series of points) or by a specific equation y = cos π/4 or cos 45° (one point).
2. If f_2 is integrated to "go back" to some original function f_1, there is again a choice of whether to use a general form or a specific (definite) form of integration. In other words, do you wish to find the general (indefinite) integral form for any cos ω or the definite form of (say) cos ω from 0 to 2π.
3. The "area under the curve" can be derived if integration is visualized as a form of summing (points). From very elementary geometry, the area of a rectangle is its height times its width or h*w. This width can be broken into a series of very small incremental rectangles, h-high and Δw-wide (h * Δw), where the Δw approaches zero. The total area is n(h * Δw), where n is the total number of small rectangles. In this simple illustration, the top of the large rectangle is assumed perfectly flat and thus with a

constant slope. If the top of the rectangle is not perfectly flat but, instead, some other definable function, then calculus is best used to find the area.

MM and GM ANTIDERIVATIVES, THE INTEGRAL AND INTEGRATION

Illustrations of antiderivatives:

1. If $f_1 = x^2$, then its derivative $\frac{d}{dx}$ shall be defined as f_2, and we already know $f_2 = 2x$, and now, going "backwards," the antiderivative or integral if f_2 is $f_1' = x^2 + C$ where the "C" stands for a constant—to be explained later. (E 14.38)

2. If $f_1 = \sin x$, then its derivative $\frac{d}{dx}$ shall be defined as f_2, and we know $f_2 = \cos x$, and now, going "backwards," the antiderivative or integral of f_2 is $f_1' = \sin x + C$ where the "C" stands for a constant—to be explained later. (E 14.39)

The area under the curve:

The following graphic and mathematical models show the relationships between [1] determining the area of a geometric figure using geometry, [2] using the elementary ideas of the sum (Σ) of "n" areas as Δx goes to zero and n goes to infinity, and [3] using the integral.

The area of a rectangle:

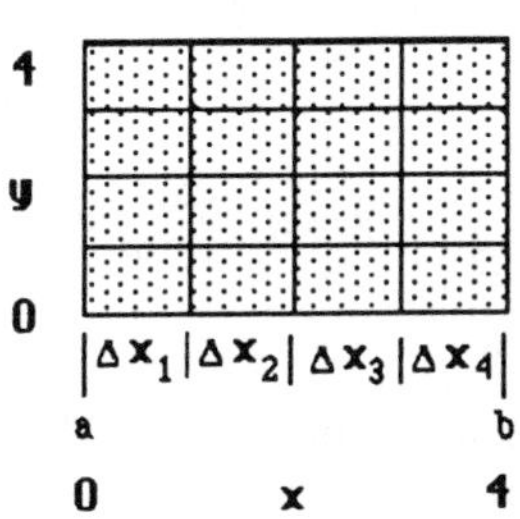

FIGURE 14.29 An x-y rectangle bounded at the top by y = f(x) = 4.

1. Using geometry, the area of Figure 14.29 = xy = 4 * 4 = 16

2. The area of each vertical column is y * Δx or 4 * 1 = 4. Using a summing process, $Area_{col} = f(x_i)\ \Delta x_i$, where the small column area $f(x)\ \Delta x = 4 * 1 = 4$. In this figure, there are four Δx's (four columns) and thus n = 4.

$$Area_{Tot} = \sum_{n=1}^{4} (f(x)\ \Delta x_1) + (f(x)\ \Delta x_2) + (f(x)\ \Delta x_3) + (f(x)\ \Delta x_4)$$

$$= (4 * 1) + (4 * 1) + (4 * 1) + (4 * 1) = 16$$

3. Now, using integration:

$$Area_{Tot} = \lim_{\Delta x \to 0} \sum_{n=1}^{\infty} f(x_i)\ \Delta x_i = \int_{a=0}^{4} x\ dx = 4 * x\Big|_0^4 = (4 * 4) - (4 * 0) = 16$$

The area of a right triangle:

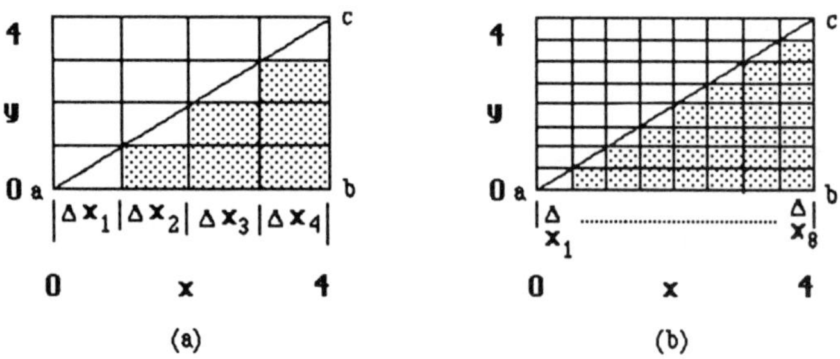

FIGURE 14.30 A right triangle bounded at the top by y = f(x) = x: (a) The approximate area of the abc triangle can be determined by summing the dotted rectangles under the line y = x. If more and more samples (Δx's) are selected, in the limit, Δx will go to zero and the (n) samples will go to infinity; (b) The approximate area of this abc triangle can also be determined by summing all the dotted rectangles under the line y = x.

1. Using geometry, the area of Figure 14.30 (a) $= \frac{1}{2} xy = \frac{1}{2}(4 * 4) = 8$

2. Using a summing process, in Figure 14.30 (a) and (b) the sum of the rectangles when n is the number of whole samples (columns) is:

$$Area_{(a)} \approx \sum_{k=1}^{n} (y * \Delta x) \approx 0 * 0 + 1 * 1 + 2 * 1 + 3 * 1 = 6$$

$$\text{Area}_{(b)} \approx \sum_{k=1}^{n} (y\ \Delta x) \approx \frac{1}{2} * \frac{1}{2} + \frac{2}{2} * \frac{1}{2} + \frac{3}{2} * \frac{1}{2} + \frac{4}{2} * \frac{1}{2} + \frac{5}{2} * \frac{1}{2}$$

$$+ \frac{6}{2} * \frac{1}{2} + \frac{7}{2} * \frac{1}{2} = \frac{28}{4} = 7$$

If the (n) samples were again increased and thus the Δx was decreased, the area would approach and, finally, in the limit, equal the true area of 8.

3. Using integration:

$$\text{Area}_{\text{Tot}} = \lim_{\Delta x \to 0} = \sum_{k=1}^{n} f(x_i)\ \Delta x_i = \int_{x=a}^{b} f(x)\ dx$$

$$= \int_{x=0}^{4} x\ dx = \frac{1}{2} * x^2 \Big|_0^4 = \left(\frac{1}{2} * 16\right) - \left(\frac{1}{2} * 0\right) = 8$$

14.8 THE FUNDAMENTAL THEOREM OF CALCULUS AND INTEGRATION

TM THE FUNDAMENTAL THEOREM OF CALCULUS

The first form of the Fundamental Theorem of Calculus formalizes the process of integration and thus defines the integral. It also infers that integration is the inverse of differentiation and that there is a profound connection between integration and the area under the curve. The rule assumes that a function f is continuous from a starting point "a" to a stated end point "b."

In the following two forms, the basic convention of $y = f(x)$ is assumed. Textbooks use variations in this and other notations. For instance, F(x) can be the same as $f(x)$, but most often it indicates the integral of $f(x)$. Likewise, sometimes $dF(x) = df(x) = d\int f(x) = f(x)\ dx$. F' or f' is usually written as the first derivative and F" or f'' the second derivative, etc. Each book usually develops its own style.

There are two structures that indicate the integration process. The lone integral sign "$\int$" indicates an *indefinite integral.* The indefinite integral has no limits specified and is used only to indicate the fundamental form or process. When the integral sign shows a bottom and top letter or number "$\int_a^b$," it is termed a *definite integral* and specifies that the integration process must only include values of the function from a to b, π to 2π, etc.

MM THE FUNDAMENTAL THEOREM OF CALCULUS

The first form of The Fundamental Theorem—the integral and integration:

$\int f(x)\,dx = F(x) = F(b) - F(a)$ an indefinite integral (here F is the $\int$ of f)

$$\int_a^b f(x)\,dx = F(x)\Big|_a^b = F(b) - F(a) \text{ a definite integral} \qquad \text{(E 14.40)}$$

The second form of The Fundamental Theorem—the derivative and differentiation:

$$f'(x) = \lim_{\Delta x \hat{}0} \frac{f(x + \Delta x) - \Delta x}{\Delta x} \text{ (here } f' \text{ is the derivative of } f) \qquad \text{(E 14.41)}$$

14.9 THE CONSTANT OF INTEGRATION

TM THE CONSTANT OF INTEGRATION

In electronics problems, the value of the constant of integration is usually a dc voltage or current value. By remembering that the derivative of a constant is zero, one can see that the integral of zero can be a constant, or that any integral can produce a constant plus the integral of the varying function. This is called a constant of integration.

Assume an output-signal wave form that is characterized as $y_{out} = \cos \omega t$. The integral of $\cos \omega t$ is $\sin \omega t + C$. This new sine function may or may not have a dc bias—its equivalent wave form may or may not be ac coupled.

Mathematically, an integrated function can have any number of "constant of integration" values. The constant value depends upon the fixed value (in electronics, this is often called the dc bias) or pedestal that the function may be riding upon.

MM and GM THE CONSTANT OF INTEGRATION

Example 14.2: If $\frac{dy}{dx}$ equals cos x, find a family of curves equal to sin x. By integration, $\int \cos x = \sin x + C = \sin x + C_1 = \sin x + C_2 = \sin x + C_n$

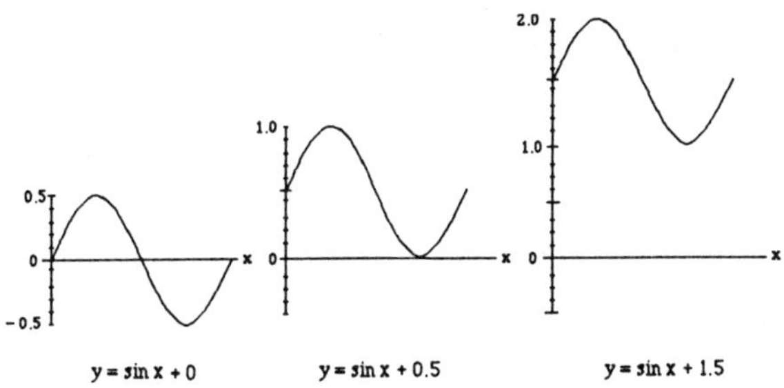

FIGURE 14.31 A family of sine figures with different constants of integration (different dc values).

14.10 SOME BASIC FORMULAS FOR INTEGRATION

TM SOME BASIC FORMULAS FOR INTEGRATION

Although E 14.40 provides the fundamental rule for integration, it is much more practical to integrate using a look up table of integration models or a mathematics computer program. The following list gives most of the integration models used in fundamental circuit analysis.

MM A BRIEF TABLE OF INTEGRALS FOR BASIC CIRCUIT ANALYSIS

This section follows the format of most books on calculus by using the letters "u" and "v" for functions and "a," "n" and "e" to represent constants.

$$\int a\, dx = a \int dx \text{ and } \int 0\, dx = C \qquad \text{(E 14.42)}$$

$$\int a^u\, du = \frac{a^u}{\ln a} + C \qquad \text{(E 14.43)}$$

$$\int e^u\, du = e^u + C \qquad \text{(E 14.44)}$$

$$\int ue^{au}\, du = \frac{e^{au}}{a^2}(au - 1) + C \qquad \text{(E 14.45)}$$

$$\int u\, dv = uv - \int v\, du + C \qquad \text{(E 14.46)}$$

$$\int \frac{du}{u^2 + a^2} = \frac{u}{a} \tan^{-1} \frac{u}{a} + C \qquad \text{(E 14.47)}$$

$$\int \frac{du}{(u^2 + a^2)^2} = \frac{1}{2a^2}\left(\frac{u}{x^2 + a^2} + \frac{u}{a} \tan^{-1} \frac{u}{a}\right) + C \qquad \text{(E 14.48)}$$

$$\int \sin u \, du = -\cos u + C \qquad \text{(E 14.49)}$$

$$\int \cos u \, du = \sin u + C \qquad \text{(E 14.50)}$$

$$\int \tan u \, du = \ln \sec u + C = -\ln(\cos u) + C \qquad \text{(E 14.51)}$$

$$\int \sin^2 au \, du = \frac{u}{2} - \frac{\sin 2au}{4a} + C \qquad \text{(E 14.52)}$$

$$\int \cos^2 au \, du = \frac{u}{2} + \frac{\sin 2au}{4a} + C \qquad \text{(E 14.53)}$$

$$\int_0^{\pi} \frac{\sin ax}{x} = \frac{\pi}{2} \text{ for } a > 0 \text{ and } -\frac{\pi}{2} \text{ for } a < 0 \qquad \text{(E 14.54)}$$

14.11 NUMERICAL INTEGRATION

Section 14.7 introduced a technique of "finding the area under the curve" by using a series of (decreasing) rectangles. The area can be calculated from the integral. This general process can be reversed to find the integral from the area. There are some "curves" that are impossible to integrate using the Fundamental Rule of Calculus. Some unique functions, usually used in higher mathematics or from particular sets of recorded data, require the technique termed numerical integration. This method of integration is also used by some calculators and computers as their fundamental algorithms for doing most, if not all, of the integration problems. The following section introduces two common techniques for numerical integration. Although the use of flat-topped rectangles is possible, trapezoidal and parabolic topped columns (algorithms) are usually used. The parabolic technique is known as Simpson's Rule.

TM THE TRAPEZOIDAL AND SIMPSON'S RULES

A trapezoid can be simply defined as a rectangular column, as illustrated in Section 14.7, with a slanted (f(x)) roof. Likewise, a Simpson column can be defined as a rectangular column with a parabolic (f(x)) roof. By using either the trapezoidal roof or the parabolic roof, a curve (function) can be quite closely approximated by a series of these columns placed one after another. The trapezoidal approach is easiest but less accurate, while the more accurate parabolic approach requires some restrictions. The parabolic Simpson's Rule approach is more common.

MM THE TRAPEZOIDAL AND SIMPSON'S RULES

For a single trapezoid:

Area = the width times the average height

$$= \Delta x * \frac{1}{2}(y_n + y_{n+1}) \text{ where } \Delta x = \frac{b-a}{n} \text{ and n = the number}$$

of intervals from a to b. (E 14.55)

For the approximate integral—The Trapezoidal Rule:

$$\int_a^b f(x)\,dx \approx \frac{1}{2}(y_0 + y_1)\Delta x + \frac{1}{2}(y_1 + y_2)\Delta x + \frac{1}{2}(y_2 + y_3)\Delta x$$

$$+ \frac{1}{2}(y_{n-1} + y_n)\Delta x$$

$$\approx \Delta x\,[\frac{1}{2}y_0 + y_1 + y_2 + \cdots + y_{n-1} + \frac{1}{2}y_n]$$

$$\approx \Delta x\,[\frac{1}{2}f(x_0) + f(x_1) + \cdots + f(x_{n-1}) + \frac{1}{2}f(x_n)] \qquad \text{(E 14.56)}$$

For the approximate integral—Simpson's Rule:

Special Note: The derivation of Simpson's Rule follows the same general pattern as the trapezoidal but is much more involved. The final result is given:

$$\int_a^b f(x)\,dx \approx \frac{\Delta x}{3}[f(x_0) + 4f(x_1) + 2f(x_2) + 4f(x_3) + \cdots$$

$$+ 2f(x_{n-2}) + 4f(x_{n-1}) + f(x_n)] \qquad \text{(E 14.57)}$$

GM THE TRAPEZOIDAL AND SIMPSON'S RULES

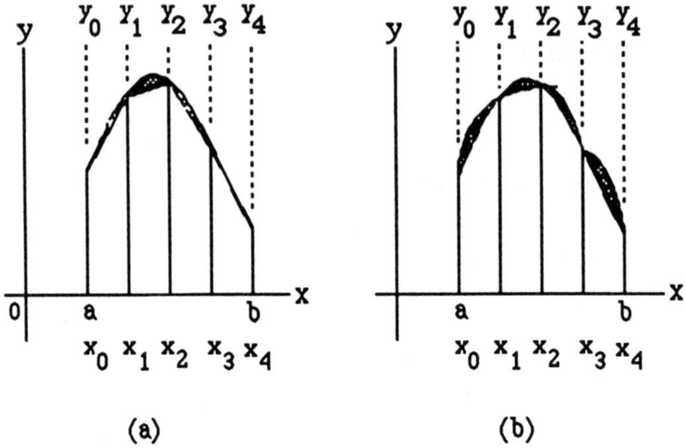

FIGURE 14.32 (a) An exaggerated graph describing the trapezoidal rule; (b) An exaggerated graph describing Simpson's Rule.

14.12 COMPUTER-AIDED CALCULUS

The personal computer is a powerful tool that can aid in the understanding of both the philosophy and the techniques of calculus. One of the most important gifts of the computer is the ability to graph functions including their integrals and their differentials. One can quickly observe the shape of the curve, its maxima and minima, or the (constant of integration) offset. Every student or practicing engineer should have a strategy to improve his or her understanding, skill and enjoyment of this beautiful creation of the mind of humankind. Here is one suggestion.

1. Have a conscious strategy! If the following one does not fit your style, make and use one of your own, but have a strategy.
2. Find and use a program that plots functions. Simple programs such as "FunPlot" (see the Zimmermann reference) cost under $25 and will prove to be among your most valuable tools. As an alternative, obtain a pocket calculator that will plot and display functions.
3. Start a reference list of books on computers and calculus. A personal library is always recommended for the professional, but don't overlook the libraries of your city and county, or nearby college or university libraries. Books like *Calculator Calculus* (see the McCarty reference) are a joy to read and fun to play with. Most instruction books for the

higher-grade pocket calculators give a good background on how the calculator does its more complex mathematics. For instance, the instruction and reference books for the Hewlett–Packard© HP–28C are excellent.

4. Keep a reference notebook of model problems. Know how to execute these problems using different computer programs or different pocket calculators. Be sure your system will integrate $y = x^2$ correctly before attempting something more exotic. If you are fortunate enough to have one of the better mathematics programs, try a new function. Then stretch yourself and see how far you can go!

Glossary

active filters Analog filters that incorporate "active" or amplifying elements such as transistors or operational amplifiers to improve the filter characteristics.

admittance The reciprocal of impedance.

algebraic sum The sum of a series of numbers that includes their algebraic or "+ and –" values.

aliasing The folding back of upper (unwanted) parts of a spectrum or signal into the lower parts of the spectrum or signal, causing false information to be added to the signal. Aliasing is usually caused by improper sampling techniques.

amplitude of spectral coefficients The magnitude of the Fourier spectrum coefficients.

antiderivatives Integrals.

arithmetic series A series of numbers following the progression a, a + d, a + 2d, a + 3d, · · · .

averager Any hardware or software device or technique that finds the average of a series of inputs (signals).

barrier potential The electric field potential in a semiconductor junction that must be overcome before significant forward current can flow. It is approximately 0.7 volts in silicon diodes and 0.3 volts in germanium diodes.

bilateral network Any network that can be turned end-for-end (look at terminals B and A rather than A and B) and still have the same characteristics.

bipolar A semiconductor device that operates using the flow of two types of charge—electrons and holes.

branches Any portion of the circuit that has one or more circuit elements in series.

bulk resistance The true or "ohmic" resistance found in the different semiconductor materials in the junctions of diodes or transistors.

Cartesian The familiar set of x-y coordinates that form the geometric z-planes, s-planes, impedance planes, etc.

characteristic equation A type of system differential equation, usually in the form $As^2 + Bs + C = 0$. The roots of this equation will indicate the character—the resonant frequency, the damping, etc.—of the system.

coefficient of coupling A ratio number, usually a decimal fraction "k," that states the ratio of the flux lines that couple a second coil or winding compared to the total of the flux lines produced in the first coil or winding.

common-mode rejection ratio (CMRR) The ratio, usually in dB, of the differential gain to the common-mode gain of a differential amplifier.

common-mode rejection The ability of a differential amplifier to reject a signal, such as ac hum, that is in phase on both of the differential inputs.

complex frequency The designation for the special use of "frequency" when it is used in a complex s-plane. The complex-plane vertical axis can be $\pm j\omega$, sometimes including the resonant frequency $j\omega_0$, and the horizontal axis can be $\pm\alpha$, usually termed the complex damping or neper-frequency axis.

complex function A function or signal that is defined in terms of both amplitude and phase.

complex numbers Numbers that are used in a two-valued mathematical system where an operator is used to define the relationship of the second number. R + jX is a complex number.

complex plane A geometric model indicating two parameters that have a ±90° relationship such as resistance and reactance.

complex variable A circuit parameter, such as voltage or current, that can (must) be defined by TWO terms such as magnitude and phase.

conductance The reciprocal of resistance—the ability to conduct. It has the units of mho or siemen.

conjugate matching The technique of connecting a complex (R±jX) generator impedance with a load impedance composed of an equal value of R but the "other" X. If the generator is R+jX, the load should be R-jX, etc. This maximizes the power transfer from the generator to the load.

constant current generator A generator that outputs one specified current regardless of the load, including, in theory, both an open and a short circuit.

constant of integration A constant, fixed term that is added to the result after an integration. Since the integration produces, in theory, an infinite number of answers (curves), a particular constant is needed to conform with the conditions of any one particular problem. In circuit calculations, the constant of integration often represents the particular dc offset.

constant voltage generator A generator that outputs one specified voltage regardless of load, including, in theory, a short circuit.

convolution sum The "sum" or result of convolving a function. This is not the same as an arithmetic or algebraic sum.

cosine wave A wave form that follows the characteristic cosine function. Usually it will be a normalized POSITIVE unity amplitude at zero reference degrees, a normalized zero amplitude at 90°, a normalized NEGATIVE unity at 180°, etc.

critical damping The damping of a circuit, usually an RLC resonant circuit, that produces the most rapid rise-time without allowing overshoot.

damping axis The horizontal or α axis in the complex s-plane.

definite integral An integral that has specific limits given.

depletion layer The region near the edges of a p-n junction that contains a concentration of recombined holes and electrons and thus lacks free holes and electrons.

difference equation Digital domain equations that describe the characteristics and operation of circuits such as digital filters. They are somewhat similar to differential equations in the analog domain.

differential equations Equations of motion, voltage, current, etc., that are characterized by the differential form used in calculus, such as $\frac{dx}{dt}$ or $\frac{d^2x}{dt^2}$.

doping The process of adding a small amount of a second material to a pure semiconductor to give it special properties, such as making it p-type or n-type. Pure or intrinsic silicon is often doped with arsenic (+), phosphorus (+), and boron (–) or gallium (–).

eddy currents Currents that are induced in the core material of a transformer by the primary flux. These currents produce one of the power losses in the transformer.

element One single part of a circuit.

envelope The outline of a wave form. The outline of an amplitude modulated wave is the most common example.

ferrite A man-made magnetic core material, with high resistivity and permeability, containing iron oxide and other materials such as manganese, nickel and zinc.

first order An equation containing only terms with an (implied) exponent of one.

flux density The magnetic flux per unit area.

flux The model used to represent magnetism or magnetic force. Flux is usually modeled as containing a line of force.

frequency axis The complex-frequency plane vertical axis where, for sine waves, $s = j\omega$.

frequency domain The mathematical model of a signal or a circuit where such parameters as magnitude, response, etc., are specified in terms of frequency.

frequency response The amplitude or output of a signal or circuit specified in terms of frequency.

frequency spectrum *See* spectrum.

function A mathematical correspondence or term that relates one idea or parameter to another. y = "sin" x relates y to x according to a sinusoid function. A general function is $y = f(x)$.

gain margin The amount the gain can change, due to temperature, aging, etc., in a feedback system without making the system unstable.

gain-bandwidth product The gain time of the bandwidth in a feedback system. For a given system, it is always a constant number. Thus, if the system gain is reduced, the bandwidth is proportionally increased.

geometric series A number series that follows the form $S = a + ar + ar^2 + \cdots + ar^{n-1} + \cdots$ and $S = \frac{a}{1-r}$.

half section One-half of an analog filter section, usually a "T" or "π" section, that was so divided to allow it to be used for improving a particular frequency or phase response or for impedance matching.

harmonic series A number series that follows the form

$$\sum_{n=1}^{\infty} \frac{1}{n} = 1 + \frac{1}{2} + \frac{1}{3} + \cdots \frac{1}{n} + \cdots.$$

homogeneous solution The solution of an equation in the form $f(x) = 0$. In circuit analysis, it is most common to find the homogeneous solution of the second order resonant-circuit equation $As^2 + Bs + C = 0$.

hybrid A result produced from combining two or more ideas or elements. Circuit analysis uses hybrid or h-parameters that incorporate both voltage and current and hybrid circuits that combine two or more circuit types or technologies.

hysteresis Something based upon history. The loop or path that the magnetization of an iron transformer core takes depends upon the previous magnetization.

H–network A five-element network configured or symbolized in the form of an H.

image impedance The impedance "seen" at a network input. It is usually desirable to have the image impedance equal the generator impedance and to have the generator impedance, the network impedance, and the load impedance all equal the characteristic impedance.

impedance The combination of resistive and reactive elements—for example, R + jX.

impulse wave or impulse function (*see* unit impulse) A very, very small width input "spike." In theory, the input pulse would have a width of zero and a height of infinity. The impulse function has meaning in theoretical analysis and the very narrow, but finite pulse has meaning as a troubleshooting tool.

indefinite integral (*see* definite integral) An integral without any specific limits.

infinite series A series of numbers that keeps going on and on to infinity—it never converges or goes to a final number.

input bias current The no-signal input current of a differential amplifier. It is sometimes defined as the average of the two input currents to a differential amplifier.

input offset current The difference between the two input currents in a differential amplifier.

input offset voltage The ADDITIONAL input (offset) that is needed to bring the no-signal output of an operational amplifier to zero.

integrating, integration The process of finding an antiderivative. The very special technique of summing numbers, areas, etc., by using mathematical functions and limits.

inverse Laplace Transform "Going back" to the original integral domain of an equation from its Laplacian form.

iron A basic, natural element with the symbol Fe.

KCL Kirchhoff current law.

Kirchhoff's current and voltage laws THE fundamental laws that explain the flow of current in all electronic circuits and the voltages that these currents produce. ALL circuit analysis starts with these laws.

KVL Kirchhoff voltage law.

ladder network A continuing network configured like the sides and the crosspieces of a carpenter's ladder.

Laplace Transform A technique for, among other things, solving differential equations.

large signal (*see* small signal) A signal with enough magnitude that it may be affected by some of an amplifier's non-linearities.

leakage inductance The concept of a series inductor representing the generated lines of force (flux) that do NOT link the windings and thus do not contribute to transformer action.

linear Straight line. An equation or a circuit where the output has a straight-line relationship with the input. A first-order function.

loops (Sometimes called a *mesh*) is any portion of a circuit that closes back on itself.

LTI Linear, time invariant.

lumped constant filters *See* passive filters.

magnetizing current The small primary current necessary to produce a secondary voltage when the secondary is open circuited.

magnetomotive force (mmf) The force required to produce magnetic flux—usually stated in ampere-turns.

Miller feedback capacitor A feedback capacitor in a tube, transistor, or operational amplifier that is a function of gain. These amplifiers are usually easier to analyze if this input-output feedback capacitor is replaced by a newly derived shunt input and a shunt output capacitor.

moving average The average of a series of numbers (usually digital) that continually changes, usually as a function of time. For instance, the daily averages for the stock market, averaged over a thirty-day period, can keep changing day by day.

mutual inductance (*see* self inductance) The ability of coil_1 (the primary or secondary) to oppose the change of current in coil_2 (the secondary or primary) is termed mutual inductance.

natural response The response of a circuit that results from only the energy stored in the circuit itself. The natural response does not include any external forcing functions.

nodes The junction or connection of two or more circuit elements.

non-recursive A type of digital filter in which the output is derived only from input values—there is no feedback.

normalize To make a term, an equation, the labels on a graphic axis, etc., conform to a simplified standard or reference. For example, values are often all divided so they are all referred to as a reference value of unity.

Norton's theorem Any two-terminal linear, bilateral network can be replaced be an equivalent circuit consisting of a current source and a shunt impedance.

Ohm's law The relationship between voltage, current and resistance in a dc circuit or voltage, current and impedance in an ac circuit.

overdamping The damping of a circuit, usually an RLC resonant circuit, that produces an excessively slow rise-time—there is no overshoot.

particular solution The solution of an equation, usually by integration, that includes finding the value of a constant by using the specified information or the particular initial conditions.

passive filters Electronic filters that do not contain "active" or amplifying elements such as transistors or operational amplifiers.

permeability The ability of a material to be magnetized. Permeability corresponds to conductance.

phase margin The amount of phase change allowed in a feedback system before the system becomes unstable.

phase response The change of output phase of a system, compared to the input, as a function of frequency.

phasors A rotating vector, in a complex plane, that indicates the projection of a sinusoidal function such as voltage or current.

Pi (π) network A three-element network configured or symbolized with a general π shape.

power bandwidth Sometimes termed large-scale bandwidth, it is the maximum peak-value frequency of a large-amplitude sinusoidal wave that an operational amplifier can amplify without slew-rate distortion.

propagation constant Usually denoted by the symbol γ, it is the natural logarithm of the ratio of the input current to the output current of a network or transmission line.

Q For a tuned circuit, Q is two π times the ratio of the maximum energy stored to the total energy lost per period. It represents the sharpness of tuning or the ratio of the resonant frequency divided by the (minus three dB) bandwidth.

Q-point The no-signal operating point on the load line for a diode, transistor, vacuum tube, etc.

quantizing The process of assigning digital values to the amplitude of a signal. There are 2^n levels with binary coding. For example, for an 8-bit code, there are 2^8 or 256 levels.

radian The angle of a circle where the arc of the angle equals the radius of the circle. There are 2π radians in the diameter of a circle. One radian equals 180/π, or about 57.3 degrees.

reactance The property of an inductor or capacitor to oppose the change of current or voltage.

recursive A type of digital filter where the output is derived from both input and previous output signal values—a feedback digital filter.

relative permeability The ratio of the permeability of a material to that of a vacuum or free space.

reluctance The property of a material that opposes the creation of a magnetic field.

rise time The time required for the rise of an input step function. Rise time is usually stated as the time from the ten percent rise point to the ninety percent rise point.

s-parameters Specialized measurements of incident and reflected waves in transmission lines, special filters and high-frequency transistors.

sampling frequency The rate of taking samples of an input wave form. If distortion-free sampling is desired, the sampling frequency must be at least twice that of the highest input signal frequency.

sampling The process of "taking a snapshot" of the instantaneous value of an input wave form.

sawtooth wave A wave form starting at a reference value and then rising as a linear function of time. At some point, the rising wave form returns rapidly to the reference point and again starts its linear rise.

second order An equation where the highest order of exponents is 2. $y = A^2 + 2B + C$ is a second-order equation.

self inductance or inductance (*see* mutual inductance) The ability of a wire or a coil of wire to oppose the change of current in that wire or coil is termed inductance.

shifting factor The amount a difference equation is moved plus or minus.

sine wave A periodic wave form whose y-axis amplitude is always the projection of a point moving uniformly around a circle.

skin effect The forcing of the current (flow), at high frequencies, into surface regions of a solid conductor. Since the current is thus more concentrated in the surface region, the i^2R loss is greater than would be predicted if the simple dc resistance value was used in the loss calculation.

slew rate The maximum signal rate-of-change that an operational amplifier can faithfully amplify. Slew rate is specified in volts per second.

slope The "tilt" of a line—usually with reference to the horizontal or zero-slope position.

small signal A signal that is small enough not to be affected by any of the non-linearities of an amplifier or a network.

spectral coefficients The amplitude and phase or the real and imaginary parts of the frequencies that make up a spectrum.

spectrum A set of frequencies that exist from some low-frequency value to some high-frequency value in a particular circuit or signal. The "light spectrum" is continuous and the spectrum of a piano (notes) is discrete.

square wave A continuous, repeating analog signal that is on (high) for fifty percent of the time and off (low) for fifty percent of the time.

steel A primarily iron-based material that also includes carbon and, sometimes, other elements that are used to change the basic properties of iron—an alloy of iron.

step wave or step function A transient test signal that, theoretically, starts at zero and immediately rises to a nominal "one" (or some predetermined value) and continues at that level forever.

substrate The underlying platform or support for an integrated circuit. The electrical (insulation) properties and the physical (thermal) properties of a substrate are very important.

superposition The concept that any cause or effect can be broken into the sum of the individual causes or effects. For electric networks, the current through any element or the voltage across that element can be considered as the sum of currents or voltages coming from each independent input source.

T-network A three-element network configured or symbolized with a general T shape.

tangent A line, next to the graph of some function, such as a sine, that has the same slope as the function where the tangent line just touches the function.

theory of sampling The collection of ideas and principles that explains the results when a signal is sampled periodically.

Thevenin's Theorem Any two-terminal linear, bilateral network can be replaced by an equivalent circuit consisting of a voltage source and a series impedance.

time constant A number somewhat like the "half-life" in physics. The time it takes the energy in a circuit to increase to about 63 percent of its final value or to decrease to about 37 percent of its original value.

time domain The area of circuit analysis or signal processing where time and time response is one of the major parameters to be considered. If a function is graphed, time is usually the abscissa or x-axis coordinate.

time response The response of a network, often to an input step, as a function of time.

topology In circuit analysis, the study of the configurations of circuits and networks.

transfer function The mathematical relationship between the input and output of a circuit or network.

transform pairs The "going forward" and "going backward" results of a mathematical transformation such as the Laplace Transform. The forward transform is usually simply called the transform where the going back operation is termed the inverse transform.

transient response The response to a sudden change of an input signal. *See also* natural response.

underdamped The damping of a circuit, such as an RLC resonant circuit, that produces a fast rise-time, always with overshoot.

unipolar A semiconductor device that operates using the flow of only one type of charge—electrons or holes.

unit circle The circle configuration, in the z-plane, that, like the jω axis in the s-plane, separates stable operation from non-stable operation.

unit impulse wave or unit impulse function (*see* impulse) An impulse function where the magnitude (area) of the pulse is defined as unity or one.

vectors Complex-plane lines, many times arrowed lines, that are used to indicate complex numbers.

weighting factor As used in difference equations, a number that scales the value of unit impulses from one.

Bibliography

GENERAL REFERENCES FOR CIRCUIT ANALYSIS

Al-Nasser, F., "Tables Speed Design of Low-Pass Active Filters," *EDN*, March 15, 1971.

Aleksandrov, A. D., Kolmogorov, A. N., and Lavrent'ev, M. A., (eds.), *Mathematics: Its Content, Methods, and Meaning*. Cambridge, MA: MIT Press, 1963. (Translated from the Russian edition.)

Angelo, E. J., Jr., *Pole-Zero Patterns in the Analysis and Design of Low-order Systems*. New York: McGraw-Hill Book Company, 1964.

Baher, H., *Analog and Digital Signal Processing*. Chichester, England: John Wiley & Sons, 1990.

Bazin, L. J., "Design of Active R-C Filters," RCA Internal Unpublished Notes, January 20, 1971.

Bennett, W. R., Jr., *Scientific and Engineering Problem-Solving with the Computer*. Englewood Cliffs, NJ: Prentice-Hall, Inc., 1976.

Bogart, T. F., Jr., *Electronic Devices and Circuits*. Columbus, OH: Merrill Publishing Company, 1986.

Boylestad, R. L., *Introductory Circuit Analysis,* 5th ed. Columbus, OH: Merrill Publishing Company, 1987.

Burington, R. S., *Handbook of Mathematical Tables and Formulas.* Sandusky, OH: Handbook Publishers, Inc., 1946.

Butterworth, S., "Effective Resistance of Inductance Coils at Radio Frequency," *Experimental Wireless and Wireless Engineering,* Vol. 3, August 1926.

Cadzow, J. A., *Discrete-Time Systems An Introduction with Interdisciplinary Applications.* Englewood Cliffs, NJ: Prentice-Hall, Inc., 1975.

Cannon, R. H., Jr., *Dynamics of Physical Systems.* New York: McGraw-Hill Book Company, 1967.

Chamberlin, H., *Musical Applications of Microprocessors.* Rochelle Park, NJ: Hayden Book Co., 1980.

Cohen, A. R., *Outline of Linear Circuits and Systems.* New York: Regents Publishing Company, Inc., 1965.

Cowles, L. G., *Transistor Circuits and Applications.* Englewood Cliffs, NJ: Prentice-Hall, Inc., 1968.

Dattorro, J., "The Implementation of Recursive Digital Filters for High-Fidelity Audio," *Journal of the Audio Engineering Society*, November 1988.

Edminister, J. A., *Theory and Problems of Electric Circuits.* New York: Schaum Publishing Company, 1965.

Engineering Staff of Texas Instruments, Inc., *Transistor Circuit Design.* New York: McGraw-Hill Book Company, 1963.

Erst, S. J., *Electronics Equation Handbook.* Blue Ridge Summit, PA: Tab Books, Inc., 1989.

Fink, D. G. and Christiansen, D., *Electronics Engineers' Handbook,* 3rd ed. New York: McGraw-Hill Book Company, 1989.

Hogben, L., *Mathematics for The Millions.* New York: W. W. Norton & Company, 1971.

Hopkins, R. S. and Hurst, R. N., "Introduction to Digital Television," RCA Internal Unpublished Notes, September 1976.

Horowitz, P. and Hill, W., *The Art of Electronics*. Cambridge, England: Cambridge University Press, 1980.

Hurley, R. B., *Junction Transistor Electronics*. Englewood Cliffs, NJ: Prentice-Hall, Inc., 1973.

IEEE, *Programs for Digital Signal Processing*. New York, IEEE Press, 1979.

ITT-Federal Telephone and Radio Company (eds.), *Reference Data for Radio Engineers*. New York: Federal Telephone and Radio Company, 1953.

Johnson, D. E. and Hilburn, J. L., *Rapid Practical Designs of Active Filters*. New York: John Wiley & Sons, 1975.

Johnson, J. R., *Electric Circuits*. Hasbrouck Heights, NJ: Hayden Book Company, 1984.

Kaufman, M. and Seidman, A. H. (eds.), *Handbook of Electronic Calculations*. New York: McGraw-Hill Book Company, 1979.

Kuhfittig, P. K. F., *Technical Calculus with Analytic Geometry*. Monterey, CA: Brooks/Cole Publishing Company, 1983.

Kulathinal, J., *Transform Analysis and Electronic Networks with Applications*. Columbus, OH: Merrill Publishing Company, 1988.

Lancaster, D., *Active Filter Cookbook*. Indianapolis, IN: Howard W. Sams and Company, 1975.

Landee, R. W., Davis, D. C., and Albrecht, A. P., *Electronic Designers' Handbook*. New York: McGraw-Hill Book Company, 1957.

Lowdon, E., *Practical Transformer Design Handbook*. Blue Ridge Summit, PA: Tab Books, Inc., 1989.

Lynn, P. A. and Fuerst, W., *Introductory Digital Signal Processing with Computer Applications*. New York: John Wiley & Sons, 1989.

Malvino, A. P., *Electronic Principles*, 4th ed. New York: McGraw-Hill Book Company, 1989.

McCarty, G., *Calculator Calculus*. Palo Alto, CA: Pace-Ficklin Publishing Company, 1975.

Mett, C. L. and Smith J. C., *Calculus with Applications*. New York: McGraw-Hill Book Company, 1985.

Michie, J. N., *Differential and Integral Calculus*. New York: D. Van Nostrand Company, 1947.

Middlebrook, R. D., *Differential Amplifiers*. New York: John Wiley & Sons, 1963.

Murakami, T., Sonnenfeldt, R. W., et al., *Color Television Lecture Course*. RCA Internal Unpublished Color Television Design Course, 1956.

Nilsson, J. W., *Electric Circuits*. Reading, MA: Addison-Wesley Publishing Company, 1986.

Oppenheim, A. V. and Schafer, R. W., *Digital Signal Processing*. Englewood Cliffs, NJ: Prentice-Hall, Inc., 1975.

Pappas, T., *The Joy of Mathematics*. San Carlos, CA: Wide World Publishing/Tetra, 1989.

Protter M. H. and Morrey, C. H., Jr., *Calculus with Analytic Geometry*. Reading, MA: Addison-Wesley, 1966.

Rabiner, L. R. and Gold, B., *Theory and Application of Digital Signal Processing*. Englewood Cliffs, NJ: Prentice-Hall, Inc., 1975.

Roddy, D. and Coolen, J., *Electronic Communications*. Reston, VA: Reston Publishing Company, Inc., 1984.

Rosenstark, S., *Feedback Amplifier Principles*. New York: Macmillan Publishing Company, 1986.

Ryder, J. D. and Thompson, C. M., *Electronic Circuits and Systems*. Englewood Cliffs, NJ: Prentice-Hall, Inc., 1976.

Schafer, R. and Mersereau, R., "What Designers Need to Know About Digital Filtering," *Communications Systems Equipment Design*, October 1964.

Sevin, L. J., Jr., *Field-Effect Transistors.* New York: McGraw-Hill Book Company, 1965.

Shea, R. S., *Principles of Transistor Circuits.* New York: John Wiley & Sons, 1953.

Shockley, W., *Electrons and Holes in Semiconductors.* Princeton, NJ: Van Nostrand, 1950.

Smith, R. P., *Circuits, Devices and Systems,* 3rd ed. New York: John Wiley & Sons, 1976.

Staff Members, Massachusetts Institute of Technology. *Applied Electronics.* Englewood Cliffs, NJ: Prentice-Hall, Inc., 1943.

Staff of Motorola, Inc., *Linear Integrated Circuits Data Book.* Phoenix, AZ: Motorola, Inc. 1971.

Staff, *The Electric Circuits Problem Solver©.* Piscataway, NJ: Research and Education Association, 1988.

Staff, *The Electronic Communications Problem Solver©.* Piscataway, NJ: Research and Education Association, 1984.

Stanley, W. D., *Operational Amplifiers with Linear Integrated Circuits.* Columbus, OH: Merrill Publishing Company, 1989.

Stanley, W. D., *Transform Circuit Analysis for Engineering and Technology.* Englewood Cliffs, NJ: Prentice-Hall, Inc., 1968.

Strawn, J., *Digital Audio Signal Processing.* Las Altos, CA: William Kaufmann, Inc., 1985.

Strum, R. D. and Ward, J. R., *Laplace Transform Solutions of Differential Equations.* Englewood Cliffs, NJ: Prentice-Hall, Inc., 1968.

Strum, R. D. and Kirk, D. E., *First Principles of Discrete Systems and Digital Signal Processing.* Reading, MA: Addison-Wesley Publishing Company, 1988.

Terman, F. E., *Radio Engineering,* 3rd ed. New York: McGraw-Hill Book Company, 1947.

Terman, F. E., *Radio Engineers' Handbook*. New York: McGraw-Hill Book Company, 1943.

Terrell, T. J., *Introduction to Digital Filters*. New York: John Wiley & Sons, 1988.

The Proceedings of the AES 7th International Conference, *Audio in Digital Times*. New York: The Audio Engineering Society, 1990.

Thomson, C. M., *Mathematics for Electronics*. Englewood Cliffs, NJ: Prentice-Hall, Inc., 1976.

TRW LSI Products, *Digital Signal Processing*. La Jolla, CA: TRW LSI Products, 1981.

Tuma, J. J., *Handbook of Numerical Calculations in Engineering*. New York: McGraw-Hill Book Company, 1989.

Ware, L. A. and Reed, H. R., *Communications Circuits*. New York: John Wiley & Sons, 1958.

Watkinson, J., *The Art of Digital Audio*. London: Focal Press, 1989.

Weinberg, L., *Network Analysis and Synthesis*. New York: McGraw-Hill Book Company, 1962.

Yunk, M., *Design of Modern Transistor Circuits*. Englewood Cliffs, NJ: Prentice-Hall, Inc., 1973.

Zverev, A. I., *Handbook of Filter Synthesis*. New York: John Wiley & Sons, 1967.

REFERENCE COMPUTER PROGRAMS FOR CIRCUIT ANALYSIS

BV Engineering, ACNAP (a program for ac circuit analysis). BV Engineering Professional Software, Riverside, CA 92507.

BV Engineering, ACTFIL (a program for active-filter design). BV Engineering Professional Software, Riverside, CA 92507.

BV Engineering, DCNAP (a program for dc circuit analysis). BV Engineering Professional Software, Riverside, CA 92507.

BV Engineering, LCFIL (a program for passive-filter design). BV Engineering Professional Software, Riverside, CA 92507.

BV Engineering, PCPLOT (a high-resolution graphics program). BV Engineering Professional Software, Riverside, CA 92507.

BV Engineering, SPP (a program for signal processing analysis and design). BV Engineering Professional Software, Riverside, CA 92507.

Drexel University—Onaral, B. and Cain, M. R., Discrete Signals and Systems I, Drexel University, Philadelphia, PA 19104.

Drexel University—Onaral, B. and Cain, M. R., Discrete Signals and Systems II, Drexel University, Philadelphia, PA 19104.

Dynacomp, Inc., Digital Filter. Dynacomp, Inc., Rochester, NY 14618.

MicroSim Corporation, PSpice (a broad-based circuit analysis and design program). MicroSim Corporation, Irvine, CA 92718.

University of the Pacific—Zimmermann, W., FunPlot. University of the Pacific, Stockton, CA 95211.

Index